AF528472

Edible Alliums

THE AUTHORS

Dr. Sanyat Misra is B.Sc. Forestry from UAS, Bangalore; M.Sc. (Ag.) Hort. from GBPUAT, Pantnagar; and Ph.D. from Dr. YSP Univ. of Horticulture and Forestry, Nauni, Solan (H.P.). He served NBPGR at Shimla & Delhi continuously for 3 years as Research Associate. As a regular Jr. Scientist *cum* Asstt. Prof., he joined B.A.U., Ranchi-834006 in 2004. At present he is working there as Head of Department. He to his credit has 45 research publications, some 70 chapters in the books published from England in 2017. He has authored 10 books whereas 2 are still in press. In the Department, he is teaching almost all the horticultural courses being offered in the University. Also as a Chairman, he guided 7 M.Sc. students. He is involved in all the research and developmental activities in the Deparment.

Mr. Sanjay Mishra is a Computer Specialist and has written 2 books on **Computer Education**, 3rd as "**Soilless Crop Production**", and 4th "**Herbs, Spices and Medicinal Plants**". His 5th **book** is already in the press which is likely to be out hardly in one month, and the 6th has already gone to the press. His efforts for bringing out these publications are superb.

Dr. **R.L. Misra**, after serving IARI, New Delhi for 40 years, superannuated in May 2008 as Project Coordinator, AICRPF *cum* Principal Scientist. His contribution in horticulture field is immense which is acclaimed worldwide. To his credit he has **30 gladiolus varieties**, and has got about 500 publications out of which there are about 200 only research publications and more than 100 book chapters. He **guided 6 M.Sc. students** and **16 Ph.D. scholars** for their thesis works. He edited **Journal of Ornamental Horticulture** continuously for **12 years**. He is Life Member in 12 professional societies. He is **Honorary Fellow of 4 Societies**, *viz.*, ISGPB, New Delhi (1980), ISOH, New Delhi (2009), HSI, New Delhi (2018) and Society for Promotion of Horticulture, IIHR, Bengaluru (2022). He was bestowed with **Urivi-Vikram Charitable Trust Award** (2003),**HSI Gold Medal** (2013),**ISOH Lifetime Achievement Award in Floriculture** (2018), and **Lifetime Achievement Award in Horticulture** from Odisha Horticultural Society, Bhubaneswar (2023). To his credit he has some 20 published books including 3 quite volumetric ones from Kruger-Brentt, Middlesex, England. These 3 books if kept together are best on flower crops world over. Even after his retirement, from 2010 to 2014, he was Adjunct Faculty in IARI. Still at the age of 80 years, he remains quite active.

Edible Alliums

Authors

Sanyat Misra

Sanjay Mishra

R.L. Misra

2024

Daya Publishing House®
A Division of
Astral International Pvt. Ltd.
New Delhi – 110 002

ISBN: 978-93-5461-981-6

Publisher's Note:

Every possible effort has been made to ensure that the information contained in this book is accurate at the time of going to press, and the publisher and author cannot accept responsibility for any errors or omissions, however caused. No responsibility for loss or damage occasioned to any person acting, or refraining from action, as a result of the material in this publication can be accepted by the editor, the publisher or the author. The Publisher is not associated with any product or vendor mentioned in the book. The contents of this work are intended to further general scientific research, understanding and discussion only. Readers should consult with a specialist where appropriate.

Every effort has been made to trace the owners of copyright material used in this book, if any. The author and the publisher will be grateful for any omission brought to their notice for acknowledgement in the future editions of the book.

Published by : **Daya Publishing House®**
A Division of
Astral International Pvt. Ltd.
– ISO 9001:2015 Certified Company –
4736/23, Ansari Road, Darya Ganj
New Delhi-110 002
Ph. 011-43549197, 23278134
e-mail: info@astralint.com
Website: www.astralint.com

Printed at : **Replika Press Pvt Ltd**

Preface

In 2012 when we were writing the book '**Commercial Ornamental Bulb Science**' which was then published from Westville Publishers, Paschim Vihar, New Delhi in 2013, under 1st Chapter 'True Bulbs' we encountered ***Allium*** as one genus containing many flowering ornamentals together with "**Edible Onions**". However, as then the book was on ornamental side so we had to skip the edible ones which have got their turn now in the form of the present publication '**Edible Alliums**'. Here we have tried to put all the edible alliums together at one platform whether cultivated or collected from the wild anywhere globally. We discussed the things together and then we started this on scratch basis but it was very encouraging when we started getting enormous literature on the subject. Many of these alliums, we found that apart from being used as spices are also used as vegetables as well as certain even quite effective against certain health problems. And now the book is before the public to judge its quality. People would be most welcome to point out the deficiencies if any in this book and together authors will be highly thankful to such critics ~ as such critical evaluations will be quite helpful in improving this book further.

This book is a ready reckoner for all those who in any way are involved acquiring knowledge about vegetables, vis-à-vis their culinary and medicinal uses; to the olericulture academicians; to working scientists who are already in this line; to all those Professors who are engaged in educating the horticultural students whether at B.Sc., M.Sc. or Ph.D. levels; to all those growers who commercially or otherwise are involved in growing of any of the alliums; vis-à-vis to all those students who in any way have opted courses in vegetables. We are aware that in

India, the most common alliums being used in our cuisine are onion and garlic but what is option for those who religiously or by any otherwise means do not use these, yes for them also there are various kinds of 'Jimbus' or 'Faran' and moreover essentially this does not require to be cultivated as people collect these directly from the wild to use fresh, dried or in powdered form as is being done in Himachal Pradesh, Uttarakhand, Jammu & Kashmir, NEH Regions and in parts of hilly West Bengal cum South India, as well as in many other countries. Also the book is quite useful to those who believe in home treatment using Ayurvedic System of Medicines as many alliums have many such properties. Systematically all the edible Allium crops as mentioned below are briefly described further.

The species of Allium described in this book are i) *Allium cepa* (**Onion**); ii) *Allium sativum* (**Garlic**); iii) *Allium ampeloprasum* var. *porrum* (syn. *A. porrum;* **Leek**), iv) *A.a. var. ampeloprasum* (**Elephant garlic**), v) *A.a. var. kurrat* (syn. *A. kurrat;* **Kurrat**), vi) *A.a. var. sectivum* ('**Pearl-Onion Group**'); vii) *Allium shoenoprasum* (**Chives**), viii) *A. ramosum* (**Chinese chives**), ix) A. *tuberosum* (**Garlic chives**), x) A. *hookeri* (**Hooker chives**); xi) *Allium fistulosum* (**Welshy onion**); xii) *Allium oschaninii* (**French grey shallot**); xiii) *Allium stipitatum* (**Persian shallot**); xiv) *Allium tricoccum* (**Rams**), xv) *A. ursinum* (**Ramsons**); xvi) *Allium scorodoprasum* (**Sand leek**), xvii) *A. oleraceum* (**Field garlic**), xviii) *A. vineale* (**Crow garlic**), xix) *A. canadense* (**Canada onion**); xx) *Allium hypsistum,* xxi) *A. negianum,* xxii) *A. przewalskianum,* xxiii) *A. stracheyi,* xiv) *A. wallichii,* xv). *A. auriculatum,* xxvi) *A. carolinianum,* xxvii) *A. consanguineum,* xxviii) *A. fasciculatum,* xxix) *A. humile,* xxx) *A. oreoprasum,* xxxi) *A. prattii,* xxxii) *A. roylei*; xxxiii) *Allium chinense* (**Chinese onion**); xxxiv) *Allium stellatum* (**Cliff onion**); xxxv) *Allium cyaneum* (**Dark blue garlic**); xxxvi) *Allium drummondii* (**Drummond's onion**); xxxvii) *Allium monanthum* (**Korean wild chive**); xxxviii) *Allium cernuum* (**Nodding onion**); xxxix) *Allium rotundum* (**Round-headed leek**); xxxx) *Allium galanthum* (**Snowdrop onion**); xxxxi) *Allium triquetrum* (**Three-cornered leek**); xxxxii) *Allium bisceptrum* (**Twincrest onion**); and xxxxiii) *Allium anceps* (**Twinleaf or Kellogg's onion**).

Authors highly acknowledge all the internet references quoted in the text after every chapter.

Sanyat Misra

Sanjay Mishra

R.L. Misra

Contents

Introduction

Allium **L.,** formerly belonging to the family of Alliaceae, followed by Liliaceae as the onions have superior ovaries but finally now these have been placed in the family of **Amaryllidaceae** due to flowers in this case appearing in umbels (Anon., 2022), and the type species being *Allium sativum* L. (common garlic). The genus *Allium* initially standing for garlic in Latin (Gledhill, 2008) was first mentioned by Linnaeus in 1753 and since then it was so frequently used with so many synonyms such as *Aglitheis* Raf., *Anguinum* Fourr., *Ascalonicum* P.Renault, *Berenice* Salisb. 1866, illegitimate homonym not Tul. 1857, *Briseis* Salisb., *Butomissa* Salisb., *Calliprena* Salisb., *Caloscordum* Herb., *Camarilla* Salisb., *Canidia* Salisb., *Cepa* Mill., *Codonoprasum* Rchb., *Endotis* Raf., *Geboscon* Raf., *Getuonis* Raf., *Gynodon* Raf., *Hexonychia* Salisb., *Hylogeton* Salisb., *Iulus* Salisb., *Kalabotis* Raf., *Kepa* Tourn. ex Raf., *Kromon* Raf., *Loncostemon* Raf., *Maligia* Raf., *Milula* Prain, *Moenchia* Medik. 1790, illegitimate homonym not Ehrh. 1783 nor Roth 1788 nor Wender. ex Steud. 1841, *Molium* (G.Don) Haw., *Moly* Mill., *Molyza* Salisb., *Nectaroscordum* Lindl., *Ophioscorodon* Wallr., *Panstenum* Raf., *Phyllodolon* Salisb., *Plexistena* Raf., *Porrum* Mill., *Praskoinon* Raf., *Raphione* Salisb., *Rhizirideum* Fourr., *Saturnia* Maratti, *Schoenissa* Salisb., *Schoenoprasum* Kunth, *Scorodon* Fourr., *Stelmesus* Raf., *Stemodoxis* Raf., *Trigonea* Parl., *Validallium* Small, *Xylorhiza* Salisb. 1866, illegitimate homonym not Nutt. 1840.

Origin and Brief History

It is a large genus of monocotyledonous flowering plants characterised by herbaceous biennial and perennial geophytic true bulbs, some of which are borne on rhizomes, with an onion or garlic odour and flavour (Quattrocchi, 1999; Gledhill, 2008). Though naturally Allium species occur in temperate climates of the Northern Hemisphere including Mexico, a few species occur in Brazil (such as *A. sellovianum*), Chile (*A. juncifolium*) and tropical Africa ~ southern Africa and Ethiopia (*A. spathaceum*) (en.wikipedia.org; Block, 2010). Majority of Allium species are native to Northern Hemisphere, being spread throughout the Holarctic region from dry subtropics to the boreal zone (Li *et al.*, 2010) but predominantly in Asia that is why these vary accordingly in their hardiness. *Allium dregeanum* Kunth is the sole known species of *Allium* endemic to sub-Saharan Africa (De Sarkar *et al.*, 1997).

Wild Allium in Behbahan (Iran)	**Selection of cultivated alliums displayed at the BBC Gardeners' World Live show**

By Photogir - Own work, CC BY-SA 4.0, https://commons.wikimedia.org/w/index.php?curid=89493262	***By Andy Mabbett - Own work, CC BY-SA 3.0,.commons.wikimedia.org/w/index.php?curid=15531749***

The genus *Allium* is much popular worldwide for its two species *A. cepa* (onion) and *A. sativum* (garlic) and is used as spice crops. *A. cepa* 'Aggregatum Group' (Shallot), *A. fistulosum* L. (Welsh onion/Stone Leek) and *A. schoenoprasum* L. (Chives/Leek) are also cultivated commercially to be used as spices, vegetables and for medicines. Domestication of wild alliums that have characteristic onion- or garlic-like odour and flavour, started since ancient times and is still continuing in the diversity rich regions, with major centres of diversity being in Central Asia and Western North America (Li *et al.*, 2010) where these are used as seasoning spices and green vegetables by different communities (Pandey *et al.*, 2021, 2022). *Wild Allium* species have long been used as a leafy vegetable, seasoning spice, and for therapeutic purposes, vis-a-vis for ornamental gardening especially non-edible ones. They are found in abundance in Indian Himalayan region and Nepal. Globally, several wild species are being commercially cultivated for vegetable, e.g. *A. komarovianum* Vved. (*A. sacculiferum* Maxim.) in North Korea, *A. canadense* L. in Cuba (Fritsch and Friesen, 2002) and *A. obliquum* L., native to Eastern Europe-Siberia which is grown there as a therapeutic plant

(Pandey *et al.*, 2022). Chinese or Japanese garlic (*A. macrostemon*), Naples garlic (*A. neapolitanum* Cirillo), ramsons (*A. ursinum* L.) and long-rooted garlic (*A. victorialis*) are some of the potential domesticated species for edible purposes. *A. przwalskianum* and recently described new species *A. negianum* collected from Niti Valley of Uttarakhand (India) have been domesticated and are now under commercial cultivation (Pandey *et al.*, 2021).

Allium species and several other related groups were until recently classified as the member of Liliaceae, but recent taxonomic revisions now place them in the family Amaryllidaceae. The decision to include a species in the genus *Allium* is taxonomically difficult as the species boundaries are not very clear (en.wikipedia.org). Still there is ambiguity regarding the estimates in the number of species under *Allium* as Knud (1998) mentions the total number as 260, Sanyat Misra and Misra (2013) up to 500, Xu and Kamelin (2000) and Mabberley (2017) up to 850, in 2013 through an internet site http://www.theplantlist.org/ it is 972, as per Li *et al.* (2010) and Govaerts *et al.* (2021) some 1100 and Leek (2008) mentions it as high as about 1250 distributed worldwide. This number may still increase as scientists are putting their concerted efforts in plant exploration worldwide. Most widely used alliums are **onions** (*Allium cepa*), **garlic** (*Allium sativum*), **leeks** (*Allium porrum*), **chives** (*Allium schoenoprasum*) and **shallots** (*Allium cepa var. aggregatum*). Out of around 200 species that grow in Asia (Keusgen *et al.*, 2006), 34 are reported growing in India alone (Dasgupta, 2006). High Himalayan mountains are the home for many species of *Allium* such as *A. auriculatum* Kunth, *A. caesioides* Wendelbo, *A. gilgiticum* F.T.Wang & Tang, *A. macranthum* Baker, A. *roylei* Stearn, *A. humile* Kunth, *A. sikkimense* Baker, *A. stracheyi* Baker and *A. wallichii* Kunth (Maity *et al.*, 2018; Mohan *et al.*, 2019). Nine species of *Allium* are reported in Sikkim alone (Srivastava, 1996; Maity *et al.*, 2018).

Also known as bulb crops, the genus *Allium* is an economically important group of crops cultivated worldwide for their use as vegetables, spices and medicines. This genus contains several species variously referred to as onions and cultivated for food such as the common onion (*Allium cepa*), the Japanese bunching onion (*Allium fistulosum*), the tree onion (*A.* × *proliferum*), and the Canada onion (*Allium canadense*) (Fritsch and Friesen, 2002).

Botany

Alliums are herbaceous geophytic perennials mostly having true tunicated bulbs, with a few exceptions which are rhizomatous (Sanyat Misra and Misra, 2013) where these ovoid-cylindrical (rarely with underground bulbils) bulbs solitarily or aggregating on rhizome (either garlic or onion) (Pandey *et al.*,

2022). Among the species, the bulbs vary in size from small (around 2–3mm in diameter) to rather large (8–10cm). The plants are perennialized by the bulbs reforming annually from the base of the old bulbs or are produced on the ends of rhizomes, though in a few species at the ends of stolons. Some species such as *A. fistulosum* develop thickened leaf-bases rather than forming bulbs as such and a few species have even tuberous roots.The bulbs are solitary or clustered and tunicated. The outer fibrous coats of bulbs are commonly brown or grey with smooth texture. Inner coats of the bulbs are usually membranous (en.wikipedia.org/wiki/Allium).

Height of alliums is highly variable, *i.e.* from 5 to 150 cm where umbellate flowers appear at the top of the leafless stalks (en.wikipedia.org/wiki/Allium). Stem is solid-hollow (Pandey *et al.,* 2022). Many alliums have basal leaves that commonly wither away from the tips downward before or while the plants flower but some species have persistent foliage, many have linear, channelled or flat leaf blades which are straight or variously coiled but some species have broad leaves including *A. victorialis* and *A. tricoccum,* however, the leaves are sessile and very rarely narrowed into a petiole (en.wikipedia.org/wiki/Allium) though species from the Old World such as *Allium vineale* (Field Garlic) generally have round hollow leaves while those of the New World have flat solid leaves (Illinoiswildflowers, 2023). They are radical or cauline, sheathing at base, filiform-linear, fistular-flat and lanceolate-oblanceolate (Pandey *et al.,* 2022).

The flowerheads on most ornamental onion varieties rise well above a clump of strap-like leaves. Foliage of early bloomers emerges very early in spring and generally dies back just as the plants come into bloom. Late-blooming alliums maintain their attractive green foliage much longer into the season. It's wise to position the early bloomers in the landscape so that surrounding plantings hide the deteriorating leaves.

Inflorescences are terminal umbels that are produced on terete or flattened flowering scapes, subtended by noticeable spathe bracts, which are commonly fused and normally have around three veins (en.wikipedia.org/wiki/Allium). Umbels have few to many flowers, rarely with aerial bulbils (Pandey *et al.,* 2022). In some species, such as *A. paradoxum*, the bulbils replace some or all the flowers. In umbels, the outside flowers bloom first and then flowering progresses to the inside, each flower with one style and six epipetalous erect or sometimes pendent stamens (as per species the anthers and pollen can vary in colouration) having six petal-like tepals produced in two whorls, and the 3-lobed (3-loculed) ovaries are superior (en.wikipedia.org/wiki/Allium).

Spathes are 1 to multi-valved, persistent or caducous. Pedicellate flowers are bisexual, regular, polyphyllous, hypogynous; tepals 6 in 2 whorls, free or connate at base, variously coloured; stamens 6 in 2 whorls, free or connate at base; filaments entire, toothed or cuspidate; anthers elliptic to oblong, dorsifixed, dehiscing longitudinally inwards; ovary tricarpellary syncarpous, trilocular subglobose or ovoid; style linear-filiform arising from base; stigma sub-globose or 3-lobed; and capsule is loculicidal. Seeds are 3, 6 or even more; testa black-darkest brown, non-lustrous, compressed-globose and winged or wingless. Whole plant may emit onion or garlic odour (Pandey *et al.*, 2022).

The capsular fruits open longitudinally along the capsule wall between the partitions of the locule, exhibiting small and rounded black seeds with very hard seed coats (en.wikipedia.org/wiki/Allium).

Genetics

Although alliums are valuable horticultural and food commodities yet genetic and genomic knowledge and its application for their genetic improvement is not sufficient enough to impact farmers' field and industry at significant level (Meena *et al.*, 2016; Jo *et al.*, 2017; Rajkumar *et al.*, 2015). *Allium* genomes are classified as giant genomes and among the largest ones in plant kingdom (Jaume *et al.*, 2014) where onions have the largest genome among all the vegetable crops (Shukla *et al.*, 2016) though not among the alliums and garlic genome is roughly equal to that of onion. The most common basic chromosome number in onion is x = 8 though other numbers may have x = 7, 9, 10, 11 with variation even at ploidy levels (Traub, 1968; Friesen *et al.*, 1992; Huang *et al.*, 1995; Xu *et al.*, 1998; Zhou *et al.*, 2007).

Chromosome numbers for economically important alliums mainly used in culinary uses and for medicinal uses are described in the following table:

Chromosome numbers in most commonly cultivated edible Allium species			
Category of Allium	Botanical name	chromosome no.(somatic)	Ploidy level
Common onion	Cepa group (*Allium cepa* L. or *A. cepa* var. cepa L.)	16	Diploid
Multiplier onion: shallot and potato onion	Aggregatum group (*Allium cepa* L. or *A. cepa* var. aggregatum G. Don)	16	Diploid
Garlic • Hard neck • Soft neck	 *A. sativum* var. ophioscorodon *A. sativum* var. sativum	16	Diploid
Other edible Allium species of global importance			
Chives	*A. schoenoprasum* L.	16, 24 & 32	Diploid, triploid & tetraploid

Leek	*A. ampeloprasum* ssp. porrum	32	Tetraploid
Japanese bunching onion/Welsh onion	*A. fistulosum*	16	Diploid
Lesser known alliums with high value in niche markets			
Tree onion/ Egyptian walking onion	*A. × proliferum*	16	Diploid
	A. × cornutum	2n=3x=24	Triploid
Great-headed garlic	*A. ampeloprasum var. ampeloprasum*	32, 48 & 64	Tetraploid, hexaploid & Octoploid
Kurrat	*A. ampeloprasum var. kurrat*	32	Tetraploid
Pran - Kashmir (India)	*A. cepa var. viviparum/ proliferum*	24	Triploid
Jaynaut - NE India	*A. hookeri*	22	Triploid
Jambu/Faran - Nepal, Northern Uttarakhand (India)	*A. consanguineum* Kunth. *A. stracheyi Baker.*		
Jimbur - Sikkim (India)	*A. wallichii* Kunth.		
Dune – Northern Uttarakhand (India)	*A. humile* Kunth, syn *A. govanianum* Wall. ex Baker.		
Bengaluru rose onion	*A. cepa* L.	16	Diploid
Rakkyo/scallion - NE India, China, Japan	*A. chinense*	16, 24 & 32	Diploid, Triploid & Tetraploid
(Gohil and Koul, 1981; Kiełkowska, 2012; Shah, 2014; Malik *et al.* 2020)			

Nutritional Values

Per cent composition of raw flesh of onions, spring onions, garlic and leeks (Source: Athar *et al.*, 2004; Hedges and Lister, 2007):

Composition	*Units*	*Onion, flesh, raw*	*Spring onion, flesh of bulb, raw*	*Garlic cloves, raw, peeled*	*Leeks, bulb, raw*
Water	G	87.9	86.8	64.3	86
Energy	kcal	40	40	97	35
Protein	G	1.27	0.9	7.9	1.9
Total fat	G	0.12	0.3	0.6	0.4

Composition	*Units*	*Onion, flesh, raw*	*Spring onion, flesh of bulb, raw*	*Garlic cloves, raw, peeled*	*Leeks, bulb, raw*
Carbohydrate, available	G	8.53	8.5	15	5.9
Dietary fibre (Englyst, 1988)	G	2.36	1.7	8	3.3
Ash	G	0.46	0.9	1.5	1
Sodium	mg	2.21	13	4	9
Phosphorus	mg	39.5	24	170	43
Potassium	mg	184	230	620	310
Calcium	mg	21.2	140	19	63
Iron	mg	0.24	1.2	1.9	1.1
Beta-carotene equivalents	µg	10	586	T	49
Total vitamin A equivalents	µg	1.7	98	T	8
Thiamin	mg	0.043	0.05	0.13	0.1
Riboflavin	mg	0.015	0.09	0.04	0.05
Niacin	mg	0.734	1	0.4	0.6
Vitamin C	mg	7.1	25	17	18
Total saturated fatty acids	G	0.023	0.058	0.122	0.061
Total monounsaturated fatty acids	G	0.02	0.049	0.015	0.006
Total polyunsaturated fatty acids	G	0.054	0.133	0.342	0.253
Dry matter	G	12.1	13.2	35.7	14
Total nitrogen	G	0.2	0.15	1.27	0.31
Glucose	G	3.6	3.5	0.4	2.4
Fructose	G	2	4.2	0.6	2.4
Sucrose	G	2.4	0.6	0.57	1.1
Total available sugars	G	8	8.3	1.6	5.9
Starch	G	0.53	0.2	13.4	0
Alcohol	G	0	0	0	0
Total niacin equivalents	mg	1.04	1.2	1.5	0.9
Soluble non-starch polysaccharides	G	1.31	0.9	5.5	1.6

Composition	Units	Onion, flesh, raw	Spring onion, flesh of bulb, raw	Garlic cloves, raw, peeled	Leeks, bulb, raw
Insoluble non-starch polysaccharides	G	1.06	0.8	2.5	1.6
Energy	kJ	166	167	402	144
Magnesium	mg	8.43	11	25	10
Manganese	μg	161	208	500	188
Copper	mg	0.059	0.13	0.06	0.1
Zinc	mg	0.25	0.1	1	0.4
Selenium	μg	0.16	0.16	2	
Potential niacin from tryptophan	mg	0.304	0.2	1.1	0.3
Vitamin B6	mg	0.036	0.1	0.38	0.25
Folate, total	μg	26.9	40	5	86
Vitamin E	mg	0.3	0.3	0.01	0.92
Source: Hedges and Lister (2007)					

Phytochemical Constituents

Allium species comprise of compounds that exhibit many *in vivo* and *in vitro* bioactivities. Both wild and cultivated species have many key metabolites and other sulphur-rich substances, such as allicin in common garlic, *i.e. Allium sativum* (O'Donnell *et al.*, 2008). *Allium cepa* (common onion) is full of many important secondary metabolites. Flavonoids, fructans, and oraganosulphur compounds have been isolated from *Allium cepa*. Both its wild and cultivated species are full of key sulphur-containing compounds.

Alkaloids

O'Donnell *et al.* (2008) have stated several alkaloids in the genus:

- Three pyridine-N-oxide alkaloids having disulfide functional groups have been isolated from the *Allium stipitatum* such as 2-(methyldithio) pyridine-N-oxide, 2- (methylthiomethyl) dithio- pyridine-N-oxide and 2,2′-dithio-bis-pyridine-N-oxide. These compounds have been reported to contain antimicrobial activities against several bacterial species.
- A thiosulfinate natural products allicin that widely occurred in genus *Allium* was resolved by the synthesis of compound 2-(methyldithio)

pyridine-N-oxide by simple method of Kitson and Loomes. These compounds were isolated as yellow and orange oils from the chloroform extracts with the different molecular formulas.

Flavonoids

Flavonoids and their derivatives isolated from several cultivated species (including *A. sativum*, *A. cepa*, *A. schoenoprasum*) and also from wild species of *Allium* (like *A. filidens*, *A. griffithianum*, *A. rosenbachianum*) have anti-oxidant, anti-tumor, anti-inflammatory and anti-mutagenic activities (Borlinghaus *et al.*, 2014). Several different flavonoids like flavone, flavonol, flavenone have been found in various *Allium* species (Timothy *et al.*, 2022).

Carbohydrates

Allium sativum contains the highest amounts of carbohydrates, which is also reported to be present in considerable amounts in other species such as *A. porrum* and *A. cepa*. *Allium cepa* contains fructo-oligosaccahrides and *Allium porrun* contains active polysaccharides like glucuronic acid, galactose, and rhamnose. The extracts were reported to have higher polyuronic and protein amount and low sugar value. *Allium sativum* mainly contains reducing sugar such as glucose, fructose, sucrose and maltose that have high biological activities to stimulate the immune system of the human body (Kratchanova *et al.*, 2010; Timothy *et al.*, 2022).

Sulphur Compounds

Oraganosulfur compounds such as di-allylsulfide, sulfinate, allylpropylsulfide and S-methyl-L-cysteine sulphoxide were isolated from *Allium cepa*. Allicin (diallylthiosulfinate), a defense molecule was isolated from *A. sativum* which had broad range of biological activities. These compounds have been used for prevention of cancer, diabetes and cardiovascular problems (Borlinghaus *et al.*, 2014; Timothy *et al.*, 2022). Cysteine sulphoxides (alliin) were isolated from number of species such as A. *filidens*, *A. griffithianum*, *A. rosenbachianum*, *A. barsczewskii* and *A. sativum*. Twenty seven volatile sulfur rich compounds were extracted from *A. tuberosum* including sulfide, disulfides, trisulfides, and tetrasulfides with ethyl, butyl, and pentyl groups (Borlinghaus *et al.*, 2014; Krest *et al.*, 2000). Thioethyl and thiopentyl compounds were also reported from

A. schoenoprasum (Timothy *et al.*, 2022). These sulfur containing compounds have maximum antioxidant and antitumor activities (Block, 2013).

Flavour Precursors

Plants of the genus *Allium* produce chemical compounds, mostly derived from flavour precursors S-alk(en)yl cysteine sulphoxides (ACSOs) found in their cells, which give them a characteristic onion or garlic taste and odour (Eric Block, 2010). The flavour precursors are in the form of S-transprop-1-enyl cysteine sulfoxide (isoalliin) and S- propenyl cysteine sulphoxide (propiin) in onion, S- propyl cysteine sulphoxide in leek, S-allyl cysteine sulfoxide (alliin) in garlic and S-methyl cysteine sulfoxide (methiin) in other alliums (Jones *et al.*, 2004; Randle and Lancaster, 2002). Some alliums belong to a methiin/alliin-type, like ramsons (*A. ursinum*), "*see Ramps and Rampsons*". The characteristic *Allium* flavour depends on the sulphate content of the soil where the plant grows. (Eric Block, 2010) In the rare occurrence of sulphur-free growth conditions, all *Allium* species completely lose their usual pungency.

The precursors are odourless, stable, non-volatile amino acids which are responsible for volatile flavour, pungency, antioxidant activities, nutraceuticals and medicinal properties, and making it as medicinal herb, when cells are damaged, the amino acids break down under the influence of alliinase enzyme and lachrymatory-factor synthase into highly reactive sulphenic acids, ammonia and pyruvate and then sulphenic acids react with other compounds to form a range of disulphides (Jones *et al.*, 2004; Kamenetsky and Rabinowitch, 2006).

Compounds in Various Alliums

Volatile phytochemical constituents (non-organosulphur compounds) of *Allium cepa*, *A. sativum*, *A. schoenoprasum*, *A. tuberosum*, *A. ampeloprasum* and *A. fistulosum*. Source: Alam *et al.* (2022). \

No.	Compounds	Sources	Plant part(s)	References
1	Dipropyl monosulfide	*A. sativum*	Bulbs	Mikail, 2010; Martins *et al.*, 2016
2	Methyl allyl sulphide	*A. sativum*	Bulbs	Mikail, 2010; Martins *et al.*, 2016
3	Diallyl sulphide	*A. schoenoprasum*	Leaves	Singh *et al.*, 2018
4	Dimethyl disulfide	*A. tuberosum*	Aerial parts	Jannat *et al.*, 2019
		A. schoenoprasum	Leaves	Singh *et al.*, 2018

No.	Compounds	Sources	Plant part(s)	References
5	Methyl-propyldisulphide	*A. schoenoprasum*	Leaves	Block, 1992; Singh *et al.*, 2018
6	Pentyl hydrodisulfide	*A. schoenoprasum*	Leaves	Block, 1992; Singh *et al.*, 2018
7	Methyl pentyldisulfide	*A. schoenoprasum*	Leaves	Block, 1992; Singh *et al.*, 2018
8	Dipropyl disulfide	*A. schoenoprasum*	Leaves	Block, 1992; Singh *et al.*, 2018
		A. cepa	Bulbs	Teshika *et al.*, 2019
		A. sativum	Bulbs	Mikail, 2010; Martins *et al.*, 2016
9	Methyl-1-propenyl disulfide	*A. schoenoprasum*	Leaves	Singh *et al.*, 2018
		A. tuberosum	Aerial parts	Jannat *et al.*, 2019
10	Allyl methyl disulfide/ Methyl-2-propenyl disulfide	*A. tuberosum*	Aerial parts	Jannat *et al.*, 2019
		A. sativum	Bulbs	Martins *et al.*, 2016
		A. cepa	Bulbs	Teshika *et al.*, 2019
11	1-propenyl propyl disulfide/Cis-propenyl propyl disulfide	*A. schoenoprasum*	Leaves	Singh *et al.*, 2018)
		A. cepa	Bulbs	Teshika *et al.*, 2019
12	Trans-propenyl propyl disulfide	*A. cepa*	Bulbs	Teshika *et al.*, 2019
13	Allyl propyl disulfide	*A. schoenoprasum*	Leaves	Singh *et al.*, 2018
14	Diallyl disulfide	*A. tuberosum*	Aerial parts	Jannat *et al.*, 2019
		A. sativum	Bulbs	Mikail, 2010; Martins *et al.*, 2016
15	Methyl-1-(methylthio) ethyl disulfide	*A. schoenoprasum*	Leaves	Singh *et al.*, 2018
16	Methyl-1-(methylthiopropyl) disulfide	*A. schoenoprasum*	Leaves	Singh *et al.*, 2018
17	Dimethyl trisulfide	*A. schoenoprasum*	Leaves	Singh *et al.*, 2018
		A. tuberosum	Aerial parts	Jannat *et al.*, 2019
18	Methyl propyl trisulfide	*A. schoenoprasum*	Leaves	Singh *et al.*, 2018
		A. cepa	Bulbs	Teshika *et al.*, 2019
19	Dipropyl trisulfide	*A. schoenoprasum*	Leaves	Singh *et al.*, 2018
		A. cepa	Bulbs	Teshika *et al.*, 2019
		A. sativum	Bulbs	Mikail, 2010; Martins *et al.*, 2016

No.	Compounds	Sources	Plant part(s)	References
20	Methyl-1-propenyl trisulfide	*A. schoenoprasum*	Leaves	Singh *et al.*, 2018
21	Allyl methyl trisulfide	*A. tuberosum*	Aerial parts	Jannat *et al.*, 2019
		A. schoenoprasum	Leaves	Singh *et al.*, 2018
22	Diallyl trisulfide (DATS)	*A. sativum*	Bulbs	Martins *et al.*, 2016
23	Di-1-propenyl trisulfide	*A. schoenoprasum*	Leaves	Singh *et al.*, 2018
24	Dimethyl tetrasulfide	*A. schoenoprasum*	Leaves	Singh *et al.*, 2018
		A. ampeloprasum	Bulbs	Anides *et al.*, 2019
25	Dipropyl tetrasulfide	*A. schoenoprasum*	Leaves	Singh *et al.*, 2018
		A. sativum	Bulbs	Mikail, 2010; Martins *et al.*, 2016
26	Allicin	*A. sativum*	Bulbs	Martins *et al.*, 2016
		A. cepa	Bulbs	Zamri and Abd, 2019
27	S-allyl cysteine	*A. sativum*	Bulbs	Martins *et al.*, 2016
		A. cepa	Bulbs	Zamri and Abd, 2019
28	S-allyl cysteine sulfoxide	*A. sativum*	Bulbs	Martins *et al.*, 2016
		A. cepa	Bulbs	Teshika *et al.*, 2019
29	S-allylmercaptocysteine	*A. cepa*	Bulbs	Zamri and Abd, 2019
30	2-vinyl-4H-1,3-dithiin	*A. sativum*	Bulbs	Martins *et al.*, 2016
31	1,2-vinyldithiin	*A. sativum*	Bulbs	Martins *et al.*, 2016
32	Allixin	*A. sativum*	Bulbs	Martins *et al.*, 2016
33	Xanthiazone	*A. cepa*	Bulbs	Zamri and Abd, 2019
		A. ampeloprasum	Bulbs	Zamri and Abd, 2019
34	Xanthiside	*A. cepa*	Bulbs	Zamri and Abd, 2019
		A. ampeloprasum	Bulbs	Zamri and Abd, 2019
35	2-hydroxyxanthiside	*A. cepa*	Bulbs	Zamri and Abd, 2019
		A. ampeloprasum	Bulbs	Zamri and Abd, 2019
36	Entadamide A-β-D-glucopyranoside	*A. cepa*	Bulbs	Zamri and Abd, 2019
		A. ampeloprasum	Bulbs	Zamri and Abd, 2019
37	Glucoerucin	*A. cepa*	Bulbs	Zamri and Abd, 2019
38	Onionin A1	*A. cepa*	Bulbs	Nohara *et al.*, 2017
		A. fistulosum	Bulbs	Nohara *et al.*, 2017
39	Onionin A2	*A. cepa*	Bulbs	Nohara *et al.*, 2017
		A. fistulosum	Bulbs	Nohara *et al.*, 2017
40	Onionin A3	*A. cepa*	Bulbs	Nohara *et al.*, 2017
		A. fistulosum	Bulbs	Nohara *et al.*, 2017
41	Garlicnin B1	*A. sativum*	Bulbs	Nohara *et al.*, 2017
42	Garlicnin B2	*A. sativum*	Bulbs	Nohara *et al.*, 2017
43	Garlicnin B3	*A. sativum*	Bulbs	Nohara *et al.*, 2017
44	Garlicnin B4	*A. sativum*	Bulbs	Nohara *et al.*, 2017

No.	***Compounds***	***Sources***	***Plant part(s)***	***References***
45	Garlicnin K1	*A. sativum*	Bulbs	Nohara *et al.*, 2017
46	Garlicnin K2	*A. sativum*	Bulbs	Nohara *et al.*, 2017
47	Garlicnin C1	*A. sativum*	Bulbs	Nohara *et al.*, 2017
48	Garlicnin C2	*A. sativum*	Bulbs	Nohara *et al.*, 2017
49	Garlicnin C3	*A. sativum*	Bulbs	Nohara *et al.*, 2017
50	Garlicnin A	*A. sativum*	Bulbs	Nohara *et al.*, 2017
51	Garlicnin I	*A. sativum*	Bulbs	Nohara *et al.*, 2017
52	Welsonin A1	*A. fistulosum*	Bulbs	Nohara *et al.*, 2017
53	Welsonin A2	*A. fistulosum*	Bulbs	Nohara *et al.*, 2017
54	Garlicnin J	*A. sativum*	Bulbs	Nohara *et al.*, 2017
55	Garlicnin G	*A. sativum*	Bulbs	Nohara *et al.*, 2017
56	Garlicnin L-1	*A. sativum*	Bulbs	Nohara *et al.*, 2017
57	Garlicnin L-2	*A. sativum*	Bulbs	Nohara *et al.*, 2017
58	Garlicnin L-3	*A. sativum*	Bulbs	Nohara *et al.*, 2017
59	Garlicnin L-4	*A. sativum*	Bulbs	Nohara *et al.*, 2017
60	E-Ajoene	*A. sativum*	Bulbs	Martins *et al.*, 2016
		A. cepa	Bulbs	Zamri and Abd, 2019
		A. ampeloprasum	Bulbs	Zamri and Abd, 2019
61	Z-Ajoene	*A. sativum*	Bulbs	Martins *et al.*, 2016
62	Garlicnin E	*A. sativum*	Bulbs	Nohara *et al.*, 2017
63	Garlicnin F	*A. sativum*	Bulbs	Nohara *et al.*, 2017
64	2-methyl-2-pentenal	*A. schoenoprasum*	Leaves	Block, 1992; Singh *et al.*, 2018
65	2-methyl-2-butenal	*A. schoenoprasum*	Leaves	Block, 1992; Singh *et al.*, 2018
66	Trans-2-Ethyl-3-methylthiophane	*A. ampeloprasum*	Bulbs	
67	(Z)-1-(Methylthio)-1-propene	*A. ampeloprasum*	Bulbs	
68	Dichloroacetic acid	*A. fistulosum*	Leaves	Ajayi *et al.*, 2019
69	1-Buten-3-yne, 1-chloro-, (Z)-	*A. fistulosum*	Leaves	Ajayi *et al.*, 2019
70	3-Ethyl-3-heptanol	*A. ampeloprasum*	Bulbs	
71	Nonane	*A. schoenoprasum*	Leaves	Singh *et al.*, 2018
72	n-Pentadecane	*A. ampeloprasum*	Bulbs	
73	n-Hexadecane	*A. ampeloprasum*	Bulbs	
74	4,6-dimethyldodecane	*A. ampeloprasum*	Bulbs	
75	α-farnesene	*A. schoenoprasum*	Leaves	Singh *et al.*, 2018
76	E-Beta farnesene	*A. schoenoprasum*	Leaves	Singh *et al.*, 2018
77	n-Heptadecane	*A. ampeloprasum*	Bulbs	
78	n-Octadecane	*A. ampeloprasum*	Bulbs	
79	n-Eicosane	*A. ampeloprasum*	Bulbs	

No.	Compounds	Sources	Plant part(s)	References
80	n-Heneicosane	*A. ampeloprasum*	Bulbs	
81	Hexadecenoic acid, ethyl ester	*A. ampeloprasum*	Bulbs	
82	Octadecanoic acid, ethyl ester	*A. ampeloprasum*	Bulbs	
83	Linoleic acid, ethyl ester	*A. ampeloprasum*	Bulbs	
84	Borneol	*A. schoenoprasum*	Leaves	Singh *et al.*, 2018
85	p-Menth-8(10)-ene-2,9-diol	*A. ampeloprasum*	Bulbs	
86	1-Isopropyl-1,2-cyclohexanediol/1,2-Cyclohexanediol, 1-(1-methylethyl)-cis-	*A. ampeloprasum*	Bulbs	
87	D-Limonene	*A. fistulosum*	Leaves	Ajayi *et al.*, 2019
88	α-Pinene	*A. fistulosum*	Leaves	Ajayi *et al.*, 2019
89	Sesquiphellandrene	*A. schoenoprasum*	Leaves	Singh *et al.*, 2018
90	Thymol	*A. fistulosum*	Leaves	Ajayi *et al.*, 2019
91	Phenol, 2,4-bis (1,1-dimethylethyl)-	*A. ampeloprasum*	Bulbs	
92	α-copaene	*A. schoenoprasum*	Leaves	Singh *et al.*, 2018
93	Caryophyllene	*A. schoenoprasum*	Leaves	Singh *et al.*, 2018
Saponins				
94	Tuberoside A	*A. tuberosum*	Seeds	Sang *et al.*, 1999; Jannat *et al.*, 2019
95	Tuberoside B	*A. tuberosum*	Seeds	Sang *et al.*, 1999; Jannat *et al.*, 2019
96	Tuberoside C	*A. tuberosum*	Seeds	Sang *et al.*, 1999; Jannat *et al.*, 2019
97	Tuberoside D	*A. tuberosum*	Seeds	Jannat *et al.*, 2019
98	Tuberoside E	*A. tuberosum*	Seeds	Jannat *et al.*, 2019
99	Tuberoside F	*A. tuberosum* +-	Seeds	Jannat *et al.*, 2019
100	Tuberoside G	*A. tuberosum*	Seeds	Jannat *et al.*, 2019
101	Tuberoside H	*A. tuberosum*	Seeds	Jannat *et al.*, 2019
102	Tuberoside I	*A. tuberosum*	Seeds	Jannat *et al.*, 2019
103	Tuberoside J	*A. tuberosum*	Seeds	Jannat *et al.*, 2019
104	Tuberoside K	*A. tuberosum*	Seeds	Jannat *et al.*, 2019
105	Tuberoside L	*A. tuberosum*	Seeds	Jannat *et al.*, 2019
106	Tuberoside M -		Seeds	Jannat *et al.*, 2019
107	Tuberoside N	*A. tuberosum*	Seeds	Jannat *et al.*, 2019
108	Tuberoside O	*A. tuberosum*	Seeds	Jannat *et al.*, 2019
109	Tuberoside P	*A. tuberosum*	Seeds	Jannat *et al.*, 2019
110	Tuberoside Q	*A. tuberosum*	Seeds	Jannat *et al.*, 2019
111	Tuberoside R	*A. tuberosum*	Seeds	Jannat *et al.*, 2019
112	Tuberoside S	*A. tuberosum*	Seeds	Jannat *et al.*, 2019

No.	Compounds	Sources	Plant part(s)	References
113	Tuberoside T	*A. tuberosum*	Seeds	Jannat *et al.*, 2019
114	Tuberoside U	*A. tuberosum*	Seeds	Jannat *et al.*, 2019
115	Ascalonicoside A1	*A. cepa*	Whole plants	Corea *et al.*, 2005
116	Ascalonicoside A2	*A. cepa*	Whole plants	Corea *et al.*, 2005
117	Ascalonicoside B	*A. cepa*	Whole plants	Corea *et al.*, 2005
118	Yayoisaponin A	*A. ampeloprasum*	Whole plants	Sata *et al.*, 1998
119	Yayoisaponin B	*A. ampeloprasum*	Whole plants	Sata *et al.*, 1998
120	Yayoisaponin C	*A. ampeloprasum*	Whole plants	Sata *et al.*, 1998
121	Aginoside	*A. ampeloprasum*	Whole plants	Sata *et al.*, 1998
122	Tropeoside A1	*A. cepa*	Whole plants	Corea *et al.*, 2005
123	22-O-methyl derivative of Tropeoside A1	*A. cepa*	Whole plants	Corea *et al.*, 2005
124	Tropeoside B1	*A. cepa*	Whole plants	Corea *et al.*, 2005
125	22-O-methyl derivative of Tropeoside B1	*A. cepa*	Whole plants	Corea *et al.*, 2005
126	Tropeoside A2	*A. cepa*	Whole plants	Corea *et al.*, 2005
127	22-O-methyl derivative of Tropeoside A2	*A. cepa*	Whole plants	Corea *et al.*, 2005
128	Tropeoside B2	*A. cepa*	Whole plants	Corea *et al.*, 2005
129	22-O-methyl derivative of Tropeoside B2	*A. cepa*	Whole plants	Corea *et al.*, 2005
130	Dioscin	*A. ampeloprasum*	Whole plants	Sata *et al.*, 1998
131	Tuberosine A [(25S)-5β-spirostan-2β,3β-diol 3-O-β-D-glucopyranoside]	*A. tuberosum*	Roots	Jannat *et al.*, 2019
132	Tuberosine B [(25S)-5β-spirostan 2β,3β,19-triol3-O-β-D-glucopyranoside]	*A. tuberosum*	Roots	Jannat *et al.*, 2019
133	Tuberosine C [(25S)-5β-spirostan-2β,3β-diol 3-O-α-L-rhamnopyranoyl-(1-4)-O-β-D-glucopyranoside]	*A. tuberosum*	Roots	Jannat *et al.*, 2019

No.	Compounds	Sources	Plant part(s)	References
134	(3β,5α,6β,25R)-6-[(β-D-glucopyranosyl) oxy]- spirostan-3-ylO-β-D-glucopyranosyl-(1?2)-O-[β- D-glucopyranosyl-(1?3)]-β-D-galactopyranoside	*A. ampeloprasum*	Bulbs	Adao *et al.*, 2011
135	(20S,25S)-spirost-5-en-3β,12β,21-triol 3-O-α-L-rhamnopyranosyl-(1–2)-β-D- glucopyranoside	*A. schoenoprasum*	Whole plants	Timite *et al.*, 2013
136	(20S,25S)-spirost-5-en-3β,11α,21-triol 3-O-α-L-rhamnopyranosyl-(1–2)-β-D- glucopyranoside	*A. schoenoprasum*	Whole plants	Timite *et al.*, 2013
137	Laxogenin 3-O-α-L-rhamnopyranosyl-(1–2)-[β-D-glucopyranosyl-(1–4)]-β-D- glucopyranoside	*A. schoenoprasum*	Whole plants	Timite *et al.*, 2013
138	Laxogenin-3-O-α-L-rhamnopyranosyl-(1–2)-β-D-glucopyranoside	*A. schoenoprasum*	Whole plants	Timite *et al.*, 2013
139	Diosgenin-3-O-α-L-rhamnopyranosyl-(1–2)-O-β-D-glucopyranoside (Prosapogenin A of dioscin)	*A. schoenoprasum*	Whole plants	Timite *et al.*, 2013
140	Diosgenin-3-O-β-D-glucopyranosyl-(1–4)-[α-L-rhamnopyranosyl-(1–2)]-β-D-glucopyranoside (deltonin)	*A. schoenoprasum*	Whole plants	Timite *et al.*, 2013
141	(25R)-5α-spirostan-3β,11α-diol3-O-β-D-glucopyranosyl-(1–3)-[β-D glucopyranosyl-(1–4)]-β-D-galactopyranoside	*A. schoenoprasum*	Whole plants	Timite *et al.*, 2013
142	(25R)-furost-5-en-3β,22α,26-triol 26-O-β-D-glucopyranosyl-3-O-α-L-rhamnopyranosyl-(1–2)-[β-D- glucopyranosyl-(1–4)]-β-D-glucopyranoside (deltoside)	*A. schoenoprasum*	Whole plants	Timite *et al.*, 2013
143	26-O-β-D-glucopyranosyl-(25 R)-3β,22x, 26-trihydroxyl-5α-furostane3-O-bchacotrioside	*A. tuberosum*	Seeds	Jannat *et al.*, 2019

No.	Compounds	Sources	Plant part(s)	References
144	26-O-β-D-glucopyranosyl-(25S)-3β,5β,6α,22x,26-pentahydroxyl-5β-furostane 3-O-α-L-rhamnopyranosyl-(1□4)-β-D-glucopyranoside	*A. tuberosum*	Seeds	Jannat *et al.*, 2019
145	3-O-α-L-rhamnopyranosyl-(1□4)-β-D-glucopyranosyl 3β,5β,6α, 16β-tetrahydroxypregnane 16-(5-O-β-D-glucopyranoyl-4(S)- methyl-5-hydroxypentanoic acid) ester	*A. tuberosum*	Seeds	Jannat *et al.*, 2019
Flavonoids				
146	Kaempferol	*A. schoenoprasum*	Leaves	Singh *et al.*, 2018
		A. fistulosum	Leaves	Vlase *et al.*, 2013
		A. ampeloprasum	Seeds and leaves	Abd and Ali, 2013
		A. cepa	Stems	Corea *et al.*, 2005
147	Quercetin	*A. schoenoprasum*	Leaves	Singh *et al.*, 2018
		A. sativum	Bulbs	Martins *et al.*, 2016
		A. cepa	Stems	Corea *et al.*, 2005; Teshika *et al.*, 2019; Fredotovic *et al.*, 2017
		A. ampeloprasum	Seeds and leaves	Abd and Ali, 2013
		A. fistulosum	Seeds and leaves	Vlase *et al.*, 2013
148	Myricetin	*A. schoenoprasum*	Leaves	Singh *et al.*, 2018
		A. cepa	Leaves	Fredotovic *et al.*, 2017
149	Luteolin	*A. schoenoprasum*	Leaves	Singh *et al.*, 2018
150	Isorhamnetin	*A. schoenoprasum*	Leaves	Singh *et al.*, 2018
		A. cepa	Whole plants	Fredotovic *et al.*, 2017
151	Naringenin	*A. schoenoprasum*	Leaves	Singh *et al.*, 2018
152	Taxifolin	*A. cepa*	Bulbs	Corea *et al.*, 2005
153	Isoquercetin	*A. schoenoprasum*	Leaves	Singh *et al.*, 2018; Teshika *et al.*, 2019
		A. fistulosum	Whole plants	Vlase *et al.*, 2013
154	Quercitrin	*A. fistulosum*	Whole plants	Vlase *et al.*, 2013

No.	Compounds	Sources	Plant part(s)	References
155	Rutin	*A. schoenoprasum*	Leaves	Singh *et al.*, 2018
156	Kaempferol 3-O-β-sophoroside	*A. tuberosum*	Aerial parts	Jannat *et al.*, 2019
157	Kaempferol 3,4-O-di-O-β-D-glucoside	*A. tuberosum*	Aerial parts	Jannat *et al.*, 2019
158	Astragalin/Kaempferol 3-β-D-glucoside	*A. schoenoprasum*	Leaves	Singh *et al.*, 2018
159	Isorhamnetin 3-β-D-glucoside	*A. schoenoprasum*	Leaves	Singh *et al.*, 2018
160	Isorhamnetin 3-O-galactopyranoside	*A. fistulosum*		Zolfaghari *et al.*, 2021
161	Spiraeoside/Quercetin-4?-monoglucoside	*A. cepa*	Stems	Fredotovic *et al.*, 2017; Teshika *et al.*, 2019
162	Quercetin-3, 4?-diglucoside	*A. cepa*	Stems	Fredotovic *et al.*, 2017; Teshika *et al.*, 2019
163	Quercetin-7,4?-diglucoside	*A. cepa*	Stems	Teshika *et al.*, 2019
164	Isorhamnetin-3,4?-diglucoside	*A. cepa*	Stems	Teshika *et al.*, 2019
165	Quercetin-3,7,4?-triglucoside	*A. cepa*	Stems	Teshika *et al.*, 2019
166	Taxifolin 7-glucoside	*A. cepa*	Bulbs	Corea *et al.*, 2005
167	Delphinidin-3,5-diglucoside	*A. cepa*	Stems	Teshika *et al.*, 2019
168	3-O-β-D-(2-O-feruloyl)-glucosyl-7, 40-di-O-β-D-glucosylkaempferol	*A. tuberosum*	Aerial parts	Jannat *et al.*, 2019
169	3-O-β-sophorosyl-7-O-β-D-(2-O-feruloyl) glucosyl kaempferol	*A. tuberosum*	Aerial parts	Jannat *et al.*, 2019
Anthocyanins				
170	Cyanidin 3-(6-malonylglucoside)	*A. schoenoprasum*	Flowers and stem	Fossen *et al.*, 2000; Singh *et al.*, 2018
		A. sativum	Leaves	Phan *et al.*, 2019
171	Cyanidin 3-(3-malonylglucoside)	*A. schoenoprasum*	Flowers and stem	Fossen *et al.*, 2000; Singh *et al.*, 2018
172	Cyanidin-3-O-glucoside	*A. schoenoprasum*	Flowers and stem	Fossen *et al.*, 2000; Singh *et al.*, 2018
173	Peonidin-3?-glucoside	*A. Cepa*	Bulbs	Fredotovic *et al.*, 2017; Teshika *et al.*, 2019
174	Malvidin-3`-glucoside	*A. cepa*	Bulbs	Fredotovic *et al.*, 2017; Teshika *et al.*, 2019
Phenolic compounds				
175	Protocatechuic acid	*A. cepa*	Stems	Teshika *et al.*, 2019
176	Vanillic acid	*A. ampeloprasum*	Seeds and leaves	Abd and Ali, 2013

No.	*Compounds*	*Sources*	*Plant part(s)*	*References*
177	Gallic acid	*A. schoenoprasum*	Roots, stalk and leaves	Parvu *et al.*, 2014; Singh *et al.*, 2018
		A. cepa	Stems	Teshika *et al.*, 2019
		A. ampeloprasum	Seeds and leaves	Abd and Ali, 2013
178	p-coumaric acid	*A. schoenoprasum*	Roots, stalk and leaves	Parvu *et al.*, 2014; Singh *et al.*, 2018
		A. fistulosum	Whole plants	Vlase *et al.*, 2013
		A. ampeloprasum	Seeds and leaves	Abd and Ali, 2013
179	Caffeic acid	*A. ampeloprasum*	Seeds and leaves	Abd and Ali, 2013
180	Ferulic acid	*A. schoenoprasum*	Roots, stalk and leaves	Parvu *et al.*, 2014; Singh *et al.*, 2018
		A. fistulosum	Whole plants	Vlase *et al.*, 2013
		A. cepa	Stems	Teshika *et al.*, 2019
181	Sinapic acid	*A. schoenoprasum*	Roots, stalk and leaves	Parvu *et al.*, 2014; Singh *et al.*, 2018
		A. fistulosum	Whole plants	Vlase *et al.*, 2013
182	N-coumaroyltyramine	*A. fistulosum*	Whole plants	Zolfaghari *et al.*, 2020
183	Moupinamide/N-feruloyl tyramine	*A. ampeloprasum*	Seeds and stems	Sadeghi *et al.*, 2013
		A. fistulosum	Whole plants	Zolfaghari *et al.*, 2020
184	N-caffeoyl tyramine	*A. ampeloprasum*	Seeds and stems	Sadeghi *et al.*, 2013
		A. fistulosum	Whole plant	Zolfaghari *et al.*, 2020
185	N-coumaroyltyrosine	*A. fistulosum*	Whole plant	Zolfaghari *et al.*, 2020
186	Persicoimidate	*A. ampeloprasum*	Seeds and stems	Sadeghi *et al.*, 2013
		A. fistulosum	Whole plants	Zolfaghari *et al.*, 2020
187	Fistuloimidate A	*A. fistulosum*	Whole plants	Zolfaghari *et al.*, 2020
188	Fistuloimidate B	*A. fistulosum*	Whole plants	Zolfaghari *et al.*, 2020
189	Chlorogenic acid	*A. cepa*	Stems	Teshika *et al.*, 2019
		A. ampeloprasum	Seeds and leaves	Abd and Ali, 2013
190	Tuberonoid A	*A. tuberosum*	Aerial parts	Jannat *et al.*, 2019

No.	Compounds	Sources	Plant part(s)	References
191	Tuberonoid B	*A. tuberosum*	Aerial parts	Jannat *et al.*, 2019
192	Tuberosine D	*A. tuberosum*	Roots	Jannat *et al.*, 2019
193	(7R, 8S)-dihydrodehydrodiconiferyl alcohol-di-9, 90-O-β-D-glucopyranoside	*A. tuberosum*	Whole plants	Jannat *et al.*, 2019
Organic acids and fatty acids				
194	Oxalic acid	*A. cepa*	Stems	Teshika *et al.*, 2019
195	Succinic acid	*A. cepa*	Stems	Teshika *et al.*, 2019
196	Malic acid	*A. cepa*	Stems	Teshika *et al.*, 2019
197	Tartaric acid	*A. cepa*	Stems	Teshika *et al.*, 2019
198	Citric acid	*A. cepa*	Stems	Teshika *et al.*, 2019
199	Ascorbic acid	*A. cepa*	Stems	Teshika *et al.*, 2019
200	Palmitic acid	*A. schoenoprasum*	Leaves	Shirshova *et al.*, 2013; Singh *et al.*, 2018
201	Linoleic acid	*A. schoenoprasum*	Leaves	Shirshova *et al.*, 2013; Singh *et al.*, 2018
		A. tuberosum	Seeds	
		A. tuberosum	Seeds	
202	Linolenic acid	*A. schoenoprasum*	Leaves	Shirshova *et al.*, 2013; Singh *et al.*, 2018
203	Oleic acid	*A. tuberosum*	Seeds	
204	Stearic acid	*A. tuberosum*	Seeds	
205	Arachidic acid	*A. tuberosum*	Seeds	
206	Docosanoic acid/Behenic acid	*A. tuberosum*	Seeds	
207	Tricosanoic acid	*A. tuberosum*	Seeds	
208	Tetracosanoic acid/ Lignoceric acid	*A. tuberosum*	Seeds	
Steroids				
209	Sitosterol	*A. schoenoprasum*	Whole plants	Shirshova *et al.*, 2013; Singh *et al.*, 2018
		A. fistulosum	Whole plants	Vlase *et al.*, 2013
210	Stigmasterol	*A. schoenoprasum*	Whole plants	Shirshova *et al.*, 2013; Singh *et al.*, 2018
		A. fistulosum	Whole plants	Vlase *et al.*, 2013
211	Cholesterol	*A. schoenoprasum*	Whole plants	Shirshova *et al.*, 2013
212	Campesterol	*A. schoenoprasum*	Whole plants	Shirshova *et al.*, 2013; Singh *et al.*, 2018

No.	Compounds	Sources	Plant part(s)	References
		A. fistulosum	Whole plants	Vlase *et al.*, 2013
Amino acids				
213	Tryptophan	*A. schoenoprasum*	Whole plants	Singh *et al.*, 2018
		A. tuberosum	Seeds	
214	Threonine	*A. schoenoprasum*	Whole plants	Singh *et al.*, 2018
		A. tuberosum	Seeds	
215	Leucine	*A. schoenoprasum*	Whole plants	Singh *et al.*, 2018
		A. tuberosum	Seeds	
216	Isoleucine	*A. schoenoprasum*	Whole plants	Singh *et al.*, 2018
		A. tuberosum	Seeds	
217	Lysine	*A. schoenoprasum*	Whole plants	Singh *et al.*, 2018
218	Methionine	*A. schoenoprasum*	Whole plants	Singh *et al.*, 2018
		A. tuberosum	Seeds	
219	Phenylalanine	*A. schoenoprasum*	Whole plants	Singh *et al.*, 2018
		A. tuberosum	Seeds	
220	Tyrosine	*A. schoenoprasum*	Whole plants	Singh *et al.*, 2018)
		A. tuberosum	Seeds	
221	Valine	*A. schoenoprasum*	Whole plants	Singh *et al.*, 2018
		A. tuberosum	Seeds	
222	Alanine	*A. tuberosum*	Seeds	
223	Histidine	*A. schoenoprasum*	Whole plants	Singh *et al.*, 2018
		A. tuberosum	Seeds	
224	Aspartic acid	*A. schoenoprasum*	Whole plants	Singh *et al.*, 2018
		A. tuberosum	Seeds	
225	Glutamic acid	*A. schoenoprasum*	Whole plants	Singh *et al.*, 2018
		A. tuberosum	Seeds	
226	Cystine	*A. tuberosum*	Seeds	
227	Arginine	*A. schoenoprasum*	Whole plants	Singh *et al.*, 2018

No.	Compounds	Sources	Plant part(s)	References
		A. tuberosum	Seeds	
228	Glycine	*A. schoenoprasum*	Whole plants	Singh *et al.*, 2018
		A. tuberosum	Seeds	
229	Proline	*A. schoenoprasum*	Whole plants	Singh *et al.*, 2018
		A. tuberosum	Seeds	
230	Serine	*A. schoenoprasum*	Whole plants	Singh *et al.*, 2018
Glutamyl peptides				
231	γ-Glutamyl-S-methyl-L-cysteine	*A. sativum*	Whole plants	Martins *et al.*, 2016
		A. cepa	Whole plants	Bora and Sharma, 2009
232	γ-Glutamyl-methionine	*A. cepa*	Whole plants	Bora and Sharma, 2009
233	γ-Glutamyl-isoleucine	*A. cepa*	Whole plants	Bora and Sharma, 2009
234	γ-Glutamyl-valine	*A. cepa*	Whole plants	Bora and Sharma, 2009
235	γ-Glutamyl-leucine	*A. cepa*	Whole plants	Bora and Sharma, 2009
236	γ-Glutamylphenylalanine	*A. cepa*	Whole plants	Bora and Sharma, 2009
237	γ-Glutamyl-tyrosine	*A. cepa*	Whole plants	Bora and Sharma, 2009
238	γ-Glutamyl-S-methyl-L-cysteine sulfoxide	*A. cepa*	Whole plants	Bora and Sharma, 2009
239	γ-Glutamyl-S-trans-(1-propenyl)-L-cysteine-sulfoxide	*A. cepa*	Whole plants	Bora and Sharma, 2009
240	Glutathione	*A. cepa*	Whole plants	Bora and Sharma, 2009
241	γ-Glutamyl-S-(2-carboxypropyl)-cysteinylglycine	*A. cepa*	Whole plants	Bora and Sharma, 2009
242	Cysteine-glutathione disulfide	*A. cepa*	Whole plants	Bora and Sharma, 2009
Vitamins				
243	Thiamin	*A. tuberosum*	Seeds	
		A. schoenoprasum	Whole plants	Singh *et al.*, 2018
244	Riboflavin	*A. tuberosum*	Seeds	

No.	Compounds	Sources	Plant part(s)	References
		A. schoenoprasum	Whole plants	Singh *et al.*, 2018
245	Vitamin C	*A. sativum*	Whole plants	Martins *et al.*, 2016
		A. schoenoprasum	Whole plants	Singh *et al.*, 2018
246	Pantothenic acid	*A. schoenoprasum*	Whole plants	Singh *et al.*, 2018
247	Vitamin B complex	*A. sativum*	Whole plants	Martins *et al.*, 2016
		A. schoenoprasum	Whole plants	Singh *et al.*, 2018
248	Vitamin B6	*A. schoenoprasum*	Whole plants	Singh *et al.*, 2018
249	Niacin	*A. tuberosum*	Seeds	
250	Vitamin A	*A. schoenoprasum*	Whole plants	Singh *et al.*, 2018
251	Vitamin E	*A. schoenoprasum*	Whole plants	Singh *et al.*, 2018
252	Vitamin K	*A. schoenoprasum*	Whole plants	Singh *et al.*, 2018
253	β-carotene	*A. schoenoprasum*	Whole plants	Singh *et al.*, 2018
Nucleosides				
254	Thymine	*A. tuberosum*	Roots	Jannat *et al.*, 2019
255	Adenine	*A. tuberosum*	Roots	Jannat *et al.*, 2019
256	Uridine	*A. tuberosum*	Roots	Jannat *et al.*, 2019
257	Thymidine	*A. tuberosum*	Roots	Jannat *et al.*, 2019
258	Guanosine	*A. tuberosum*	Roots	Jannat *et al.*, 2019
259	Adenosine	*A. tuberosum*	Roots	Jannat *et al.*, 2019
260	Deoxyadenosine	*A. tuberosum*	Roots	Jannat *et al.*, 2019

Uses

Irrespective of the differing aroma in various species, many of the *Allium* species are used as food plants. In most cases, both bulbs and leaves are edible which are rich in carbohydrates, vitamins and minerals (Pandey *et al.*, 2021). The commercial commodity obtained from *Allium* vegetables is routinely consumed afresh as well as processed in almost all the cultures of the world (Malik *et al.*, 2020). Cooking and consumption of plant parts are due to the large variety of textures and flavours whether strong or weak. Of the various alliums that have been cultivated from the earliest times, about a dozen species are economically important as crops or garden vegetables together with an increasing number of species and hybrids as important ornamental plants.

In Garhwal and Kumaon Himalayan regions of Uttarakhand of India, fresh young leaves and/or bulbs/pseudo-stems of many wild species are consumed raw, in soup, cooked like vegetable or pickled. Bulbs/pseudo-stem of *A. griffithianum*, *A. tuberosum*, *A. pratii* and *A. victorialis* are consumed raw, cooked or pickled; and the cloves/bulbs of *A. ampeloprasum* as well as *A. chinense* are pickled. In Pithoragarh region of Kumaon, tender leaves of *A. stracheyi* and *A. negianum* are consumed as pot-herb. Fleshy-fibrous roots of *A. hookeri* are consumed as vegetable in NEH region or in soups and pickles of *A. stracheyi* in the NW Himalaya, *A. fasciculatum* leaves and scapes (young inflorescence in Bhutan) are usually used as vegetable, in salad and in soups (Pandey *et al.*, 2021).

Sulphur compounds released from their breakdown along with other bioactives compounds such as flavonols and other phenols exhibit prophylactic and curative properties against many human ailments such as microbial infections, carcinogenesis and degenerative diseases (Liguori *et al.*, 2017; Nicastro *et al.*, 2015; Bisen and Emerald, 2016). In Charaka Samhita, garlic and onion are recorded as diuretics and anti-rheumatic agents as well as remedies for cardiac, gastrointestinal and ophthalmic disorders (Alam *et al.*, 2022). The characteristic smell of alliums is due to sulphur compounds which have beneficial effects on the circulatory, digestive and respiratory systems. Garlic is the most pungent of all alliums and highest in therapeutic values. To some extent, most alliums are antibiotic apart from containing some iron and vitamins.

On account of numerous ethnomedicinal records of *Allium* vegetables such as chives, garlic, leeks, onions and scallions being used in cancer, a clinical investigation was performed to ascertain their protective effects against prostate cancer. The study revealed that consumption of these vegetables (at least 10g per day) reduced the prevalence of prostate cancer compared to subjects who took less than 2.2g per day. The occurrence of stomach and esophageal cancer was also demonstrated to be minimized following the intake of *Allium* vegetables (Gao *et al.* 2018; You *et al.* 2019).

Toxicity

There are many poisonous species that are somewhat similar in appearance such as 'death camas' (*Toxicoscordion venenosum*) of North America for which dogs and cats are highly susceptible but poisonous species do not have distinctive scent of onions or garlic. Some people are allergic to garlic and certain other species of *Allium*.

Propagation

While most *Allium* species reproduce sexually with high degree of out-breeding, asexual reproduction is a norm in several species such as garlic, multiplier onion, tree onion, pran and many others (Brat, 1965) giving rise to their clones. Some alliums reproduce by forming little bulbs or 'offsets' around the old one as in case of 'multiplier onions', as well as by seed. In some cases white bulbs can be seen being pushed out of the soil. In late autumn or very early spring these bulbs can be carefully lifted and gently peeled off the offsets to replant straight into the soil. Several species can form bulbils in the flowerhead as in case of so-called tree onion or Egyptian onion (*A.* ×*proliferum*) where the bulbils may be a few but are large enough to be pickled (en.wikipedia.org/wiki/Allium).

Seed propagation is common in onion, leek, Japanese bunching onion, chives as well as in Chinese chives which produces apomictic seeds. The two basic seed-production systems in the biennial alliums are seed-to-seed and the seed-to-bulb-to seed cycles. The latter allows for roguing off-types thus assuring production of true-to-type seed. Seeds of almost all the alliums are short-lived therefore only fresh seeds should be used for germination. Only proper growth, harvest and storage conditions enable long-time storage. The main good storage practices include low temperatures, low humidity, minimum contamination with storage pests and gentle handling.

Cultural practices

Allium seeds are planted from mid-September through October. Like most spring-flowering bulbs, alliums require a period of cold weather in order to bloom. When grown for bulbs, they are long-season crops and therefore are harvested about nine months after seeding. These seeds are small and have very hard seed coats so they are slow to germinate. A well prepared moist seed-bed is important to obtain a good yield. Seed may be planted in drills (rows) on a garden row and allowed to mature in place, or they may be transplanted into a permanent row a few months later. Soaking seed for several hours in warm water will promote good germination.

Alliums thrive in locations having open sun with well-drained rich soil though will tolerate partial shade. Most alliums prefer to grow in dry sandy loam soils rather than the soggy one but will grow in clay provided it is amended to improve drainage. Good drainage is particularly important in the winter. The growers should turn the soil and prepare the beds high enough or on raised beds to facilitate good drainage. A complete fertilizer such as 8-24-24 NPK @ 1kg per 18m of row should be mixed into the bed.

When seeds are properly germinated these are thinned to proper spacing if the seedlings are directly planted into the garden bed. Since alliums grow less well when they have to fight for nutrients with their fellow plants, it is best to plant them 15-20cm apart in high-density plantings. Transplants are commonly planted for onions and leeks. Plants should be about the size of a thin pencil or smaller so that they are large enough to withstand the winter cold and shock of planting. Too large a plant going through winter may be more easily induced to bolt, that is, to send up a seed stalk prematurely which split the bulbs. Transplants should be planted in well developed beds. It is recommended to use a complete fertilizer like 8-8-8 NPK at the rate of 1kg and calcium nitrate 450g as side-dressing per 15m of row should be administered followed by immediate irrigation. Complete fertilizer is preferable in poor soils.

Dry bulb sets or cloves are planted and grown much like transplants. The standard method for calculating the ideal depth is to dig a hole three to four times of its polar diameter and place the bulb at the bottom with its pointed end up, *i.e.* root side down. It is recommended to water and fertilize as the plant begins to grow in early spring. On heavy or sticky soils, a little soil may be carefully loosened or pulled away from the plant to reduce restriction of the bulb. Grower should be careful not to dig or cultivate deeply and injure the shallow roots.

An organic mulch cover applied in late winter will help control weeds, conserve moisture and keep the soil loose around the bulb shoulders. Cooler spring soil of mulched alliums can also help produce larger bulbs. When growing leeks, it is advised to rake up soil or mulch heavily around the plant's neck to encourage a thick white stem 10-13 cm long for cooking.

Since alliums are shallow-rooted and grow slowly, they will be subject to weed pressure. Growing alliums with black plastic mulch is very effective in larger home gardens. The black film should cover the prepared row a week or more before planting. Black plastic mulch controls most weeds around the plants, *vis-a-vis* helps warming the soil which accelerate the growth naturally. In smaller gardens, pine leaves, hay, straw and dried leaves are effective mulching materials. Options for chemical weed control are limited when growing alliums. Therefore, shallow cultivation during developmental stage is best. Bermuda and crab grasses can be controlled as these emerge with the active ingredient sethoxydim (Poast, Hi-Yield Grass Killer) without injuring alliums. Broadleaf weed control options are limited to pre-emergence control only with trifluralin (Treflan or Miracle Gro Weed Preventer). This herbicide should be administered 5cm deep into the soil prior to transplanting allium bulbs.

Harvest

Shallots and young onions can be harvested for scallion use any time they reach sufficient size. With multiplying onions and shallots, the clumps may be pulled out, one stalk may be broken and then replanted. This stalk will reestablish and begin to divide again if the weather is not too hot. Leeks are harvested before bulbing but at this time their white and thick necks should be big enough to use, usually during spring. Use them in soups, stews or for general onion use. When weather becomes hot enough, to some extent all these alliums have formed some kind of bulbs. Allium bulbs are ready for harvest when they have stopped swelling. At that time the leaves lose their healthy colour, necks become soft and most of the tops lay over. Garlic bulbs will have cloved at harvest time. When two-thirds of the tops lie over, the plants can be pulled and the bulbs should be stored in a dark location with good airflow.

The species

Due to their medicinal properties, food flavourings and importance in vegetable making, several alliums have been cultivated since the time immemorial. In flavour and pungency, species differ greatly from each other though a few may be odourless so for naming such species the spelling of *Allium* has exactly been reverted back and named Muilla. Some of most economic species of the genus other than ***A. cepa*** (Onion; in Hindi: Pyaj) and ***A. sativum*** (garlic; in Hindi: Lahsun) are ***A. ampeloprasum*** (leek), ***A. ampeloprasum*** var. ***ampeloprasum*** (Levant garlic, Round-headed garlic), ***A. schoenoprasum*** syn. ***A. sibiricum*** (Chives), ***A. fistulosum*** (Scallion, Spring onion, Welsh onion), ***A. tuberosum*** (Cuchay, Chinese chives, Garlic chives), ***A. oschaninii*** (French shallots), ***A. tricoccum*** (Ramps, Wild leek), ***A. ursinum*** (Ramsons), ***A. scorodoprasum*** (Rocambole, Sand leek), ***A. hypsistum & A. przewalskianum*** (Jimbu), and various other ***Allium*** species that are called scallions.

References

Abd, F.R.A. and R.F.M Ali, 2013. Proximate compositions, phyto-chemical constituents, antioxidant activities and phenolic contents of seed and leaf extracts of Egyptian leek (*Allium ampeloprasum var. kurrat*). *Europ. J. Chem.*, **4**(3):185–90. doi: 10.5155/eurjchem.4.3.185-190.711.

Adao, C.R., B.P. da Silva and J.P. Parente, 2011. A new steroidal saponin with antiinflammatory and antiulcerogenic properties from the bulbs of *Allium ampeloprasum var. porrum. Fitoterapia*, **82**(8):1175–80. doi: 10.1016/j.fitote.2011.08.003.

Ajayi, G.O., M.A. Akinsanya, A.T. Agbabiaka, K.S. Oyebanjo, T.D. Hungbo and J.A. Olagunju, 2019. D-Limonene: A major bio- active constituent in *Allium fistulosum identified by GC-MS analysis. J. Phytopharmac.*, **8**(5):257–9. doi: 10.31254/ phyto.2019.8509.

Alam, A., Abdullah Al Arif Jahan, Md. Sazzadul Bari, Labony Khandokar, Md. Hasan Mahmud, Muhammed Junaid, Mohammed Salahuddin Chowdhury, Mohammad Forhad Khan, Veronique Seidel and Md. Areeful Haque, 2022. *Allium* vegetables: Traditional uses, phytoconstituents, and beneficial effects in inflammation and cancer. *Crit. Rev. Food Sci. & Nutr.*, pp. 1-32. DOI: 10.1080/10408398.2022.2036094.

Anon. 2022. *Allium ampeloprasum* - Wild Leek. In: First-Nature. https://www.first-nature.com/flowers/allium-ampeloprasum.php. Retrieved 22 December, 2022.

Anon. 2023. Cliff Onion. Illinois wild flowers. https://www.illinoiswildflowers.info/prairie/plantx/cl_onionx.htm. Retrieved 30 January, 2023.

Athar, N., G. Taylor, J. McLaughlin and J. Skinner, 2004. FOODfiles 2004. In, New Zealand Institute for Crop & Food Research Limited and New Zealand Ministry of Health. Source:https://www.researchgate.net/publication/265192188_The_nutritional_attributes_of_Allium_species.

Bisen, P.S. and M. Emerald, 2016. Nutritional and Therapeutic Potential of Garlic and Onion (*Allium sp.*). *Curr. Nutr. Food Sci.,* **12:** 190. DOI: 10.2174/1573401312666160608121954.

Block, E. 1992. The organosulfur chemistry of the genus *Allium* – Implications for the organic chemistry of sulfur. *Angewandte Chemie International Edition in English,* **31** (9):1135–78.

Block, E. 2010. Garlic and Other Alliums. The Royal Society of Chemistry, Cambridge, U.K., p. 12. https://en.wikipedia.org/wiki/Allium_dregeanum.

Block, E. 2013. Fifty years of smelling sulfur. *J. Sulfur Chem.*, **34**(1-2): 158-207. (https://www.sciencedirect.com/science/article/pii/S2667142522000811).

Bora, K. and A. Sharma, 2009. Phytoconstituents and therapeutic potential of *Allium cepa* Linn. - A review. *Pharmacognosy Reviews,* **3** (5):170.

Borlinghaus, J., F. Albrecht, M.C. Gruhlke, I.D. Nwachukwu and A.J. Slusarenko, 2014. Allicin: Chemistry and biological properties. *Molecules,* **19**(8): 12591-12618. (https://www.sciencedirect.com/science/article/pii/S2667142522000811).

Brat, V. 1965. Genetic Systems in *Allium* III. Meiosis and Breeding Systems. *Hered.,* **5** (20): 325–339. DOI: 10.1038/hdy.1965.47.

Corea, G., E. Fattorusso, V. Lanzotti, R. Capasso and A.A. Izzo, 2005. Antispasmodic saponins from bulbs of red onion, *Allium cepa* L. *J. agric. Food Chem.,* **53**(4):935–40. doi: 10.1021/jf0484040.

Dasgupta, S. 2006. Alliaceae. In: Fascicles of Flora of India, no. 23 (eds Singh, N.P. and M. Sanjappa). BSI, Calcutta, India, pp. 18-9. https://www.animalmedicalresearch.org/Vol.10_Issue-2_December_2020/FOLK-MEDICINAL%20USES%20OF.pdf.

De Sarker, D., M.A.T. Johnson, A. Reynolds and P.E. Brandham, 1997. Cytology of the highly polyploid disjunct species, *Allium dregeanum* (Alliaceae), and of some Eurasian relatives. *Bot. J. Linn. Soc.,* **124** (4): 361–73.

Eric Block. 2010. Garlic and Other Alliums. *The Lore and the Science.* Royal Society of Chemistry, pp. 100-106. ISBN 978-0-85404-190-9. https://en.wikipedia.org/wiki/Allium.

Fossen, T., R. Slimestad, D.O. Ovstedal and O.M. Andersen, 2000. Covalent anthocyanin-flavonol complexes from flowers of chive, *Allium schoenoprasum. Phytochem.,* **54**(3):317–23. doi: 10.1016/ s0031-9422(00)00102-3.

Fredotovic, Z., M. Sprung, B. Soldo, I. Ljubenkov, I. Budic-Leto, T. Bilusic, V. Cikes-Culic, and J. Puizina, 2017. Chemical composition and biological activity of *Allium cepa L. and Allium × cornutum* (Clementi ex Visiani 1842) methanolic extracts. *Molecules,* **22**(3):448. doi: 10.3390/molecules22030448.

Friesen, N., P. Hanelt, K. Hammer and H. Knupffer, 1992. The genus *Allium*: taxonomic problems and genetic resources. Systematics of the Siberian polyploid complex in subgenus *Rhizirideum* (*Allium*). Germany Institut für Pflanzengenetik und Kulturpflanzenforschung, Gatersleben. Proc. of intern. Symp., pp. 55-66, held at Gatersleben, Germany, 11–13 June 1991.

Fritsch, R.M. and N. Friesen, 2002. Evolution, Domestication and Taxonomy. In: *Allium* Crop Science: Recent Advances (eds Rabinowitch, Haim D. and Lessley Currah), CABI Pub., Wallingford, UK. doi:10.1079/9780851995106.0005. ISBN 0-85199-510-1. OCLC 228168061. S2CID 189956991.

Gao, Q., X.-B. Li, J. Sun, E.-D. Xia, F. Tang, H.-Q. Cao, and H. Xun, 2018. Isolation and identification of new chemical constituents from Chinese chive (Allium tuberosum) and toxicological evaluation of raw and cooked Chinese chive. *Food and Chem. Toxicol.,* **112**:400–11. doi: 10.1016/j.fct.2017.02.011.

Gledhill David, 2008. The Names of Plants. Cambridge University Press, pp. 43. ISBN 9780521866453.

Gohil, R.N. and A.K. Koul, 1981. Cytology of the Tetraploid *Allium chinense* G. Don. *Caryologia,* **34**(1): 73–81. DOI: 10.1080/00087114.1981.10796874.

Govaerts, R., S. Kington, N. Friesen, R. Fritsch, D.A. Snijman, R. Marcucci, P.A. Silverstone-Sopkin and S. Brullo, 2021. World Checklist of Amaryllidaceae. Facilitated by the Royal Botanic Gardens, Kew. http://apps.kew.org/wcsp/ [accessed 21 May 2021]

Hedges, L and C. Lister, 2007. The nutritional attributes of *Allium species. Crop & Food Research Confidential Report No. 1814.* New Zealand Institute for Crop & Food ResearchLimited.10.13140/2.1.4265.4402.https://www.researchgate.net/publication/265192188_The_nutritional_attributes_of_Allium_species.

https://en.wikipedia.org/wiki/Allium. en.wikipedia.org.

http://www.theplantlist.org/

Hu, G., Y. Lu and D. Wei, 2006. Chemical characterization of Chinese chive seed (*Allium tuberosum Rottl.*). *Food Chem.,* **99** (4):693–7. doi: 10.1016/j.foodchem.2005.08.045.

Huang, R.F., J.M. Xu and H.A. Yu, 1995. A study on karyotypes and their evolutionary trends in *Allium* Sect. *Bromatorrhiza Ekberg* (Liliaceae), *Cathaya,* **7**: 133-45.

Jannat, K., T. Rahman and M. Rahmatullah, 2019. Traditional uses, phytochemicals and pharmacological properties of *Allium tuberosum* Rottler ex spreng. *J. Med. Pl. Stud.,* **7**:214–20.

Jaume, P.J., L.J. Kelly, I.J. Leitch, W.J. Zomlefer and M.F. Fay, 2014. A Universe of Dwarfs and Giants: Genome Size and Chromosome Evolution in the Monocot Family Melanthiaceae. *New Phytol.,* **20**:1484–97. DOI: 10.1111/nph.12617.

Jo, J., P.M. Purushotham, K. Han, H.R. Lee, G. Nah and B.C. Kang, 2017. Development of a genetic map for onion (*Allium cepa* L.) using reference-free genotyping-by sequencing and SNP assays. *Front. Plant Sci.,* **8**: 1606. DOI: 10.3389/fpls.2017.01606.

Jones, M.G., J. Hughes, A. Tregova, J. Milne, A.B. Tomsett and H.A. Collin, 2004. Biosynthesis of the flavour precursors of onion and garlic. *J exp Bot.,* **55**:1903–18. http://www.kiran.nic.in/pdf/IJHF/Vol_30-2/8.pdf.

Kamenetsky, R. and H.D. Rabinowitch, 2006. The genus Allium: a developmental and horticultural analysis. *Hort. Rev.,* **32**:329–78. http://www.kiran.nic.in/pdf/IJHF/Vol_30-2/8.pdf.

Keusgen, M,, R.M. Fritsch, H. Hisoriev, P.A. Kurbonova and F.O. Khassanov, 2006. Wild *Allium* species (Alliaceae) used in folk medicine of Tajikistan and Uzbekistan. *J. Ethnobio Ethnomed.,* **2**:18-26. https://www.animalmedicalresearch.org/Vol.10_Issue-2_December_2020/FOLK-MEDICINAL%20USES%20OF.pdf.

Kiełkowska, A. 2012. Meiotic Irregularities in Interspecific Crosses within Edible Alliums. In Meiosis-Molecular Mechanisms and Cytogenetic Diversity (ed.

Swan, A.). IntechOpen. https://www.intechopen.com/books/meiosis-molecular-mechanisms-and-cytogeneticdiversity/meiotic-irregularities-in-the-interspecific-crosses-within-edible-alliums (accessed Aug 09, 2018).

Knud, R. 1998. Alliaceae. In: The Families and Genera of Vascular Plants, vol. III (ed. Klaus Kubitzki), pp. 70-78. Springer-Verlag: Berlin;Heidelberg, Germany. ISBN 978-3-540-64060-8.

Kratchanova, M., M. Nikolova, E. Pavlova, I. Yanakieva and V. Kussovski, 2010. Composition and properties of biologically active pectic polysaccharides from leek (*Allium porrum*). *J. Sci. Food. Agri.,* **90**(12): 2046-51. https://www.sciencedirect.com/science/article/pii/S2667142522000811).

Krest, I., J. Glodek and M. Keusgen, 2000. Cysteine sulfoxides and alliinase activity of some *Allium* species. *J. Agri. Food Chem.,* **48**(8): 3753-60. (https://www.sciencedirect.com/science/article/pii/S2667142522000811).

Leek. 2008. New World Encyclopedia. Retrieved 04:27, January 4, 2023 from https://www.newworldencyclopedia.org/p/index.php?title=Leek&oldid=787109.

Li, Q.Q., S.D. Zhou, X.J. He, Y. Yu, Y.C. Zhang and X.Q. Wei, 2010. Phylogeny and biogeography of *Allium* (Amaryllidaceae: Allieae) based on nuclear ribosomal internal transcribed spacer and chloroplast rps 16 sequences, focusing on the inclusion of species endemic to China. *Ann. Bot.,* **106** (5): 709–733. https://doi.org/10.1093/aob/mcq177.

Liguori, L., R. Califano, D. Albanese, F. Raimo, A. Crescitelli and M.D. Matteo, 2017. Chemical composition and antioxidant properties of five white onion (*Allium cepa* L.) landraces. *J. Food Qual.* Article ID 6873651, 9. DOI: 10.1155/2017/6873651.

Mabberley, D.J. 2017. Mabberley's Plant–book: A Portable Dictionary of Plants, their Classification and Uses (4th edn.). Cambridge University Press, Cambridge, England. https://www.animalmedicalresearch.org/Vol.10_Issue-2_December_2020/FOLK-MEDICINAL%20USES%20OF.pdf.

Maity, D., G.G. Maiti and A.S. Chauhan, 2018. Flora of Kanchenjunga Biosphere Reserve, Sikkim. Botanical Survey of India, Kolkata, pp. 668-70. https://www.animalmedicalresearch.org/Vol.10_Issue-2_December_2020/FOLK-MEDICINAL%20USES%20OF.pdf.

Malik, G., Ajmer Singh Dhatt and Ajaz Ahmed Malik, 2020. A Review of Genetic Understanding and Amelioration of Edible *Allium* Species. Food Reviews International. DOI: 10.1080/87559129.2019.1709202.

Martins, N., S. Petropoulos and I.C. Ferreira, 2016. Chemical composition and bioactive compounds of garlic (*Allium sativum* L.) as affected by pre- and post-harvest conditions: A review. *Food Chem.,* **211**:41–50. doi: 10.1016/j.*foodchem*.2016.05.029.

Meena, L.K., S.L. Bairwa, M. Kumari and M.K. Wadhwani, 2016. Performance of Onion in Bihar - an Economic Analysis. *Econ. Aff.,* **61**(2): 299–304. DOI: 10.5958/0976-4666.2016.00038.3.

Mikail, H. 2010. Phytochemical screening, elemental analysis and acute toxicity of aqueous extract of *Allium sativum* L. bulbs in experimental rabbits. *J. Medic. Pl. Res.,* **4** (4):322–6.

Mohan, M., S.Z. Haider, S.S. Gautam and P. Singh, 2019. *Allium stracheyi* Baker, the endangered and traditional medicinally important herb of Uttarakhand Himalaya, India: a review. *Res. Pharm.*, **9**: 01-05. https://www.animalmedicalresearch.org/Vol.10_Issue-2_December_2020/FOLK-MEDICINAL%20USES%20OF.pdf.

Nicastro, H.L., S.A. Ross and J.A. Milner, 2015. Garlic and Onions: Their Cancer Prevention Properties. *Cancer Prev. Res.* (Phila.), **8**(3): 181–9. DOI: 10.1158/1940-6207,CAPR14-0172.

Nohara, T., Y. Fujiwara, M. El-Aasr, T. Ikeda, M. Ono, D. Nakano and J. Kinjo, 2017. Antitumor allium sulfides. *Chem. & Pharmac. Bull.,* **65**(3):209–17. doi: 10.1248/cpb.c16-00844.

O'Donnell, G., R. Poeschl, O. Zimhony, M. Gunaratnam, J.B. Moreira, S. Neidle, D. Evangelopoulos, S. Bhakta, J.P. Malkinson and H.I. Boshoff, 2008. Bioactive Pyridine-N- oxide Disulfides from *Allium stipitatum. J. Nat. Prod.,* **72**(3): 360-65. (https://www.sciencedirect.com/science/article/pii/S2667142522000811).

Pandey, A., P.K. Malav, K.M. Rai and S.P. Ahlawat, 2021. Neodomesticates of the Himalayan *Allium* spices (*Allium* species) in Uttarakhand, India and studies on eco-geography and morphology. *Gen. Res. and Crop Evol.,* **68**: 2167-79.

Pandey, A., Pavan Kumar Malav, K.M. Rai and S.P. Ahlawat, 2022. Genus *Allium* L. of the Indian Region: A Field Guide for Germplasm Collection and Identification, 109p.. *ICAR-National Bureau of Plant Genetic Resources*, New Delhi, India. http://www.nbpgr.ernet.in/Downloadfile.aspx?EntryId=9197.

Parvu, A., M. Parvu, L. Vlase, P. Miclea, A. Mot and R. Silaghi-Dumitrescu, 2014. Anti-inflammatory effects of *Allium schoenoprasum* L. leaves. *J. Physiol. and Pharmac.,* **65**(2):309–15.

Phan, A.D.T., G. Netzel, P. Chhim, M.E. Netzel and Y. Sultanbawa, 2019. Phytochemical characteristics and antimicrobial activity of Australian grown garlic (*Allium sativum* L.) cultivars. *Foods,* **8**(9):358. doi: 10.3390/foods8090358.

Quattrocchi, U., 1999. CRC World Dictionary of Plant Names: Common Names, Scientific Names, Eponyms, Synonyms and Etymology. CRC Press, Boca Raton, Florida, USA, Vol. 1, p. 91. ISBN 978-0-8493-2673-8. https://en.wikipedia.org/wiki/Allium.

Rajkumar, H., R.K. Ramagoni, V.C. Anchoju, R.N. Vankudavath and A.U.Z. de novo Syed, 2015. Transcriptome analysis of *Allium cepa* L. (onion) bulb to identify allergens and epitopes. *PLoSONE,* **10**(8), e0135387. DOI: 10.1371/journal.pone.0135387.

Randle, W.M. and J.E. Lancaster, 2002. Sulphur Compounds in Alliums in Relation to Flavour Quality. In: Allium Crop Science: Recent Advances (eds Rabinowitch, H.D. and L. Currah), pp. 330. CABI publishing, U.K.. DOI: 10.1079/9780851995106.0329.

Sadeghi, M., B. Zolfaghari, M. Senatore and V. Lanzotti, 2013. Antifungal cinnamic acid derivatives from Persian leek *(Allium ampeloprasum* subsp. *persicum). Phytochem. Letters,* **6**(3):360–63. doi: 10.1016/j.phytol.2013.04.007.

Sang, S., A. Lao, H. Wang and Z. Chen, 1999. Furostanol saponins from *Allium tuberosum. Phytochem.,* **52**(8):1611–5. doi: 10.1016/ S0031-9422(99)00312-X.

Sanyat Misra and R.L. Misra, 2013. *Allium.* In: Commercial Ornamental Bulb Science, pp. 2-10. Westville Publishing House, New Delhi.

Sata, N., S. Matsunaga, N. Fusetani, H. Nishikawa, S. Takamura and T. Saito, 1998. New antifungal and cytotoxic steroidal saponins from the bulbs of an elephant garlic mutant. *Biosci., Biotechn., and Biochem.,* **62**(10):1904–11. doi: 10.1271/ bbb.62.1904.

Shah, N.C. 2014. Status of Cultivated & Wild *Allium* Species in India: A Review. *Scitech. J.,* **1**(9): 28–36.

Shirshova, T., I. Beshlei, V. Deryagina, N. Ryzhova and N. Matistov, 2013. Chemical composition of *Allium schoenoprasum* leaves and inhibitory effect of their extract on tumor growth in mice. *Pharmaceut. Chem. J.,* **46** (11):672–5. doi: 10.1007/ s11094-013-0867-8.

Shukla, S., M.A. Iquebal, S. Jaiswal, U.B. Angadi, S. Fatma, N. Kumar, R.S. Jasrotia, Y. Fatima, A. Rai and D. Kumar, 2016. The Onion Genomic Resource: A Genomics and Bioinformatics Driven Resource for Onion Breeding. *Plant Genet.,* **8**: 9–15. DOI: 10.1016/j.plgene.2016.09.003.

Singh, V., G. Chauhan, P. Krishan and R. Shri, 2018. *Allium schoenoprasum* L.: A review of phytochemistry, pharmacology and future directions. *Nat. Prod. Res.,* **32**(18):2202–16. doi: 10.1080/14786419.2017.1367783.

Srivastava, R.C. 1996. Liliaceae. In: Flora of Sikkim, Vol 1 (Monocotyledons) (eds Hajra, P.K. and D.M. Verma). Botanical Survey of India, Calcutta, pp. 142-66. https://www.animalmedicalresearch.org/Vol.10_Issue-2_December_2020/ FOLK-MEDICINAL%20USES%20OF.pdf.

Teshika, J.D., A.M. Zakariyyah, T. Zaynab, G. Zengin, K.R. Rengasamy, S.K. Pandian and M.M. Fawzi, 2019. Traditional and modern uses of onion bulb (*Allium cepa L.*): A systematic review. *Crit. Rev. Food Sci. & Nutrit.,* **59** (sup1):S39–S70. doi: 10.1080/10408398.2018.1499074.

Timite, G., A.C. Mitaine-Offer, T. Miyamoto, C. Tanaka, J.F. Mirjolet, O. Duchamp and M.A. Lacaille-Dubois, 2013. Structure and cytotoxicity of steroidal glycosides from *Allium schoenoprasum. Phytochem.*, **88**:61–6. doi: 10.1016/j.phytochem.2012.12.001.

Timothy Prince Chidike Ezeorba, Chinonso Anthony Ezema, Rita Onyekachukwu Asomadu, Christian Chiazor Chibuogwu, Wisdom Favour Chinedu Ezeorba, Chinenye Patience Ezeorba, Parker Elijah Joshua, 2022. Allistatin: A potent yet uncharacterized therapeutic nugget in *Allium sativum*. *Pharmacol. Res. - Modern Chinese Med., Vol.* 3, pp. 100-121. ISSN 2667-1425. https://doi.org/10.1016/j.prmcm.2022.100121; (https://www.sciencedirect.com/science/article/pii/S2667142522000811).

Traub, H.P. 1968. The subgenera, sections and subsections of *Allium L. Plant Life*, **24**: 147-63.

Vlase, L., M. Parvu, E. Parvu and A. Toiu, 2013. Phytochemical analysis of *Allium fistulosum* L. and *A. ursinum* L. *Digest J. Nanomaterials & Biostructures* (DJNB), **8**(1):457–67.

Xu, J and R.V. Kamelin, 2000. *Allium* Linnaeus. In: Flora of China. Vol. 24 (Flagellariaceae through Marantaceae) (eds Wu, X.Y. and P.H. Raven). Science Press, Beijing and Missouri Botanical Garden Press, St. Louis, pp. 165-202. https://www.animalmedicalresearch.org/Vol.10_Issue-2_December_2020/FOLK-MEDICINAL%20USES%20OF.pdf.

Xu, J.M., L. Yang and X.J. He, 1998. A study on karyotype differentiation of *Allium fasciculatum* (Liliaceae), *Acta Phytotaxon. Sinica*, **36**: 346-52.

You, B. R., J.M. Yoo, S.Y. Baek and M.R. Kim, 2019. Anti-inflammatory effect of aged black garlic on 12-O-t etradecanoylphorbol-13-acetate-induced dermatitis in mice. *Nutrit. Res. and Pract.*, **13** (3):189–95. doi: 10.4162/nrp.2019.13.3.189.

Zamri, N. and H. Abd Hamid, 2019. Comparative study of onion (*Allium cepa*) and leek (*Allium ampeloprasum*): Identification of organosulphur compounds by UPLC-QTOF/MS and anticancer effect on MCF-7 cells. *Plant Foods for Human Nutrit.*, **74**(4):525–30. doi: 10.1007/ s11130-019-00770-6.

Zhou, S.D., X.J. He, Y. Yu and J.M. Xu, 2007. Karyotype studies on twenty-one populations of eight species in *Allium* section *Rhiziridium*. *Acta Phytotax. Sinica*, **45**: 207-16.

Zolfaghari, B., Z. Yazdiniapour, M. Sadeghi, M. Akbari, R. Troiano and V. Lanzotti, 2020. Cinnamic acid derivatives from Welsh onion (*Allium fistulosum*) and their antibacterial and cytotoxic activities. *Phytochem. Anal.*, **31**: 84–90.

Zolfaghari, B., Z. Yazdiniapour, M. Sadeghi, M. Akbari, R. Troiano and V. Lanzotti, 2021. Cinnamic acid derivatives from Welsh onion (*Allium fistulosum*) and their antibacterial and cytotoxic activities. *Phytochem. Anal.*: PCA **32** (1):84–90. doi: 10.1002/pca.2924.

1

Onion (*Allium cepa* L.)

Subgenus: *Cepa*; **section**: *Cepa*.

Common names: Bulb onion, Common onion, Garden onion, Onion; in Hindi: *Pyaj*.

Synonyms: *Allium angolense* Baker, *A. aobanum* Araki, *A. ascalonicum* Auct., *A.a.* f. *rotterianum* Voss ex J.Becker, *A.a.* var. *condensum* Millan, *A.a.* var. *fertile* Millan, *A.a.* var. *sterile* Millan, *A. cepa* var. *aggregatum* G.Don, *A.c.* var. *anglicum* Alef., *A.c.* var. *argenteum* Alef., *A.c.* var. *bifolium* Alef., *A.c.* var. *crinides* Alef., *A.c.* var. *flandricum* Alef., *A.c.* var. *globosum* Alef., *A.c.* var. *hispanicum* Alef., *A.c.* var. *jamesii* Alef., *A.c.* var. *lisboanum* Alef., *A.c.* var. *luteum* Alef., *A.c.* var. *multiplicans* L.H.Bailey, *A.c.* var. *portanum* Alef., *A.c.* var. *praecox* Alef., *A.c.* var. *rosum* Alef., *A.c.* var. *sanguineum* Alef., *A.c.* var. *solaninum* Alef., *A.c.* var. *tripolitanum* Alef., *A.c.* var. *viviparum* (Metzg.) Alef., *A. cepaeum* St.-Lag., *A. commune* Noronha, *A. cumaria* Buch.-Ham. ex Wall., *A. esculentum* Salisb., *A. napus* Pall. ex Kunth, *A. nigritanum* A.Chev., *A. pauciflorum* Willd. ex Ledeb., *A. salota* Dostal, *Ascalonicum sativum* P.Renault, *Cepa alba* P.Renault, *C. esculenta* Gray, *C. pallens* P.Renault, *C. rubra* P.Renault, *C. vulgaris* Garsault, *Kepa esculenta* Raf. and *Porrum cepa* (L.) Rchb.

Introduction and origin

Allium in Latin is the ancient name for 'garlic' and the specific name *cepa* the 'onion'. Some authorities believe that onions might have been the first vegetable to have been domesticated (Dhaliwal, 2017). The word 'onion' comes from the Middle English unyun, from the French 'oignon' which itself originated from the Latin 'unio' meaning 'onion' (Aggie). *Allium cepa* having sharp-flavoured bulbs and other plant parts is believed to have originated in Central Asia (Dhaliwal, 2017). It is the most widely cultivated species of the genus *Allium* (Brewster, 1994b) which also contains several other cultivated edible species variously referred to as onions such as the **Japanese bunching onion** (*Allium fistulosum*), the **tree onion** (*A.* × *proliferum*) and the **Canada onion** (*Allium canadense*). The name 'wild onion' is applied to a number of *Allium* species but *A. cepa* is exclusively known from cultivation and its ancestral wild original form is not known, though escapes from cultivation have become established in some regions (McNeal *et al.,* 2002), and the related wild species occur in Central Asia *cum* Iran.

By Stephen Ausmus released by USDA ARS, with the ID D 723 -18 (next) ., common wikimedia. org/windex.php?curid=2140112

The most closely related wild species include *A. vavilovii* Popov & Vved from Turkmenistan and NE Iran, and *A. asarense* R.M. Fritsch & Matin being endemic to the NC region of Iran (Zohary and Hopf, 2000; Fritsch *et al.,* 2001; Anon., 2023a,b) where both are harvested from the wild for local use as a food. However, Zohary and Hopf (2000) warn that **there are doubts whether the *vavilovii* collections tested represent genuine wild material or only feral derivatives of the crop**. Closely or distantly related cultivated alliums include **chives** *(Allium schoenoprasum)*, **garlic** (*A. sativum* and *A. scordoprasum*), **leeks** (*A. porrum*), **Chinese onion** (*Allium chinense*), *etc.* Several other plants in the genus *Allium* also have the common name of onion such as the **Welsh onion** *(A. fistulosum)*, but when used without qualifiers, onion usually refers to *Allium cepa*.

Onion is one of the second most important commercial crops of India which is next to potato. India is the second largest onion growing country after China and it ranks third in export of onions after Netherlands and Mexico. The major onion producing Indian states are Maharashtra, Madhya Pradesh, Karnataka, Gujarat, Rajasthan, Bihar, Andhra Pradesh, Haryana, West Bengal, Uttar Pradesh, Chhattisgarh, Odisha, Tamil Nadu, Jharkhand and Telangana (vikaspedia.in). Because of the diverse climatic conditions, Indian onions are available round the year and are famous for their pungency.

For consumption, many onion varieties are cultivated and used around the world ~ from sweet to highly pungent *cum* red to white via yellow. It is hardy enough to be planted during winter season whether it is tropical, sub-tropical or temperate region. They are pungent when chopped as contain certain chemical substances which may irritate the eyes.

They are usually served cooked, as a vegetable or part of a prepared savoury dish, but can also be eaten raw or used to make pickles or chutneys. Onion plants are simple to grow and provide a number of benefits through companion planting ~ giving the flora and vegetable garden all-natural, organic defenses against pests and other detrimental effects.

Botany

Diploid chromosome number in *Allium cepa* is 2n=2x=16. It is a hardy cool-season perennial or biennial plant that commonly sets seeds during its second year but usually grown as an annual and harvested in its first growing season for bulbs though in its second growing season for seed production only through bulbs. Indian onions have two crop cycles, first harvesting starts in November to January and the second harvesting from January to May. Modern varieties typically grow to a height of 15-45cm (en.wikipedia.org/wiki/Onion) though certain can grow up to 100cm tall. The onion plant bears a fan of fleshy, cylindrical and hollow and yellowish to bluish-green leaves. They are at their broadest about a quarter of the way up beyond which they taper towards a blunt tip.

The base of each leaf is flattened with usually white sheath that grows out of the basal plate of a bulb. From the underside of the bulb plate, a bundle of fibrous roots extends for a short way into the soil. In the maturing process, the food reserves begin to accumulate in the leaf bases so bulb of the onion swells. Its green-white umbellate flowers are bisexual and star-shaped. The seeds are glossy black and triangular in cross section. The average pH of an onion is around 5.5. In case of overwintering of onions, in autumn or during spring the foliage dies down and the outer layers of the bulb become more dry, papery and brittle.

As mentioned through efloras.org, 2023, the traits of cultivated onion are:

Parts	*Description*
Bulbs	1–3, not rhizomatous, mostly depressed-globose, varying in size from cultivar to cultivar, 5–8 × 3–10cm, and composed of concentric, fleshy and enlarged leaf bases or scales. Outer leaf bases become scaly due to loosing the moisture while inner leaves thicken with the development of bulbs and these store carbohydrate and other nutrients
Outer coats	Enclosing 1 or more bulbs, white, yellow, red brown, yellowish brown, red, pink or green, membranous, without reticulation
Inner coats	White to pink, cells obscure to quadrate
Leaves	Persistent, 4–10, sheathing proximal 1/6–1/4 scape
Blades	fistulose, usually ± semicircular in cross section, 10–50cm × 4–20mm
Scape	Persistent, solitary, erect, fistulose, inflated below middle, 30–100cm × 3–20mm
Umbel	Erect, compact, persistent, globose, up to 500-flowered, bulbils occasionally found
Spathe bracts	2-3, caducous, 3–4-veined, ovate, ± equal, apex acute to acuminate
Flowering	Jun-Aug
Flowers	Stellate to campanulate to urceolate, 3–7 mm
Tepals	Erect to ± spreading, white to pink with greenish midveins, withering in fruit, margins entire, apex obtuse or acute, outer ovate, inner oblong
Stamens	Exserted
Anthers	White
Pollen	White
Ovary	Crestless
Style	Linear, ± equalling stamens
Stigma	Capitate, unlobed
Pedicel	1–5cm

Indian onion breeding programme

In India, systematic breeding programme was started at Pimpalgaon Baswant (Nashik, Maharashtra) in 1960 and later at ICAR-IARI, New Delhi. Two early varieties that were developed through selection, *viz*., 'N-2-4-1' and 'Pusa Red' still dominate the production of onion. This breeding programme was further intensified under a coordinated project by institutes of the Indian Council of Agricultural Research (ICAR) at National Horticultural Researc and Development Foundation and State Agricultural Universities (SAUs). As a result, 45 onion varieties (including two F_1 hybrids and 6 varieties of multiplier onion) have been developed and released (Lawande *et al.*, 2009). Some of these varieties are 'Agrifound Dark Red', 'Agrifound Light Red', 'Arka Bindu', 'Arka Kalyan', 'Arka

Kirthiman', 'Arka Lalima', 'Arka Niketan', 'Arka Pragati', 'Baifeng', 'Baswant 780', 'Bhima Dark Red', 'Bhima Light red', 'Bhima Red', 'Bhima Subhra', 'Bhima Super', 'Bhima Sweta', 'CO 1', 'CO 2', 'CO 3', 'CO 4', 'Daekwanhwang', 'DOGR Hyb.-1', 'DOGR Hyb.-2', 'DOGR Hyb.-3', 'DOGR Hyb.-4', 'DOGR Hyb.-5', 'DOGR Hyb.-7', 'DOGR Hyb.-50', 'Early Globe', 'Hisar 2', 'Hybrid-35', 'Hybrid 63', 'Hybrid Nun-3001', 'Hybrid optima', 'Hybrid Orient', 'Kalyanpur Red Round', 'Nasik Red', 'N 2-4-1', 'N 53', 'N 257-9-1', 'Punjab Selection', 'Pusa Madhav', 'Pusa Ratnar', 'Pusa Red', 'Pusa White Round', 'Pusa White Flat', 'Quer-Rich', 'S 48', 'Udaipur 101', 'Udaipur 102', 'Udaipur 103 ', 'VL 1', 'VL 3', *etc.*

Onion | **Onion flowers**

Callus in onion is the best explant source for genetic transformation, followed by shoot tips and root tips. Afterwards a new technique in 2006 was devised for onion transformation using *Escherichia coli* gene that encodes phosphomannose isomerase (pmi). Biennial habit of onion, coupled with longer time taken for breeding and difficulties in attaining/maintaining genetic uniformity (due to high degree of natural cross-pollination and rapid inbreeding depression) have made this crop unattractive to breeders for further improvement (Lawande *et al.*, 2009)

Major groups

Onions can be classified into three main groups, of which two are cultivars of *A. cepa* ~ Cepa group or common onions (*A. cepa* var. *cepa*) and Aggregatum group or multiplier onions (*A. cepa* var. *aggregatum*), and the third group comprises hybrids with *A. cepa* percentage ~ tree onion and triploid onion.

Cepa or common onion group (*A. cepa* var. *cepa* with chromosome number 2n(2x)=16**):**

These are grown from seed or seed-grown sets, have a single large bulb and contain vast majority of *A. cepa* cultivars. These are usually referred to as onions. This group includes majority of cultivated varieties grown for dry bulbs, as salad or for pickling (Fritsch and Friesen, 2002). They range from the pungent varieties used for dried soups and onion powder to the mild and hearty sweet onions such as the 'Vidalia' from Georgia (USA) or 'Walla Walla' from Washington that can be sliced and eaten raw on a sandwich. The cultivars in this group may vary in photoperiodic requirements as day-length triggers bulbing, storage life, flavour and skin colour (Brewster, 1994a).

Common onion varieties are normally available in yellow/Spanish brown, red and white colours. In India, generally red and pungent types are preferred whereas in Japan, Europe and North America ~ yellow and brown types are preferred. Onion varieties differ not only in bulb colour but also for their photoperiodic requirements for which they are classified as long day, short day and day-neutral cultivars.

Yellow or brown onions

Yellow or **brown onions** (in some of the European countries the brown ones are called red onions) are sweeter with many cultivars bred specifically to demonstrate this sweetness ('Bermuda', 'Cevennes', 'Vidalia', 'Walla Walla', *etc.*) and all these turn rich dark-brown when caramelised and give French onion soup its sweet flavour. Some yellow/Spanish brown Indian varieties include ~ 'Arka Pitambar', 'Early Grano', 'Phule Swarna', 'Spanish Brown', *etc.* (vikaspedia. in; nhb.gov.in). Very little work has been done in India for development of yellow onion varieties particularly for export. Only two varieties were developed, *viz.*, 'Phule Swarna' from MPKV, Rahuri (Maharashtra) and 'Arka Pitambar' from IIHR, Bangalore and were released at the state/institute level but with low yield than red commercial varieties (Lawande *et al.*, 2009).

Most of such works was done in European countries and USA. A new yellow-onion variety 'NuMex Starlite', developed by Corgan and Holland (1993) was resistant to bolting and *Pyrenochaeta terrestris*. The short-day type was obtained by 5 recurrent selections from 'Texas Grano 502 PRR'. Among 12 short-day onion cultivars assessed in Mexico at Hermosillo (Warid and Loaiza, 1996), all the yellow cultivars had high yields with jumbo-sized (7.5-10cm bulb diameter). 'Seville, 9003C', 'Bravo', 'Quest' and 'Sweet Perfection' gave highest marketable yields each with numerous bulbs of the 30 yellow cultivars tested over a 2-year trial (Shock *et al.*, 2000). 'Texas Grano 1015 Y', a mildly pungent, sweet, short-day yellow onion variety was developed by Pike *et al* (1988) through original single-bulb selection from 'Texas Early Grano 951' through 5 generations of selections.

Similarly, 'Texas Grano 1030 Y' was developed from F_2 selections of 'Texas Early Grano 502' x 'Ben Shemen' by Pike *et al* (1988) which is a mildly pungent late maturing short-day onion variety.

Red or purple onions

Red or **purple onions** are known for their sharp pungent flavour and are the onions of choice for everyday use in Asian cuisine ~ used raw and in grilling. Some popular red onion varieties of India include ~ 'Agrifound Dark Red', 'Agrifound Light Red', 'Arka Lalima', 'Arka Pragati', 'Bhima Dark Red', 'Bhima Raj', 'Bhima Red', 'Bhima Super', 'Punjab Selection' and 'Pusa Red' (vikaspedia.in; nhb.gov.in).

Yellow 'Bermuda' onion

White onions

White onions which are traditionally used in classic Mexican cuisine are milder in flavour and have a golden colour when cooked, and a particularly sweet flavour when sauteed. Some Indian varieties of white onion are 'Agrifound White', 'Bhima Safed', 'Bhima Shubra', 'Bhima Shweta', 'Pusa White Flat' and 'Pusa White Round' (vikaspedia.in; nhb.gov.in; Lawande *et al.*, 2009).

Performance of some white onion varieties developed in India

Sl. No.	*Variety*	*Source*	*TSS(%)*	*Average yield (t/ha)*
1.	Pusa White Round	IARI, New Delhi	11.13	30.0 - 32.5
2.	Pusa White Flat	IARI, New Delhi	10.00	32.5 - 35.0
3.	Udaipur 102	RAU, Udaipur	10.06	30.0 - 35.0
4.	Agrifound White	NHRDF, Nashik	10.76	20.0 - 25.0
5.	Phule Safed	MPKV, Rahuri	10.13	25.0 - 30.0
6.	PKV White	PDKV, Akola	09.55	25.0 - 30.0
7.	Gujarat White Onion	JAU, Junagadh	—	30.0 - 32.5
8.	N-257-9-1	Agril. Deptt., M.S.	10.00	25.0 - 30.0
9.	Punjab-48	PAU, Ludhiana	11.00	30.0 - 32.5
10.	V-12	Jain Food Park, Jalgaon	15.00	35.0 - 40.0

Sl. No.	Variety	Source	TSS(%)	Average yield (t/ha)
11.	Nimar Local	Land Race, M.P.	12.50	25.0 – 30.0
12.	Talaja Local	Land Race, Bhavnagar	12.00	25.0 – 30.0
Source: Lawande et al. (2009)				

Aggregatum group (*A. cepa* var. *aggregatum*)

Aggregatum group is also referred to as multiplier onions or eveready onion. The individual bulbs of aggregatum group of onions are smaller than most conventional onions of Cepa group (cultivariable.com). Another distinguishing feature of both groups is that multiplier onions are more commonly grown by replanting the bulbs than by starting from seed whereas the conventional biennial onions must be started from seed. Moreover, the multiplier onions can be replanted indefinitely and harvested annually which makes them easier to manage than biennial onions that must be grown from seed each year.

A single plant forms an aggregate cluster of several small lateral or daughter bulbs from a master or mother bulb. It is mainly propagated vegetatively by dividing and planting underground bulbs. The plant rarely produces flowers and in those rare cases when flowers are produced the pollen is fertile so seeds are produced (Jasna, 2013). Sexually propagated shallots are found to be more closely related to common onion than the vegetetively propagated shallots, which is grouped separately. It was once classified as a separate species ~ *A. ascalonicum*. Aggregatum onions have smaller bulbs as compared to those of common onions (Fritsch and Friesen, 2002).All taxa of *A. cepa* Aggregatum group are diploids (2n=2x=16).

On a global scale, Aggregatum group is a minor *Allium* crop. However, in SE Asia and some African countries where growing season is too short for the production of bulb onion and where onion seed is hard to produce, these vegetatively propagated onions are cultivated as an important substitute for bulb onions (Rabinowitch and Kamenetsky, 2002). In India, it is extensively cultivated mainly in Himalayas, NE hills, and in parts of S&E India (Pandey *et al.*, 2022).

Morphological characteristics of *A. cepa* and Aggregatum group

Morphological characteristics of *A. cepa* and Aggregatum group of onions grown in Croatia

Morphological characters	Cepa group	Aggregatum group
Bulbs	round, flattened at the poles, 5–10cm in diameter	2-5cm diameter, semicylindrical, well developed and gathered in clusters

Morphological characters	*Cepa group*	*Aggregatum group*
Stem	up to 1m tall, hollow, conical, inflated at the base and 3cm diameter	up to 70cm tall, hollow, conical, inflated at the base, 1–1.5cm diameter
Leaves	up to 40cm long, diameter up to 2cm, semicircular and flattened on upper side	up to 40cm long, diameter up to 1cm, semicircular with flattened upper side
Spathe	shorter than inflorescence, divided into 2–4 parts, mostly triform	shorter than inflorescence, divided into 2–4 parts
Inflorescences	hemispherical with 4-9cm diameter	dense, hemispherical with 3-5cm diameter
Perianth	star-like flower shape, segments white with a green stripe, segment size 3–4.5 × 2– 2.5mm	flower shape star-like, segments white with a green stripe, segment size 3×2 mm
Flower pedicels	equal, up to 4cm long	equal up to 2cm long
Stamens	filaments 4.5mm, outer simple, interior with extended toothed base.	filaments 3–4mm, outer and inner simple
Pistils	lower than the stamens	lower than the stamens
Source: Jasna (2013)		

Indian varieties of Multiplier onion

As per eagri.org1; vikaspedia.in, some Indian varieties of Multiplier onion are:

- **Co 1 (TNAU, Coimbatore):** 7-9 large and pink bulbs per clump; yield 9-10t/ha in 85 days; propagation generally through bulbs.
- **Co 2 (TNAU, Coimbatore):** 7-9 large and crimson bulbs per clump; yield 12t/ha in 65-70 days; resistant to purple blotch (ecoursesonline.iasri.res.in); propagation generally through bulbs.
- **Co 3 (TNAU, Coimbatore):** 8-10 pink bulbs per plant, average weight about 75g per lump; resistant to thrips (ecoursesonline.iasri.res.in), has good storage quality; yield 15.8t/ha in 65 days; propagation generally through bulbs.
- **CO 4 (TNAU, Coimbatore):** 9-13 attractive light pink bulbs per plant; average weight 90g per lump; yield 19t/ha in 65 days; moderately resistant to thrips (ecoursesonline.iasri.res.in); propagation generally through bulbs.
- **CO 5 (TNAU, Coimbatore):** free-flowering and seed setting type; propagation through seed and bulbs.

- **MDU 1** (TNAU, Coimbatore)
- **Agrifound Red (NHRDF, Dindigul)**: 5-6 light red bulblets; each clump weighing 65g with yield 18-20 t/ha in 65 days having TSS 15-16%.

Classification of Aggregatum group

Aggregatum group contains shallots and potato onions. Shallots are more common as compared to potato onions. As compared to shallots, potato onions are larger, splitting into fewer bulbs, and remain enclosed within the skin of the seed bulb longer. Intermediate forms exist (Fritsch and Friesen, 2002).

Shallots, French red shallots or Dutch shallots (in Hindi: Gandhana pyaj)

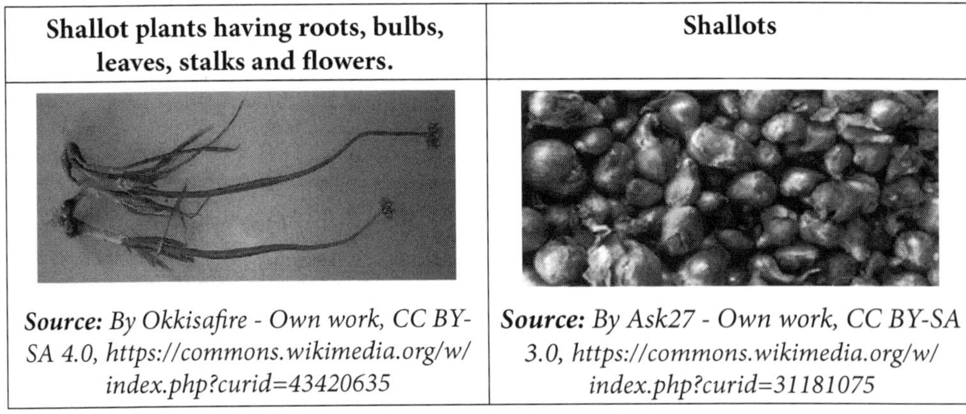

Shallot plants having roots, bulbs, leaves, stalks and flowers.	Shallots
Source: *By Okkisafire - Own work, CC BY-SA 4.0, https://commons.wikimedia.org/w/index.php?curid=43420635*	*Source:* *By Ask27 - Own work, CC BY-SA 3.0, https://commons.wikimedia.org/w/index.php?curid=31181075*

Introduction and origin

Shallots have probably originated in Central or SE Asia before making their way to India and the Mediterranean (masterclass.com, 2022). The ancient Greeks discovered shallots while trading in a Palestinian port now known as Ashkelon in Israel, while the ancient Egyptians used them as medicinal remedies (masterclass.com, 2022). Shallots are the most important subgroup within the Aggregatum group and comprise the only cultivars cultivated commercially. They form aggregate clusters of small, narrowly ovoid to pear-shaped bulbs and its clusters may contain as many as 15 bulbs (Moradi *et al.*, 2013). The skin colour of shallots can vary from golden-brown to grey to rose red, and their off-white flesh is usually tinged with green or magenta.

They are smaller in size and have milder flavour than the common onion. Like garlic, shallots are formed in clusters of offsets with a head composed of multiple cloves (en.wikipedia.org/wiki/Shallot). The biggest distinction between shallots and other onions, apart from size, shape, flavour and number of bulbs per plant, is the cellular structure; shallots breaking down much more easily than their larger counterparts when cooked, allowing for a meltier level of caramelization, or a more subtle touch for creating a foundation in things like sauces. Shallots generally have milder flavour and poor storage characteristics, while potato onions have a sharper flavour and store much better (cultivariable.com).

Etymology

The name 'shallot' originates from Ashkelon ~ an ancient Canaanite city where classical-era Greeks believed shallots originated (Green, 2004). The term **shallot** is usually applied to the **French red shallot** (*Allium cepa* var. *aggregatum* or the *A. cepa* Aggregatum Group) (en.wikipedia.org/wiki/Shallot) which is also called Dutch shallot (Iannotti, 2022). It is also used for the **Persian shallot** or **musir** (*A. stipitatum*) from the Zagros Mountains in Iran and Iraq, and the **French grey shallot** (*A. oschaninii*) which is also known as 'griselle' or 'true shallot' (Green, 2004). French grey shallot is a specific form of shallot long cultivated in France and Italy. It has greyish hue in its outer skin and is the ultimate culinary variety (masterclass.com, 2022).

French red shallots are more widely grown, easier to peel, and may be started from seed; whereas, the French grey shallots are almost always propagated by bulbs and are not easy to peel (sonomamg.ucanr.edu; kitchengardenseeds.com). The name shallot is also used for a scallion in New South Wales (Australia) and among English-speaking people in Quebec while the term French shallot refers to the plant which is described in this topic (en.wikipedia.org/wiki/Shallot). The term shallot in southern Croatia is additionally used for diploid and triploid viviparous onions, *viz. A.* × *proliferum* and *A.* × *cornutum* along with the onions in *A. cepa* Aggregatum group.

Indian Shallots

Shallots also called as Madras or *Sambar* onions in South India where these are used extensively in cooking (Fritsch and Friesen, 2002). Indian shallots are usually oval-shaped with the outer skin covered in an easy to peel paper thin layer. Once the outer skin is removed, the inner flesh of the shallot is tinged with red and purple hues (food.ndtv.com). Mainly used as flavouring for curries, the Indian shallots are like baby red onions but are more oval in shape and often have a pair of smaller bulbs inside the brown skin.

Unlike onions where each plant normally forms a single bulb, shallots, similar to garlic, form clusters of offsets. The flesh is usually pinkier than that of brown-skinned onions. They have a mild and sweeter flavour as compared to onions. Shallots pickled in red vinegar are often served in Indian restaurants along with *papads* (a salted, spiced and crisp thin circular cake made from ground pulse, sago, potato, *etc.*) and green chutney. These grow similar to garlic in a cluster and usually grow in pairs of bulbs inside the skin (food.ndtv.com). Fresh green shallots are available in the spring whereas dry shallots, which are dry skinned and have a moist flesh, are available throughout the year (food.ndtv.com).

Classification

Though there are several shallot varieties, these can be classified broadly on their colours and sizes.

- **Brown Shallots** are the most commonly found, especially in India.
- **Jersey shallot** (Pink or Jersey shallots) have a rosier colour and are the commonest ones found in U.S. Grocery Stores and markets (masterclass.com, 2022).
- **Echalion or banana shallots** are the cross between a regular shallot and a common onion having larger size of an onion and the milder flavour of a shallot (masterclass.com, 2022). These are the largest varieties and are named for their size (food.ndtv.com).

Varieties (as per Krishna Kumar *et al.*, 2015)

- **Red Sun**: French-type with golden-red skin and reddish interior rings, maturing in 80 days.
- **Gray Shallots**: 54 to 65 bulbs/kg.
- **Prisma**: Hybrid, red-to-the-core bulb having strong onion flavour with high-gloss finish, and long-day type, maturing in 100 days.
- **Bonilla**: Hybrid, round, straw-coloured, thick-skinned, very pungent, white-fleshed, 5cm bulbs, roducing 4 or 5 separate bulbs by mid-summer. Bulbs store well in spring and early summer. It matures in 100-105 days.
- **Olympus**: Hybrid, producing white-skinned *cum* white-fleshed 4 bulbs each with 5cm bulb size. It matures in 90 days and stores well.
- **Armador**: It produces only one deep red bulb with about 7.5cm diameter.
- **Ambition**: The bulb of this hybrid is a large French-style half-long teardrop-shaped with skin colour being reddish-copper though flesh creamy-white, which matures in 100 days. It is suitable for long-term storage through spring. Best grown in northern climate.
- **Holland Red**: Short, flat and plumpy bulbs bearing reddish skin and white-flesh tinged with purple. It stores well.
- **Matador**: It matures in 90 days bearing reddish-brown, round, thick-skinned and white-fleshed bulbs having 5cm bulb diameter.
- **French Red**: Here single bulb multiplies into 10 to 15 reddish-purple shallots in just one season of 100 days. It is easy to peel.

- **Mirage:** Large but half-long French style shallots with reddish-copper skin and white flesh which mature in 92 days. Bulbs are quite firm and can store through spring. It is adaptable to 35-65° latitude.
- **French Gray:** Plants produce purple-fleshed pear-shaped bulbs with tough, thick, grey-blue and wrinkled skins. It matures in 100 days.
- **Picador:** It is similar to var. 'Ambition'. It is a large half-long French-styled hybrid maturing in 105 days, producing single reddish-brown bulbs with white flesh. It stores well. It is adapted for 35-65° latitude.
- **Pikant:** It produces large French-styled mahogany bulbs with reddish flesh which matures in 80 days and stores well. It is adapted to 35-65° latitude.
- **Saffron:** This hybrid maturing in 100 days, produces bright copper-skinned bulbs with pale-yellow flesh which stores well. Well adapted for growing at 35-65° latitude.
- **Conservor:** This hybrid, similar to 'Ambition', produces single, reddish-brown bulbs, maturing in 110 days which stores well. It is adapted to 35-65° latitude.
- **Ed's Red:** This matures in 100 days producing 2.5-5cm red-skinned and red-fleshed bulbs, maturing in 100 days.

Potato onions

Multiplier, potato, Egyptian or underground onions (en.wikipedia.org/wiki/Potato_onion) differ from shallots in forming larger but fewer bulbs per cluster, and having a flattened (onion-like) shape. Shallots flower rarely while multiplier or potato onions ~ especially the white types flower annually or at irregular intervals though shallots on an average have two eyes per asymmetrical bulb, while multiplier onions ~ regardless of their size usually have only one eye whereas some varieties are intermediate between these two groups of onion (Krishna Kumar *et al.,* 2015).

They are tolerant to draught conditions. Fall planted potato onions usually get bigger than those planted in spring. These onions generally have sharper flavour and store much better as compared to shallots. Potato onions though usually have 1.2-5cm bulb diameter but some newer varieties produce even larger bulbs. It is a group of varieties which some authorities call *Allium* × *proliferum* but has also been classed in the Aggregatum Group of *Allium cepa*, similar to the shallot which sometimes produces round or irregular-shaped bulbs which in some old English varieties may be large (en.wikipedia.org/wiki/Potato_onion).

Onion

Sources: *finegardening.com, cultivariable.com*

I'itoi onion

I'itoi (pronounced 'EE-toy') onion is easy to grow a small-bulb type prolific multiplier onion adapted to the desert southwest conditions, having a shallot-like flavour. It is ideal for hot and dry climates. Its bulbs are separated and planted in fall 2.5cm below the surface and 30cm apart. Its bulbs multiply into clumps which can be harvested throughout the cooler months. Though tops die back due to summer heat but may return with heavy rains. The bulbs can remain in the ground or be harvested and stored in a cool dry place for planting in the fall. The plants rarely flower. It is propagated through division.

Bulbil forming onions - hybrids with *A. cepa* parentage

Several hybrids with *A. cepa* parentage have been cultivated including the diploid tree onion or Egyptian onion (*A.* × *proliferum*), and the triploid onion (*A.* × *cornutum*). In contrast to *A. cepa* Aggregatum group which has normal meiosis and produces fertile seed, hybrid onions *A.* × *proliferum* and *A.* × *cornutum* are sterile, and reproduce exclusively vegetatively by underground bulbs or bulbils from the inflorescence. The major dfference between Aggregatum group of onions and these two hybrid onions is the formation of bulbils in the umbel instead of flowers. Based on several morphological, cytogenetic and molecular studies, both these diploid and triploid hybrids are now recognized as a separate *Allium* crops and the new taxonomic units (Jasna, 2013).

Even though these diploid and triploid hybrids are now recognized as separate *Allium* crops and new taxonomic units, we find worth mentioning them here because they are closely related with the common onion, especially with *A. cepa* Aggregatum group. Moreover the karyotype of diploid *A.* × *proliferum* consists of one haploid *A. cepa* and one haploid *A. fistulosum* set of chromosomes, while triploid *A.* × *cornutum* consists of two sets from *A. cepa* and the third set from an unknown parent. A common morphological feature of *A. cepa* Aggregatum group and these two hybrids is the vegetative propagation using underground bulbs and a close relationship with the common onion.

Comparision of morphological characters of hybrids with parents

Morphological Characters	*A. cepa*	*A. × cornutum*	*A. × proliferum*	*A. fistulosum*
Bulbs	round, flattened at the poles, 5–10cm in diameter	elongated, pear-shaped, developed and gathered in clusters, 2–5cm diameter	elongated, less developed, 2–4cm in diameter	elongated, less developed, 1–2cm in diameter
Stem	up to 100 cm tall, hollow, conical, inflated at the base, 3cm in diameter	up to 1m tall, hollow, conical, slightly inflated at the base, diameter 3cm	80–150cm tall, hollow, conical, inflated at the base, 3–5cm in diameter	12–70cm tall, inflated in the middle, 1–2cm in diameter
Leaves	up to 40cm long, diameter up to 2cm, semicircular and flattened on the upper side	up to 40cm long, diameter up to 2cm, tubular, circular in cross-section	up to 50cm long, diameter up to 3cm, semicircular and flattened on the upper side	from 7 to 30cm long, diameter 0.5–1.5cm, cylindrical, circular in cross-section
Spathe	shorter than inflorescence, divided into 2–4 parts, mostly triform	shorter than inflorescence, usualy triform	significantly longer than the inflorescence, 1–2 parts	almost the same length as the inflorescence, 1–2 parts
Inflorescences	hemispherical, diameter 4–9cm	spherical to semispherical, with bulbils, diameter 4–5cm	hemispherical, numerous bulbils but flowering rare, diameter 3–5cm	hemispherical, diameter 1.5–5cm, dense
Perianth	flowers starry, segments white with a green stripe, segment size 3–4.5 × 2–2.5mm	Star-like flowers, segments white with a green stripe, segment size 3.5–4 × 3mm	flower shape more campanulate than starlike, segments white with a green stripe, segment size 5–7 × 3mm	closed shape of a flower, campanulate, segments yellowish-white, segment size 5–7 × 2mm
Flower pedicels	equal, up to 4cm long	Equal, 1–2cm long	Unequal, 0.3–7cm long	unequal, 0.3–2cm long

Stamens	filaments 4.5mm, outer simple, interior with extended toothed base.	filaments 3–4mm, outer simple filaments 4.5mm, inner simple, with a toothed inner base	filaments 4–7mm, outer and inner simple	filaments 8–12mm long, the outer and inner simple
Pistils	lower than the stamens	taller than the stamens	taller than the stamens	lower than the stamens
Source: Jasna (2013)				

Tree onion or Walking onion [*Allium* × *proliferum* (Moench) Schrad. ex Willd.]

Other common names: Egyptian onions, Top onions, Topsetting onions and Turfed stone leek.

Synonyms: *Allium cepa* var. *proliferum* (Moench) Regel, *Allium fistulosum* var. *viviparum* Makino, *Allium fistulosum* f. *viviparum* (Makino) M.Hiroe, *Allium multitabulatum* S. Cicina, *Allium multitabulatum* S. Cicina, *Allium* × *wakegi* Araki and *Cepa* × *prolifera* Moench

Introduction

Although some sources may still treat the tree onion as *A. cepa* var. *proliferum* or *A. cepa* 'Proliferum Group', genomic evidence has conclusively shown that tree onions are a diploid hybrid of the *A.cepa* and the Welsh onion (*A. fistulosum*) (Friesen and Klaas, 1998; Yamashita *et al.*, 2001; Brewster, 2008). Several karyological studies indicated its hybrid (allodiploid) nature and showed that the karyotype consists of one haploid *A. cepa* and one haploid *A. fistulosum* set of chromosomes (Bozzini, 1964; McCollum, 1974; Fiskesjo, 1975; Schubert *et al.*, 1983; Puizina and Papes, 1999). It has been postulated that the name 'Egyptian onion' has been derived from Romani people (Ruttle, 2011) bringing tree onions to Europe from the Indian subcontinent. It is described as a shallot which can be grown in tropical conditions (Jaime and Fer nando, 2007) and may be cultivated commercially for bulbs, bulbils and for foliage (Haim and Lesley, 2002; Vincent and Mas, 2012). It has been grown for centuries in Japan and China for use as a salad onion (Fritsch and Friesen, 2002).

Botany

Tree onions have chromosome numbers 2n=2x=16. These medium to large plants can grow up to 1m tall or even more in height with bright green, hollow, round, juicy and crunchy leaf stalk that can be harvested all year round. It is a hardy perennial that not only produces underground, slender, shallot-like cluster of bulbs attached to the common parental disc, but also develops a cluster of grape-sized, round to ovoid bulblets (also known as topsets or bulbils) at the top of stalk where a normal onion would have flowers. However, it sometimes bears small white sterile flowers in the spring. These bulblets are encased in a red or brown papery skin when young, eventually shedding the papery coverings with maturity (specialtyproduce.com). Depending on the variety being grown and growing conditions, the size of bulblets can be between 0.5-3cm in diameter, although the sizes may differ out of the specified range from time to time (sites.google.com). One leaf stalk can grow as many as 30 topsets (specialtyproduce.com). These bulbils start out tiny and green, but eventually grow larger with reddish to brown skins (hort.extension.wisc.edu). Tree onion bulblets will sprout and grow while still on the original stalk, forming a second, and sometimes even a third tier. The stalk may bend down under the weight of the new growth and fall over, creating new plants where the bulblets hit the ground, giving rise to the name, walking onion (en.wikipedia.org/wiki/Tree_onion; specialtyproduce.com). In this way the plant can 'walk' to establish plants in a broader area.

Bulblets forming (first and second tiers).	**Onion stalks with bulblets, or miniature onions grown at the top of the stalk.**

Source: *terrepromise.ca*	***Source:*** *en.wikipedia.org*

They are unlikely to produce bulbs or bulblets in their first year but will develop a good root system ready for producing bulbs and bulblets the following summer (sites.google.com). All parts are crispy and have a sharp, pungent and very strong spicy flavour; however, some relatively mild and sweet cultivars are also available (Ruttle, 2011).

The red/purple to white underground bulbs are tough-skinned, firm, crisp and can divide many times to form new clusters of bulbs (specialtyproduce.com). They are usually more pungent than bulblets and can be quite elongate or in some types may form bulbs up to 5cm across.

A sort of gigantism that characterizes this cultivar is its most striking feature, in which it also differs from other varieties of onions. Due to the rapid vegetative development, these perennial plants can grow up to the height of 130cm, and their flower pedicels can be 3–5cm long. The shape and structure of the stamen's filaments of tree onions more closely resemble *A. fistulosum*, especially the shape of the flower that is more campanulate and closed as in *A. fistulosum*, rather than stellar and open as in *A. cepa*. Stamen filaments are long, thin, with or without barely noticeable expansion on the basis of inter nal stamens. Anthers are often rudimentary or sterile. The leaf of tree onions resembles *A. cepa*, flattened on the inner side. A special feature of tree onion is the long spathe that envelops the inflorescence which it surmounts considerably. This novel feature is not characteristic of *A. cepa* or *A. fistulosum* (Jasna, 2013).

Uses

The stalks, bulbs and topsets are all edible and can be used in place of regular onions. The young leaves, shoots and stalks can be used similarly to scallions in soups, stews, omelets, stir-fries or as a finishing herb. The small size of the topsets makes them ideal for pickling. They can also be roasted whole, drizzled with oil and salt and roasted in their skin to seal in natural sugars. The bulbs can be chopped, sliced or minced and cooked with vegetables, roasted meat or brown rice and they can be fried as a crunchy side dish. The flowers can also be crumbled over salads, eggs or mushrooms for added flavour (specialtyproduce.com). Its medicinal values may be more or less same as of common onions.

Nutritional Value

Walking onions contain vitamins A, C, and K, folate, potassium, calcium and phosphorus. Moreover, it may have more or less same nutritional composition and phytochemicals as those present in common onion and/or Welsh onion.

Triploid onion (*Allium × cornutum* Clementi ex Vis. with chromosome numbers 2n=3x=24)

Triploid *A. × cornutum* which probably originated in Kashmir region of norther n India is a rather widespread garden crop in SE Asia, Europe and other parts of the world (Jasna, 2013). Based on several morphological, cytogenetic and molecular studies, it is now recognized as a separate *Allium* crop and a new taxonomic unit (Jasna, 2013). It was previously named *A. cepa* var. *viviparum*; however, taking into account its hybridogenic origin, later it was named *A. × cornutum* which is a hybrid species with three sets of chromosomes ~ two sets from *A. cepa* and one set from an unknown parent (Fritsch and Friesen, 2002). Several clones of the triploid onion are being grown locally in different regions such as 'Ljutika' in Croatia whereas 'Pran' in Poonch and Srinagar in the India-Kashmir region. There are very small genetic differences between 'Pran' and the

Croatian clone 'Ljutika', implying a monophyletic origin for this species (Friesen and Klaas, 1998). 'Pran' is grown extensively in Jammu and Kashmir.

Unlike the most flowering species of *Allium* in which the leaves are already dying back at flowering time, the leaves of triploid *A.* × *cornutum* are perennials remaining green and suitable for use throughout the year. Similar to majority of other alliums, it blooms during May and June though plants being sterile do not set seeds. Hence these are propagated only vegetatively by underground bulbs and aerial inflorescence bulbils (Jasna, 2013).

These sterile triploids do not exhibit the same effect of gigantism as tree onions. It is even slightly smaller than the diploid *A. cepa*. Although its bulbs and leaves are very tasty and used in the diet, there is a problem of their prolonged storage after harvesting as these quickly produce new leaves and roots thus losing weight immediately. Because of this reason the bulbs are usually kept by preserving in vinegar (Jasna, 2013).

Hybrids

Hybrids with *A. fistulosum* (Welsh onion)

Several old hybrids between *Allium fistulosum* as female parent and *A. cepa* 'Aggregatum Group' (tropical shallots) exist. These are known as *Allium* × *proliferum*, *Allium* × *wakegi* or *Allium fistulosum* var. *caespitosum* such as Indonesian shallot 'Sumenep' which is well adapted to tropical conditions which though never flowers but produces top onion in which the inflorescence is composed of bulbils. In the vegetative stage these plants resemble *A. fistulosum*.

Recent commercial hybrids between *A. fistulosum* as female parent and *A. cepa* as male are grown mainly in the United States and Europe for their green tops and these include 'Beltsville Bunching' (a seed-producing tetraploid), 'Louisiana Evergreen' (a sterile diploid) and 'Delta Giant' (a sterile triploid originating from a backcross of the hybrid with *A. cepa*). Another one is 'Yakura Negi Group' of hybrids grown in Japan. Plants of this group tiller abundantly in spring and summer though remain dormant in winter. Their inflorescences produce no flowers but only bulbils which are used for propagation apart from division of underground bulbs (Oyen and Messiaen, 2004).

Zimmu (*Allium cepa* L. × *Allium sativum* L.)

Interspecific hybrids between *A. cepa* (as female) and a fertile clone of *A. sativum* as male parent were obtained and this hybrid is known as Zimmu. The hybridity of the plants obtained was examined by morphological observation, chromosome analysis and ribosomal RNA gene analysis. The analyses proved that the plants were mature sexual hybrids between *A. cepa* and *A. sativum*. Each hybrid plant had keels though with fistulose leaves and formed a bulb resembling to that of *A. cepa*. The hybrids produced not only S-propenyl-L-cysteine sulfoxide which is the major flavour precursor in *A. cepa* but also S-allyl-L-cysteine sulfoxide (alliin) which is characteristic of *A. sativum* (Ohsumi *et al.*, 1993). Of the various plant extracts, the leaf extract of zimmu (*Allium cepa* × *Allium sativum*) showed maximum antifungal activity against *Rhizoctonia solani* and recorded an inhibition zone of 12mm.

Leaf extract of zimmu was also effective in inhibiting the growth of other agronomically important fungal and bacterial pathogens, *viz. Aspergillus flavus, Curvularia lunata, Alternaria solani, Xanthomonas oryzae* pv. *oryzae*, *X. campestris* pv. *malvacearum* and *X. axonopodis* pv. *citri*. The antimicrobial compound was dissoluble in methanol, and the methanolic extract showed the absorption maxima at 210 nm and 230 nm. Phenolic compounds were present in greater amounts in methanol extract of zimmu. TLC analysis showed the appearance of two blue spots at R_f =0.65 and R_f =0.90. The compounds eluted at R_f =0.65 and R_f =0.90 by preparative TLC exhibited strong antifungal activity against *R. solani* (Satya *et al.*, 2007).

Cultivars and varieties

Numerous varieties of onions have been developed with a diversity of flavours (mild to pungent), colours (white, yellow, brown, red), sizes (small, medium, large) and shapes (globose, flat, turbinate, discoidal, pear-shaped, oblate and conical). Some of them have bulbils in the inflorescence as '**tree onion**'. Onions are also classified according to their origin such as Spanish, French, Italian, English, American, Dutch and Japanese varieties ~ **Spanish onions** are large, sweet and juicy with colour ranging from yellow to red; **Italian** or **cipollini onions** are flat and red, white or yellow and often have a mild, sweet flavour; **Bermuda onions** are large and flat with white or yellow colouration and fairly mild taste (britannica.com). Though Bermuda onion with varieties 'Yellow Bermuda', 'White Bermuda' and 'Crystal Wax' gained its popularity from the island of the same name but seeds were originally brought over from the Canary Islands before 1888.

The varieties used in Spain are also classified by their earliness ~ among the **early varieties** there are 'Babosa', 'Blanca Francesa' and 'Sangre de Buey'; among the **mean-early varieties** are 'Blanca del País', 'Amarilla de Cabeza', 'Grande Achatada' and 'De Liria'; and among the **late varieties** are 'Blanca Grande Tardía', 'De Grano' and 'Valenciana Tardía Recas' (frutas-hortalizas.com). The English and the Americans distinguish the onions by their flavour ~ the '**soft type**' are characterized by their sweet taste whereas the '**strong type**' has a hot taste. The 'soft types' are usually early varieties and with less contents in soluble solids whereas the 'strong types' are late varieties that provide more nutrients. The onions are moreover classified by their use such as **green onions**, **dry onions**, **small pickler onions**, *etc* (frutas-hortalizas.com).

Globe-shaped onions may be white, yellow or red (britannica.com). Depending on the matured size and the purpose for which the onion is used, these may be referred to as **pearl**, **boiler** or **pickler onions** but differ from true pearl onions which are a different species (*Allium ampeloprasum* var. *sectivum* or *A. ampeloprasum* 'Pearl-Onion Group') (Thompson, 1995). Some Indian varieties of 'Small Common Onion' include **Agrifound Rose** (NHRDF, Chickballapur, Karnataka), **Arka Bindu** (IIHR, Bangalore), *etc.* (vikaspedia.in; nhb.gov.in). Onions may be bred and grown to maturity at smaller sizes. The majority of onions grown for pickling are common onions (Fritsch and Friesen, 2002) that are grown to a small size suitable for pickling by planting at a high density (Brewster, 1994d).

Depending on the maturity level at which they are harvested, the onions can be classified as **green onions** or **green shallots** ~ immature onions without a bulb, **spring onions/scallions/salad onions** ~ when immature onions start to bulb ~ usually white in most varieties and **dry onions** ~ mature onions with a juicy flesh covered with dry, papery skin (Herbst, 2001). The mild-flavoured green onions and spring onions can be eaten raw or lightly cooked. They are sold in bunches and have long green leaves. The scallion is a term which is also associated with various members of the genus *Allium* that lacks a fully-developed bulb. The terms scallion and green onion are particularly used to refer to the Welsh onion (*Allium fistulosum*) which does not produce dry bulbs. These green and spring onions may be grown from the Welsh onion (*A. fistulosum*), as well as from *A. cepa*. Young plants of *A. fistulosum* and *A. cepa* look very similar but may be distinguished by their leaves which are circular in cross-section in *A. fistulosum* rather than flattened on one side (Brewster, 1994c).

Onion varieties developed by different organizations in India

S. No.	Organization	Onion variety	Bulb colour
1.	Agril. Dept., Maharashtra State	N-2-4-1	Red
		N-53	Red
		N-257-9-1	White
2.	Mahatma Phule Krishi Vidyapeeth (MPKV), Rahuri, Maharashtra	Baswant-780	Red
		Phule Safed	White
		Phule Swarna	Yellow
		Phule Samarth	Red

S. No.	Organization	Onion variety	Bulb colour
3.	Indian Agricultural Research Institute (IARI), New Delhi	Pusa Red	Red
		Pusa White Flat	White
		Pusa White Round	White
		Pusa Ratnar (fair in storage)	Red
		Pusa Madhavi (good storage)	Red
		Early Grano	Yellow
		Brown Spanish (Long day)	Brown
4.	Indian Institute of Horticultural Research (IIHR), Bangalore (Karnataka)	Arka Niketan	Red
		Arka Kalyan	Red
		Arka Bindu	Red
		Arka Pragati	Red
		Arka Pitambar	Yellow
		Arka Lalima (F1 hybrid)	Red
		Arka Kirtiman (F1 hybrid)	Red
5.	Haryana Agricultural University (HAU), Hissar	Hissar 2	Red
		HOS-1	Red
6.	National Horticultural Research and Development Foundation (NHRDF), Nashik, Maharashtra	Agrifound Dark Red	Red
		Agrifound Light Red	Red
		Agrifound White	White
		Agrifound Rose (pickling type, suitable for export)	Red
		Agrifound Red (Multiplier)	Red
		L-28	Red
7.	VPKAS, Almora	VL-1 (Long day)	Red
		VL-3 (long day)	Red
8.	Rajasthan Agricultural University (RAU), Rajasthan	Udaipur 101	Red
		Udaipur 102	White
		Udaipur 103	Red
9.	Chandrashekhar Azad University of Agriculture and Technology (CSAUAT), Kanpur (UP)	Kalyanpur Red Round	Red
10.	Punjab Agricultural University, (PAU), Ludhiana (Punjab)	Punjab Selection	Red
		Punjab Red Round	Red
		Punjab Naroya	Red
		Punjab-48	White
		Punjab White	White

S. No.	*Organization*	*Onion variety*	*Bulb colour*
11.	Tamil Nadu Agricultural University (TNAU), Coimbatore (Tamil Nadu)	CO-1, (Multiplier)	Red
		CO-2	Red
		CO-3	Red
		CO-4	Red
		MDU-1	Red
12.	Regional Agricultural Research Station (RARS), Durgapura (Rajasthan)	Rajasthan Onion -1	Red
		Aprita (RO-59)	Red
13.	Punjabrao Deshmukh Krishi Viswavidyalya, Akola (Maharashtra)	Akola White	White
14.	DOGR, Rajgurunagar	Bhima Raj	Red
		Bhima Super	Red
		Bhima Red	Red
Source: Lawande *et al.* (2009)			

AICRP recommended zones for some Indian varieties

Sl.No.	*Variety*	*Organization*	*Recommended zones**	*Year of identification*
1	Punjab Selection	PAU, Ludhiana	IV, VII & VIII	1975
2	Pusa RedI	ARI, N. Delhi	IV, VII, VIII	1975
3	Pusa Ratnar	IARI, N. Delhi	IV & VI	1975
4	S-131	IARI, N. Delhi	-	1977
5	N-257-9-1	Agrl. Dept., M.S.	-	1985
6	N-2-4-1	Agrl. Dept., M.S.	-	1985
7	Line-102	IARI, N. Delhi	I, IV, VI, VII	1987
8	Arka Kalyan	IIHR, Bangalore	IV, VI, VII, VIII	1987
9	Arka Niketan	IIHR, Bangalore	IV,VII, VIII	1987
10	Agrifound Dark Red	NHRDF, Nashik	IV	1987
11	VL-3	VPKAS, Almora	I	1990
12	Agrifound Light Red	NHRDF, Nashik	VI, VIII	1993
13	Punjab Red Round	PAU, Ludhiana	IV	1993
14	PBR-5	PAU, Ludhiana	VI	1997
15	L-28	NHRDF, Nashik	IV & VII	2006
16	HOS-1	HAU, Hissar	VI	2006
17	Bhima Raj	NRCOG, Rajgurunagar	VI	2007
18	Bhima Red	DOGR, Rajgurunagar	VII	2009

19	PKV White	PDKV, Akola	VI	2009
20	RHOR-S1	MPKV, Rahuri	VI, VIII	2009

*Details of Zones under AICRP Vegetables: **Zone I** = Himachal Pradesh & U.P. Hills, **Zone II** = West Bengal & Assam, **Zone III** = Sikkim, Meghalaya, Manipur, Nagaland, Mizoram, Tripura, Arunachal Pradesh and Andaman & Nicobar Islands, **Zone IV** = Punjab, Terai region of U.P. & Bihar, **Zone V** = Chhattisgarh, Orissa & Andhra Pradesh, **Zone VI** = Rajasthan, Gujarat, Haryana & Delhi, **Zone VII**= Madhya Pradesh & Maharashtra, and **Zone VIII** = Kar nataka, Tamil Nadu & Kerala.
Source: *Lawande et al. (2009)*

Rabi or (*rabi cum kharif*/late *kharif*) onion varieties

Source: Dhaliwal (2017); nhb.gov.in; ikisan.com/ka-onion-varieties.html; krishijagran.com (a)

Onion cultivars are also classified according to their growing seasons, *viz. rabi* season and *kharif* season cultivars.

- **Advance line-849**: The bulb of this line is globular-round-shaped with attractive red colouration maturing in 105-115 days after transplanting, producing 375-425q/ha. It is suitable to grow in *rabi* season in Central India. It has about 6-months storage capacity and moderately tolerant to *Stemphyllium* blight and purple blotch disease.
- **Advance Line-857**: Silvery white bulb of this line is round-globular which matures in 110-120 days after transplanting with an average yield of 350-400q/ha, having good storage capacity and with TSS (total suspended solids) of 14-15%. It is suitable for cultivation in rabi season in Central India.
- **Agrifound Light Red (NHRDF, Nasik):** It has been developed through selection from a local material. Its light red bulbs are 4-6cm in size, globular in shape with tight skin, containing 13% TSS (Dhaliwal, 2017), 14-15% dry matter contents and 12.20micromole/g pyruvic acid [krishijagran.com (a)]. Its average yield is 30-32.5t/ha, has good keeping quality with a fewer incidences of pre-mature bolting. Average yield is 30-32.5t/ha. It is recommended for growing during rabi all over the country. It is notified by Govt. of India vide notificastion number 1135(E), dated 10-02-19996 [krishijagran.com (a)] which can also be grown in late *kharif* season in Nasik (Maharashtra) (Dhaliwal, 2017).
- **Agrifound Red** (NHRDF, Dindigul): It is a light red variety with per cluster average bulb size of 7.15cm, number of bulblets per clump 5.79 with bulblet size of 3.94cm *cum* single bulblet weight being 8.85g (nhb.gov.in) and bulb weight of 65-67g [krishijagran.com (a)], TSS 15-16%, dry matter 18-20% and pyruvic acid 10.13micromole/g [krishijagran.

com (a)]. It matures in 65-67 days with 18-20t/ha bulb yield (nhb.gov.in). It is recommended for cultivation in both the seasons, *i.e. Kharif* and *Rabi* in Tamil Nadu, Karnataka and Kerala.

- **Agrifound Rose** (NHRDF, Chickballapur, Karnataka): It is a small pickling type common onion variety suitable for *Kharif* season in Cuddapeh district though in all the three seasons, *i.e. Rabi, Kharf* and late *Kharif* in Karnataka (nhb.gov.in), maturing in 95-110 days with average bulb yield of 19-20t/ha which is exclusively grown for dehydration [krishijagran.com (a)] and for export (nhb.gov.in). Its deep scarlet red bulbs are flattish round in shape and 2.5-3.5cm in size with TSS 15-16% (nhb.gov.in), dry matter 17-19% and pyruvic acid 10.27micromole/g.
- **Agrifound White** (NHRDF, Nasik): This tight-skinned and attractive silvery-white bulb is globular in shape having good keeping quality with an average yield of 20-25t/ha and maturing in 160-165 days (nhb.gov.in). It is suitable for growing during late *Kharif* and *Rabi* seasons and is recommended for Maharashtra, Madhya Pradesh and Gujarat [krishijagran.com (a)]. Its bulbs contain 14-15% TSS (nhb.gov.in), 15-16% dry matter and 12micromole/g pyruvic acid [krishijagran.com (a)]. It is a good variety for dehydration and export purposees (nhb.gov.in).
- **Arka Kirthiman F_1** (IIHR, Bangalore): It is a a F_1 hybrid of MS 65 × Selection 13-1-1 which is suitable for cultivation in both *Rabi* and *Kharif* seasons, producing medium-sized red globular bulbs with thin neck and 4-5 months long shelf life. It is usually harvested in about 120-130 days with average yield of 47t/ha. It is tolerant to purple blotch, basal rot and thrips (Dhaliwal, 2017).
- **Arka Lalima/Laolima** (IIHR, Bangalore): It is a F_1 hybrid of MS-18 × Sel. 14-1-1 and is suitable for cultivation in both *Kharif* and *Rabi* seasons. Its medium-sized red and globe-shaped bulbs have 4-5 months storage life and are red coloured, globe-shaped, medium-sized and firm-textured. It matures in about 130-140 days with an average yield of 50t/ha having 120-130g bulb weight. It is tolerant to purple blotch, basal rot and thrips (Dhaliwal, 2017).
- **Arka Niketan** (IIHR, Bangalore): It has been developed through mass selection from a local collection IIHR – 153 which is adaptable in VII & VIII climatic zones of India, *i.e.* Maharashtra, Madhya Pradesh, Karnataka and Tamil Nadu, maturing in 145 days with an average yield of 34t/ha. Plant grows to a height of 60-76cm producing attractively coloured thin-necked globular bulbs 4-6cm in size having good keeping

quality and containing 12-14% TSS with high pungency and dry matter content. Though, it is recommended for planting mainly in *Rabi* season but in Maharashtra it can be grown in late *Kharif* also (Dhaliwal, 2017; nhb.gov.in).

- **Arka Pitamber** (IIHR, Bangalore): It was developed through selection from the cross UD 102 × IHR 396 which is suitable for *Kharif* and *Rabi* season growing. Attractive yellow and medium-sized bulbs are globular with thin neck, mildly pungent and with good keeping quality (5 months storage life). It matures in 140 days with an average yield of 35t/ha where a single bulb weighs about 80g. It is bred for export purpose. TSS content of the bulb is 11%. It is moderately resistant to purple blotch, basal rot and thrips (Dhaliwal, 2017; nhb.gov.in).
- **Arka Pragati** (IIHR, Bangalore): It is recommended for growing in all the zones during *rabi* season. Its globular bulbs have 3-5cm size, maturing in 140-145 days with average bulb yield of 20t/ha. This contains 12-14% TSS. It has good keeping quality.
- **Co 1** (TNAU, Coimbatore): It is recommended for cultivation for zone VIII during *rabi* as well as *kharif* growing. Per clump it forms medium-sized 7-8 bulblets, maturing in 90 days with average yield being 10t/ha. It contains medium TSS.
- **Co 2** (TNAU, Coimbatore): It is recommended for zone VIII to be grown during *rabi* as well as *kharif* seasons. It matures in 65 days, producing some 7-9 bigger-sized bulblets per plant with an average yield of 12t/ha. It has good keeping quality. It contains high TSS content. It is moderately resistant to *Alternaria* blight and thrips.
- **Co 3** (TNAU, Coimbatore): It is recommended for growing during *rabi* and *kharif* seasons for zone VIII. It matures in 65 days with average bulblet yield of 16t/ha, producing 8-10 bigger-sized bulblets per clump. It has good keeping quality and contains 13% TSS. It is moderately resistant to thrips.
- **Co 4** (TNAU, Coimbatore): It is recommended for growing in zone VIII during *rabi* as well as *kharif* seasons. It matures in 60-65 days with average yield of 18t/ha, producing 8-10 bigger-sized bulblets per plant. It is moderately resistant to thrips.
- **Co-on 5** (TNAU, Coimbatore): It is recommended for growing under zone VIII during *rabi* and *kharif* seasons. Its propagation is through bulblets and seeds. It matures in 60-65 days producing bold, bigger and attractively pink bulblets some 18.9t/ha.

- **Early Grano** (IARI, New Delhi): It produces yellow, globular and quite pungent bulbs having 7-8cm size. Crop matures in 95-110 days with an average yield of 50-60t/ha though it does not store well. It contains 6-7% TSS, good for salad (nhb.gov.in) and is suitable for cultivation in the plains during *Rabi* and *Kharif* seasons.
- **Kalyanpur Red Round** (CSAUAT, Kalyanpur): It has been developed through mass selection from a local material and is recommended for growing in Uttar Pradesh, maturing in 120 days with average yield of 250-300q/ha (Dhaliwal, 2017). Its bronze-red globular bulbs are moderately pungent with 13-15% TSS content.
- **N-257-9-1** (MPKV, Rahuri, MH): Its white and globular bulbs have good keeping quality and is suitable for growing during *Rabi* season in onion growing season of Maharashtra. It matures in 140 days with average yield of 25t/ha (nhb.gov.in).
- **NHRDF-Red (LINE-28)**: Its attractively dark red bulbs having thick reddish inner scales are globular round in shape, maturing in 110-120 days with an average yield of 250-300q/ha. The variety being dark red with better storage performance, is quite popular among the farmers in North India and is notified by Govt. of India vide notification number SO-2035(e), dated 28/10/2001 for cultivation in Northern, Central and Western India in *Rabi* season [krishijagran.com (a)]. Its bulbs have medium keeping quality. Its bulbs contain 13-14% TSS, 14-15% dry matter and 12micromole/g pyruvic acid.
- **NHRDF-Red-2 (L-355)**: Its attractively red, globular and tight-skinned bulbs have thin necks with 5-6cm in diameter and mature in 100-120 days where bulbs have good keeping quality and bulb yield of 350-375q/ha. It is notified by Govt. of India notification No. SO-2363(E), dated 01/10/2012 for growing in *Rabi* season in zone III (Delhi, Uttar Pradesh, Haryana and Gujarat) and zone VI (Maharashtra, Karnataka and Andhra Pradesh) [krishijagran.com (a)]. Its bulbs contain 13-14% TSS, 14-15% dry matter and 12micromole/g pyruvic acid.
- **NHRDF Red 3 (L-652)**: Its bulbs being light bronze, globular round, thin-necked and 5.5-6cm in diameter have good keeping quality which mature in 120-130 days with average yield of 350-400q/ha. It is recommended for growing during *Rabi* season in Maharashtra, Madhya Pradesh, Gujarat, Delhi, Uttar Pradesh, Bihar, Punjab, Haryana and Rajasthan [krishijagran.com (a)]. Its bulbs have 12-13% TSS, 13-14% dry matter and 12.5micromole/g pyruvic acid.

- **NHRDF Red-4**: Its bulbs being dark red, globular round, thin-necked and 5.5-6.25cm in diameter with good keeping quality which mature in 110-120 days with an average yield of 350-400q/ha. This is a notified variety by Govt. of India vide notification number SO-3666 (E), dated 06/12/2016 [krishijagran.com (a)] for growing in *Rabi* season in zone III (Uttar Pradesh, Bihar, West Bengal and Imphal). It contains 12-14% TSS, 13-14% dry matter and 13micromole/g pyruvic acid.
- **PRO-6** (PAU, Ludhiana): This variety was developed through mass selection from an exotic material EC 383494. Its deep red, medium to large and round bulbs with good keeping quality and having tight *cum* thin neck are harvested in about 125 days with about 400q/ha (Dhaliwal, 2017). The bulbs have 12% TSS content.
- **Punjab Naroya** (PAU, Ludhiana): Developed through mass selection from a local material, its red bulbs are round with tight neck and have medium to large size. It is harvested usually after about 125 days with an average yield of 325q/ha. It is tolerant to purple blotch and thrips (Dhaliwal, 2017).
- **Punjab Selection** (PAU, Ludhiana): It was developed through mass selection from among the indigenous materials and is adapted to IV, VII, VIII, (UP, Bihar, Punjab, Maharashtra Karnataka and Mandya Pradesh). Its plant height is 55-65cm which produces quite firm, red, globular, 5-6cm in diameter and with 50-60g weight having good keeping quality where average bulb yield is 20-30t/ha (Dhaliwal, 2017; nhb.gov.in). Bulbs contain 14% TSS.
- **Punjab White** (PAU, Ludhiana): It was developed through mass selection from exotic material English White Spanish. This mildly pungent variety produces white, medium to large and round bulbs with tight neck and is quite suitable for dehydration. Average yield of this variety is 337q/ha (Dhaliwal, 2017). The bulbs contain 15% TSS.
- **Pusa Madhvi** (IARI, New Delhi): This was developed through selfing and massing from a local material collected from Muzzaffar Nagar (U.P.). It is suitable for growing in the Indo-Gangetic plains. Plants grow to about 55cm high, producing light red, medium to large and flattish-round bulbs having 11-12% TSS content. For maturity it requires 130-135 days and the bulbs have good keeping quality, *vis-à-vis* is suitable for export. Average bulb yield is 30-35 t/ha (Dhaliwal, 2017).
- **Pusa Ratnar** (NBPGR, New Delhi): This has been developed through selection from segregating population of an F_1 hybrid 'Granex', introduced from USA. It is adapted to IV and VII zones. It grows to

around 30cm tall, producing deep red, large and obovate to flat bulbs having drooping necks and are less pungent though have good keeping quality. During maturity the bulbs are quite exposed to ground level. It is usually harvested after about 125 days after transplanting with an average yield of 300-400q/ha (Dhaliwal, 2017).

- **Pusa Red** (IARI, New Delhi): This has been developed through selection from indigenous collection which requires 140-145 days for maturity with average bulb yield of 25-30t/ha. Its plant grow some 55-65cm high. Its bronze-red bulbs are medium, flat to globular, with medium pungency, good keeping quality (nhb.gov.in) and per bulb average weight being 70-90g. It can be grown both in late *Kharif* and *Rabi* seasons in Maharashtra. It is suitable for export. Bulbs contain 12-13% TSS. It is suitable for cultivation under zones IV, VII and VIII.
- **Pusa Riddhi** (IARI, New Delhi): This is suitable for growing under both the seasons, *i.e. Rabi* and *Kharif.* Its medium-sized round bulbs with tight neck are purplish-red, moderately pungent and with good keeping and worth export. It is usually harvested after about 125-130 days. Per bulb weight is 70-100g and average yield is 315q/ha (Dhaliwal, 2017).
- **Pusa White Flat**: For cultivation, it is recommended for zones IV, VI and VII during *rabi* season. It matures in 130-135 days after planting producing medium to large flattish-round bulbs to a tune of 20-30t/ha. It contains 12-14% TSS. Its bulbs are suitable for dehydration.
- **Pusa White Round**: It is recommended for cultivation in zones IV, VI and VII during *rabi* season. It matures in 130-135 days with average bulb yield of 30-32.5t/ha, producing medium to large flattish-round bulbs which contain 12-13% TSS. Its bulbs are suitable for dehydration.
- **S-48** (PAU, Ludhiana): It is suitable for growing in Punjab, UP and Bihar with about 25-30t/ha bulb yield (ikisan.com/ka-onion-varieties.html). Plants grow 40-60cm high with bulb weight of about 80g. Its bulbs have good keeping quality with 12-13% TSS.
- **Udaipur 101** (RAU, Bikaner, Udaipur): This variety is recommended for growing in Rajasthan and other adjoining states during *Rabi* season (nhb.gov.in). This matures in 150 to 160 days with 20-30t/ha of bulb yield. Bulbs are deep red, flattish-globular and sweet with little pungency. TSS is 12-14%.
- **Udaipur 103**: It is recommended for growing under zone VI as *rabi* season cropping. It matures in 100-110 days with an average bulb yield of 25-30t/ha, producing medium to large-sized oblate-globular bulbs with 10-5-13% TSS. It has medium keeping quality.

Kharif onion varieties

Source: Dhaliwal (2017); nhb.gov.in; ikisan.com/ka-onion-varieties.html; krishijagran.com (a)

- **Advance line-863:** This variety is recommended for cultivation in *Kharif* and late *Kharif* seasons [krishijagran.com (a)] which matures in 80-85 days with an average bulb yield of 280-300q/ha. Its dark red bulbs having 4.5-5cm diameter are round with medium keeping quality. Bulbs contain 12-13% TSS, 11-12% dry matter and 12.5-13micromole/g pyruvic acid.
- **Advance Line- 883:** This variety is recommended for early season growing during *Kharif* season all over India. Bulbs mature in 85-90 days with average bulb yield of 300-325q/ha. Its dark and shining red bulbs are round with 4.5-5.5cm diameter. The bulbs contain 12-13% TSS, 13-14% dry matter and 12micromole/g pyruvic acid [krishijagran.com (a)].
- **Agrifound Dark Red/ADR (NHRDF, Nasik)**: It was developed through mass selection from a local stock of *Kharif* onion grown at Nasik and is adapted for growing to zone VI (U.P., Bihar, Punjab). Its bulbs mature for harvesting in 95-110 days with an average bulb yield of 30t/ha (Dhaliwal, 2017). Its deep red, medium-large and globular bulbs with tight skin are moderately pungent with TSS content of 12-13%. This variety is quite suitable for export. It has the best storage quality among all the Kharif season varieties.
- **Agrifound Rose:** It is recommended for growing for zones V and VII, i.e. all the three seasons in Karnataka, and for *kharif* season in Cuddapah and Andhra Pradesh. It matures in 95-110 days with an average bulb yield of 19-20t/ha, producing flattish-round bulbs of 2.5-3.5cm size. Its bulbs contain 15-16% TSS.
- **Arka Bindu (IIHR, Bangalore):** It is recommended for zones V and VIII, i.e. Cuddapeh, Andhra Pradesh during *kharif* season and all the three seasons in Karnataka. It matures in 100 days with an average bulb yield of 20t/ha. Its 2.5-3.5cm bulbs are flattish-globular containing 14-16% TSS.
- **Arka Kalyan (IIHR, Bangalore):** It has been developed through mass selection from a local material IIHR–145 and is adapted to be grown in zones IV, VI, VII, VIII (Bihar, Punjab, Gujarat, Haryana, Maharashtra, Karnataka and Tamil Nadu). This matures in 140 days with average bulb yield of 47t/ha. The plant attains the height of 70-90cm, producing dark red, globular and average bulbs with per bulb weight of 100-190g having moderate keeping quality with 11-13% TSS content. It is moderately resistant to purple blotch (Dhaliwal, 2017; nhb.gov.in).

- **Baswant-780 (MPAU, Rahuri):** This variety is recommended for *Kharif* growing in Maharashtra (nhb.gov.in) though bulb keeping quality is poor. Its red and medium to large bulbs are flattish-round in shape, mildly pungent with TSS of 11-12%. It matures in 90-100 days with average bulb yield of 25t/ha.
- **N-53 (NIPHAD):** This variety was recommended by Department of Agriculture during 1960 and released at national level during 1987. It is a very popular variety for growing during *Kharif* season all over the country. Its red and medium to large bulbs having poor keeping quality are flattish-round and mildly pungent with 11-12% TSS. It is suitable for export (nhb.gov.in). It becomes ready for harvest in 90-100 days with average bulb yield of 25t/ha.
- **Phule Suvarna [MPKV, Rahuri (M.S.)]:** Its low pungent and medium-large yellow bulbs having 4-6 months keeping quality are quite amenable for year round cultivation and most suitable for export to Europe, Australia and America. It can be grown during late Kharif and Rabi seasons. Average yield of bulbs is 24t/ha (nhb.gov.in) with 11.5% TSS.

Improved Bhima group of onion varieties

Directorate of Onion and Garlic Research (DOGR) has recommended some improved onion varieties for release (icar.org.in/node/323; nhb.gov.in):

- **Bhima Dark Red**: This DOGR variety of onion with attractive dark red colouration and globe-shaped flat bulbs has been identified for release in areas of Chhattisgarh, Delhi, Gujarat, Haryana, Karnataka, Madhya Pradesh, Maharashtra, Odisha, Punjab, Rajasthan and Tamil Nadu for growing during kharif season. This matures for harvesting in 95-100 days (icar.org.in/node/323; nhb.gov.in) giving an average yield of 20-22t/ha. Its storage life is 1-1.5 months.
- **Bhima Kiran (RO597) (IC No. 572766):** It was identified at National level for release during 1st AINRPOG group meeting at CITH, Srinagar during 2010 for growing during *rabi* season in two zones, viz. III (Delhi, U.P., Haryana, Bihar and Punjab) and VI (Maharashtra, Karnataka and Andhra Pradesh). This matures for harvesting in 130 days with marketable yield potential up to 41.5t/ha though average national yield is 30.3t/ha (nhb.gov.in). Immediately after harvesting the bulbs attain light red colouration. Its thin-necked bulbs being oval to round have very less number of bolters though during rabi season some 5% out of these may double. The bulbs of this variety keep well which can be stored for 5-6 months. TSS of bulbs is around 12%.

- **Bhima Light Red:** It has been identified for release for *rabi* season planting in Karnataka and Tamil Nadu. It matures in 110-120 days with an average marketable yield of 36-40t/ha. It can be stored for 5-6 months (nhb.gov.in).
- **Bhima Raj (B780522, IC No. 561257):** Its bulbs are dark red, oval-shaped with single centre and thin-necked. This variety is suitable for growing during *kharif* and late *kharif* seasons in the states of Maharashtra, Karnataka and Gujarat though can also be grown during rabi season as has been identified in XXVth AICRP (VC) group meeting held at Hissar for zone VI (Rajasthan, Gujarat, Haryana and Delhi). This matures in 120-125 days with absolutely no bolters and with an average yield of 25-30t/ha with high percentage of marketable bulbs during rabi though highest yield potential is 40-45t/ha (nhb.gov.in). It stores well for 2-3 months. Its TSS range is 10-11% (nhb.gov.in).
- **Bhima Red (B780531, IC No. 561258):** It has been developed through bulb-to-row selection method (nhb.gov.in). Though this variety is already recommended for *rabi* season growing in Maharashtra and Madhya Pradesh, but DOGR also recommended this for release for kharif growing in areas of Delhi, Gujarat, Haryana, Karnataka, Maharashtra, Punjab, Rajasthan and Tamil Nadu. It can also be grown in late *kharif season in Gujarat, Karnataka and Maharashtra.* It matures in 105-110 days during *kharif* and 110-120 days during late *kharif* and *rabi* seasons having average marketable yield of 19-21t/ha in *kharif* season with 1-1.5 months storability, 48-52t/ha in late *kharif* season with 2-3 months storability and 30-32t/ha in *rabi* season with 3 months storability (icar.org.in/node/323). It produces round bulbs with 10-11% TSS.
- **Bhima Safed:** It has been identified for release for *kharif* season in Chhattisgarh, Gujarat, Karnataka, Madhya Pradesh, Maharashtra, Odisha, Rajasthan and Tamil Nadu. It attains maturity within 110-120 days with average marketable bulb yield of 18-20t/ha. It can be stored for 1–1.5 months (nhb.gov.in; icar.org.in/node/323).
- **Bhima Shakti** (1156) (IC No. 572769): It was identified for release at national level during IInd AINPROG Group Meeting held at Ooty (TNAU, Coimbatore) during 2011 for four zones, viz. zone III (Delhi, UP, Haryana, Bihar and Punjab), zone IV (Rajasthan and Gujarat), zone V (MP, Chhattisgarh and Orissa) and zone VI (Maharashtra, Karnataka and Andhra Pradesh) for late *kharif* as well as rabi cropping. Round bulbs of this variety mature in 130 days during late *kharif* and *rabi* seasons and have excellent storage life with 11.8% TSS. Immediately after harvesting

the bulbs attain attractive red colouration. This contains quite a less number of bolters, i.e. 2.15-5.69% and 0.36-1.38% during late *kharif* and *rabi* seasons, respectively. Marketable yield during late *kharif* is 45.9t/ha and during rabi 42.7t/ha in Maharashtra with national average yield of 29.18t/ha during *rabi* season (nhb.gov.in). The bulbs have small to medium neck thickness with uniform neck fall during rabi whereas more than 70% neck fall during late *kharif* seasons. The bulbs can be stored for 5-6 months. It is tolerant to thrips.

- **Bhima Shubra** (W009; IC No. 572763): This thin-necked white onion variety with oval to round bulbs was released at institute level during 2010 (nhb.gov.in) for cultivation in Chhattisgarh, Gujarat, Karnataka, Madhya Pradesh, Maharashtra, Odisha, Rajasthan and Tamil Nadu regions for *kharif* season (nhb.gov.in), vis-a-vis for late *kharif* in Maharashtra (nhb.gov.in). This has no bolters during kharif but only up to 6% during late kharif. Bulbs contain 10.4% TSS in kharif and 11.7% TSS in late *kharif* crop. It is amenable to tolerate environmental fluctuations hence can be cultivated in all the three seasons. It matures in 110-115 days during *kharif* and 120-130 days in late *kharif seasons.* Average marketable yield during *kharif* is 18-20t/ha whereas 36-42t/ha during late *kharif crop* (icar.org.in/node/323). Its storability is 2-3 months.
- **Bhima Shweta** (White El. Comp. Selection/ NRCWO2) (IC No. 572761): It was identified at national level for release during IInd AINPOG group meeting at Ooty (TNAU, Coimbatore) during 2011 for three zones, viz. zone III (Delhi, U.P., Haryana, Bihar and Punjab), zone V (M.P., Chhattisgarh and Orissa) and zone VI (Maharashtra, Karnataka and Andhra Pradesh) for rabi season. It produces attractively white bulbs with thin necks that are round with quite a less number of bolters. The bulbs maturing in 110-115 days during rabi have up to 3 months keeping quality. Average marketable yield during rabi season is 35.9t/ha with national average yield of 28.2t/ha where the bulbs contain 11.5% TSS. It is tolerant to thrips. (nhb.gov.in). This variety is also recommended as *kharif* crop in Chhattisgarh, Gujarat, Karnataka, Madhya Pradesh, Maharashtra, Odisha, Rajasthan and Tamil Nadu where this matures in 110-120 days. Its usual average marketable yield during *kharif* season is 18-20t/ha and in rabi 26-30t/ha (icar.org.in/node/323).
- **Bhima Super** (IC No. 561259): It has been developed from a base population of cultivar B780 by exercising selection pressure for single centered bulbs by following eight cycles of selections and attaining 95% of single centered bulbs (nhb.gov.in). It is a red onion variety from DOGR which has been identified for release for *kharif* season in the region of

Chhattisgarh, Delhi, Gujarat, Haryana, Karnataka, Madhya Pradesh, Maharashtra, Odisha, Punjab, Rajasthan and Tamil Nadu. It can also be grown in late *kharif in Gujarat, Karnataka and Maharashtra*. It is reported to have an average yield of 20-22t/ha in *kharif* and 40-45t/ha in late *kharif*. Bulbs attain maturity in 100-105 days in *kharif* and 110-120 days in late *kharif*. Mostly it produces single-centered bulbs (icar.org.in/node/323). Storability in kharif season is 1-1.5 months though in late kharif season 2-3 months.

Some other Indian varieties

{**source**: krishijagran.com; nhb.gov.in; ikisan.com/ka-onion-varieties.html; krishijagran.com (a)}

- **Akra Pragati** (IIHR, Bangalore): It was developed through mass selection which attains plant height of 50-60cm. It is attractively pink variety with uniform globe-shaped bulbs with thin neck, high pungency and early maturity. It matures in 140-145 days with bulb yield of about 20t/ha (nhb.gov.in).
- **Arka Bindu** (IIHR, Bangalore): It is a common small onion (2.5-3.5cm) of deep pink colouration having flattish globe-shaped bulbs. Highly pungent and splitting bulbs are free from premature bolters. Bulbs mature for harvesting in 100 days with average bulb yield of 25t/ha. The bulbs contain 14-16% TSS. It is suitable for export (nhb.gov.in).
- **Co-1** (TNAU, Coimbatore): It is a red multiplier onion producing 7-8 medium-sized bulbets per plant where per clump average weight comes

to 55-60g with a total per hectare yield of 10 tonnes. Plants mature in about 90 days where bulbs are fairly pungent with medium TSS (nhb.gov.in). It is cultivated throughout Tamil Nadu.

- **Co-2** (TNAU, Coimbatore): A crimson variety producing moderately bigger 7-9 bulblets per plant with average bulb weight of 60-65g per clump. It matures in 65 days with average bulb yield of 12t/ha where bulbs are pungent with high TSS. It is cultivated throughout Tamil Nadu. Good for storage. It is moderately resistant to thrips and *Alternaria* blight (nhb.gov.in).
- **Hisar–2** (HAU, Hissar): It was developed through mass selection of indigenous materials and is well adapted for growing in Haryana and Punjab. Its flattish-globular bulbs are dark red or brown having good keeping quality. It matures in 175 days with an average bulb yield of about 30t/ha (krishijagran.com). Its bulbs do not taste spicy.
- **MDU-1** (TNAU, Coimbatore): It is a multiplier onion which produces uniform-sized bright red *cum* large bulbs some 10-11 per clump that weigh about 75g, and thus per hectare yield comes to about 15t/ha. It is adapted in southern districts of Tamil Nadu. Its bulbs for harvesting mature in 60-75 days which have better keeping quality. It is tolerant to lodging due to thick erect leaves (nhb.gov.in).
- **N 2-4-1** (NIPHAD; Mahatma Phule Krishi Vishwa Vidyalaya, Rahuri, MH): It has been adapted to zones IV, VI and VII (Uttar Pradesh, Gujarat, Haryana and Maharashtra). Its average plant height is around 46cm and the plants mature for harvesting in 140-145 days with an average bulb yield of 30-35t/ha. Its brick-red, globular and pungent bulbs having good keeping quality are suitable for export (nhb.gov.in). The bulbs contain 11-13% TSS. It is tolerant to *Alternaria* blight.
- **Phule Safed** (MPKV, Rahuri): Bulbs are white with globular shape and suitable for dehydration. Its average yield is 25-30t/ha (nhb.gov.in) with 13% TSS.
- **Pusa White Flat** (IARI, New Delh): It is suitable to be grown in zones IV, VI and VII (U.P., Bihar, Punjab, Rajastan, Gujarat, Haryana and Maharashra). Bulbs are attractively white, flattish round and medium to large with 12-14% TSS with drying ratio 9:1. Plants mature for harvesting in 130-135 days (nhb.gov.in) with around 28-30t/ha bulb yield (krishijagran.com). Its shelf life is good and stores well.
- **Pusa White Round** (IARI, New Delhi): It was developed by local collection (106) and is suitable to be grown in zones IV, VI and VII (U.P.,

Bihar, Punjab, Rajastan, Gujarat, Haryana and Maharashtra). Its white bulbs are roundish-flat with 12-13% TSS having drying ratio of 8:1 and good storability (nhb.gov.in). Bulbs mature for harvesting in 130-135 days with an average yield of 30-32.5t/ha.

- **Spanish Brown** (IARI R.S., Katrain, Kullu Valley): It is a long day variety with attractive colouration having mild pungency. Bulbs mature for harvesting during August-September in 160-180 days with an average bulb yield of 28-30t/ha (nhb.gov.in). It remains good in storage at hills. Its bulb have 13-14% TSS.
- **Udaipur 102** (RAU, Bikaner Campus): Its white bulbs are round to flat with low percentage of small bulbs. Bulb size being 4.5-6.5cm. It matures in 120 days with average bulb yield of 30-35t/ha (nhb.gov.in). Bulbs contain about 12% TSS.
- **VL-3** (Vivekananda Parvatiya Krishi Anusandhana Shala, Alonora, then UP): It is adapted to humid western Himalayan regions. Its plants grow around 50-70cm with globular and pungent bulbs. Bulbs mature in 145 days with an average bulb yield of 25t/ha (ikisan.com/ka-onion-varieties.html).

Desi varieties

Main desi varieties are 'Bellery Red', 'Jhulia', 'Large Red', 'Nasik Red', 'Patna Red', 'Red Globe', 'White Globe', 'White Patna' and 'Yellow Globe' (agropedia.iitk.ac.in/content/onion-varieties).

Imported varieties

Main imported varieties include: 'Australian Brown', 'Red Italian', 'Silver Skin', 'Sweet Spanish' and 'White Portugal' (agropedia.iitk.ac.in/content/onion-varieties).

Selections from Punjab varieties

A number of varieties released by Maharashtra state such as 'N-53', 'N-404', 'N-491', 'All Red type' and many kharif varieties are nearly selections from Punjab varieties (agropedia.iitk.ac.in/content/onion-varieties).

American or pungent type

They generally produce bulbs of small size, dense-textured, strong flavour and better keeping quality. The most important among these are 'Bregham Yellow Globe', 'Early Yellow Globe', 'Yellow Globe Danvera', *etc*. 'Ebenezer' is a common yellow variety. 'Red Wethers Field' and 'Southport Red Globe' are red varieties. 'Southport White Globe' has high pungency (agropedia.iitk.ac.in/ content/onion-varieties).

Nutritional value

Though onions are not particularly high in most nutrients, they do contain anti-inflammatory, anti-cholesterol, anti-cancer and antioxidant components such as quercetin and are rich in vitamin C, as well as in chromium ~ a trace mineral tied to insulin response. Similar to garlic, onion is rich in sulphur-containing compounds that yield its pungent odour and taste and is responsible for many of its health effects (GMF, 2007). In homoeopathy, *Allium cepa* is used for rhinorrhea and hay fever (Morrison, 1993). The quercetin found in onions reduces the size and number of precancerous lesions in the human intestinal tract (GMF, 2007).

Per 100 g edible part of onion bulb contains 86.6% moisture, 11.1g carbohydrates, 1.2g protein, 0.1g fat, 0.08mg thiamine, 0.01mg riboflavin, 11mg vitamin C, 47mg calcium and 0.7mg iron (Dhaliwal, 2017).

Most onion cultivars contain about 89% water, 9% carbohydrates (including 4% sugar and 2% dietary fibre), 1% protein and negligible fat (see table below). Onions contain low amounts of essential nutrients and have an energy value of 166kJ (40 kilocalories) in a 100g amount. Onions contribute savoury flavour to dishes without contributing significant caloric content (History of onions, 2011).

Raw onion bulbs ***Table source:*** *en.wikipedia.org/wiki/Onion*	
Nutritional value per 100g	
Energy	**166kJ (40kcal)**
Carbohydrates	9.34g
Sugars	4.24g
Dietary fibre	1.7g
Fat	**0.1g**
Saturated Monounsaturated polyunsaturated	0.042g 0.013g 0.017g
Protein	1.1g
Vitamins	Quantity (%DV†)
Thiamine (B1)	4% (0.046mg)
Riboflavin (B2)	2% (0.027mg)
Niacin (B3)	1% (0.116mg)
Pantothenic acid (B5)	2% (0.123mg)
Vitamin B6	9% (0.12mg)

Folate (B9)	5% (19μg)
Vitamin C	9% (7.4mg)
Vitamin E Vitamin K	0% (0.02mg) 0% (0.4μg)
Minerals	Quantity (%DV†)
Calcium	2% (23mg)
Iron	2% (0.21mg)
Magnesium	3% (10mg)
Manganese	6% (0.129mg)
Phosphorus	4% (29mg)
Potassium	3% (146mg)
Zinc	2% (0.17mg)
Other constituents	Quantity
Water	89.11g
Fluoride	1.1μg
Units: μg = micrograms; mg = milligrammes; IU = International units	
†Percentages are roughly approximated using US recommendations for adults. ***Source:*** USDA FoodData Central	

Nutrient composition of onions (Krishna Kumar** et al.**, 2015)		
Characters	**Common onion**	**Aggregatum onion**
	A. cepa var. *cepa*	*A. cepa* var. *aggregatum*
Moisture (%)	86.6	84.3
Proteins (g)	1.2	1.8
Fat (g)	0.1	0.1
Minerals (g)	0.4	0.6
Fibre (g)	0.6	0.6
Carbohydrates (g)	11.1	12.6
Energy (K cal)	50	59
Calcium (mg)	47	40
Phosphorus (mg)	50	60
Iron (mg)	0.7	1.2
Carotene (mg)	0	15
Thiamine (mg)	0.08	0.08
Riboflavin (mg)	0.01	0.02
Niacin (mg)	0.4	0.5
Vitamin C	11	2

Composition of Raw shallots, taken from USDA (2006) (Nutrient values and weights are for edible portion, 12% refuse (skins) Table source: Hedges and Lister (2007)	
Nutrient (Aproximate)	**Value per 100g**
Water	79.8g
Energy	302kJ (72 kcal)
Protein	2.5g
Total lipid (fat)	0. g
Fatty acids, total saturated	0.017g
Fatty acids, total monounsaturated	0.014g
Fatty acids, total polyunsaturated	0.039g
Phytosterols	5mg
Ash	0.8g
Carbohydrate, by difference Sugars Dietary fibre	16.8g 7.87g 3.2g
Minerals	
Calcium, Ca	37mg
Iron, Fe	1.2mg
Magnesium, Mg	21mg
Phosphorus, P	60mg
Potassium, K	334mg
Sodium, Na	12mg
Zinc, Zn	0.4mg
Copper, Cu	0.088mg
Manganese, Mn	0.292mg
Selenium, Se	1.2mcg
Vitamins	
Vitamin C, total ascorbic acid	8mg
Thiamin (B1)	0.06mg
Riboflavin (B2)	0.02mg
Niacin (B3)	0.2mg
Pantothenic acid (B5)	0.29mg
Vitamin B-6	0.345mg
Folate, total	34mcg
Vitamin A, IU	1190IU
Vitamin A, RAE	60mcg_RAE
Retinol	0 mcg
Amino acids	
Tryptophan	0.028g

Threonine	0.098g
Isoleucine	0.106g
Leucine	0.149g
Lysine	0.125g
Methionine	0.027g
Phenylalanine	0.081g
Tyrosine	0.072g
Valine	0.11g
Arginine	0.181g
Histidine	0.043g
Alanine	0.113g
Aspartic acid	0.231g
Glutamic acid	0.517g
Glycine	0.124g
Proline	0.165g
Serine	0.113g

Phytochemicals

Onions are particularly not a rich source of core nutrients, with vitamin C being the most important, although the frequency in which onions are consumed makes their nutrients a valuable contribution to the diet. They contain a range of phytochemicals including the **flavonoids** (flavonols kaempferol and quercetin glycosides responsible for yellow flesh and brown skins of many varieties, and anthocyanins imparting a red/purple colouration to some varieties) present in both the bulbs and leaves of onions and in spring onions, **fructans** (energy store for plants), **sulphur-containing compounds** (including the cepaenes, and thiosulfinates including the cepaenes that are used in plant defence when stressed) and **saponins** (for protection against potential pathogens), with an array of biological effects including antioxidant activity, reduction in risk of thrombosis (blood clotting), anti-carcinogenic effects, anti-bacterial effects and reduction in risk of atherosclerosis/coronary heart disease. Many of the health benefits have been attributed to their antioxidant activity (Hedges and Lister, 2007).

Flavonoids

The flavonoids found in onion include quercetin (responsible for yellow pigmentation), isorhamnetin and kaempferol derivatives in varying proportions (Bilyk *et al.,* 1984). These derivatives include the 4'-glucoside, the 7,4'-diglucoside, the 3-4'-diglucoside, the 3-glucoside, the 7-glucoside, the 3,7-diglucoside, the

3-rutinoside (rutin), the 3-rhamnoside (quercitrin), the 7,4'-, and 3-glucosides of kaempferol, plus isorhamnetin 4'-glucoside, however, there are no less than eight quercetin glucosides where the predominant compounds are quercetin 4'-glucoside and quercetin 3-4'-diglucoside though with differences in flavonol composition and levels depending on variety (Hedges and Lister, 2007).

Various anthocyanins are found in onions. With early studies showing the presence of predominantly cyanidin 3-glucoside, with lesser amounts of cyanidin 3-laminaribioside and other minor unidentified cyanidin, peonidin and pelargonidin glycosides (Hedges and Lister, 2007). Terahara *et al.* (1994) reported presence of cyanidin 3-glucoside, cyanidin 3-laminaribioside and their 6"-malonyl derivatives in Japanese cultivar 'Kurenai'. Four major and six minor anthocyanins are reported in the cultivars 'Red Baron', 'Tropea' and 'Comred' (grown in Norway) including the 3-malonylglucoside, 3-dimalonylglucoside and 3,5-diglucoside derivatives of cyanidin, peonidin 3,5-diglucosides and two 3-glycosylated derivatives of pelargonidin (Fossen *et al.*, 1996). Donner *et al.* (1997) found that in red onion cultivars grown in Canada and the USA, especially the 'Mambo', 'Red Jumbo', 'Red Bone' and 'Red Granex', the main anthocyanins were cyanidin 3-glucoside, cyanidin 3-laminaribioside, cyanidin 3-(6"-malonylglucoside) and cyanidin 3-(6"- malonyllaminaribioside). Minor anthocyanins were shown to be cyanidin 3 (3"-malonylglucoside), peonidin 3-glucoside, peonidin 3-malonylglucoside and cyanidin 3-dimalonyllaminaribioside (Hedges and Lister, 2007), and the differences in the anthocyanin composition may be attributed due to varietal differences (genetic basis) and/or locations..

Onion also contains allistatin I and allistatin II bioflavonoids (Arshad *et al.*, 2017; Alamgir, 2017) that were reported to have been isolated first from garlic, followed by *A. stracheyi* (Etewa and Abaza, 2011; Ranjan *et el.*, 2014) as well as other plants like *Zingiber officinale* (Helen and Job, 2020). Many scholars stated allistatin as a non-sulphur-containing bioflavonoid (Arshad *et al.*, 2017; Farooqui and Farooqui, 2017; Gregrova *et al.*, 2013; Ebadi *et al.*, 2014) while a few researchers relate it to well-studied allicin (a sulphur-containing allyl compound) (Cundell, 2014).

Fructans

Fructans are polymers based on fructose and are indigestible ingredients that are fermented in the body and help maintain the health of the gut and colon (Gibson, 1998). Onion bulbs may contain a high concentration (35-40% of dry weight) of fructans including oligofructans or fructooligosaccharides (FOS), which constitute a major portion of the water-soluble carbohydrates, and have been associated with storage life of bulbs (Hedges and Lister, 2007); FOS in onion on fresh weight basis being 2.8% though in garlic it is 1%, in rye 0.7% and in bananas 0.3%.

Sulfur compounds

Another main group of phytochemicals found in onions is the organosulfur compounds such as cepaenes and thiosulfinates (Dorsch and Wagner, 1991; Goldman *et al.*, 1996), which are formed when an onion is cut and the cell walls are disrupted. Allinase enzymes produce sulfenic acids *via* S-alk(en)yl cysteine sulfoxides (ACSOs) that rearrange to various compounds such as thiosulfinates, cepaenes and onion lachrymatory factor (Block *et al.*, 1997; Lancaster *et al.*, 1998).

Saponins

Saponins, a diverse group of biologically active glycosides (Curl *et al.*, 1985), are divided into two main groups, triterpenoids and steroid saponins (Amagase, 2006). They comprise a carbohydrate portion attached to the triterpenoid or steroid aglycone base and are able to form stable soap-like solutions with water. They possess both beneficial and harmful bioactive qualities and are often bitter–tasting. A number of different saponins have been identified in *Allium* species, with processing giving rise to different saponins again (Corea *et al.* 2005; Amagase 2006; Lanzotti 2006).

Carotenoids (spring onions only)

The carotenoids are a group of yellow-orange-red pigments found in a variety of fruits and vegetables as well as in algae, fungi and bacteria, but their colours are often masked by chlorophyll. Carotenoids consist of a long-chain hydrocarbon molecule with a series of central, conjugated (alter nating) double bonds that confer colour and the compound's antioxidant properties. These appear to act synergistically with other carotenoids and other antioxidants. These pigments assist in the light-capturing process in photosynthesis and protect against damage from visible light (Hedges and Lister, 2007).

In humans, one of the various benefits of carotenoids is believed to be protecting the skin and the macula lutea of the eye against photoxidative damage (Sies and Stahl, 2003). There are two general classes of carotenoids ~ the carotenes and their oxygenated derivatives and the xanthophylls. The body can convert α-carotene, β-carotene and β-cryptoxanthin into retinol (vitamin A) whereas lycopene and the xanthophylls, lutein and zeaxanthin have no vitamin A capacity and due to their structural similarity they are difficult to separate for analytical purposes and amounts of the latter two compounds are often reported as a combined total (Hedges and Lister, 2007).

Factors affecting phytochemicals

Genetic and environmental conditions

There is a substantial difference in phytochemical contents between onion varieties (Bilyk *et al.*, 1984) particularly for polyphenols. Shallots have the highest level of polyphenols, *i.e.* six times the amount found in Vidalia onions (Slimestad *et al.*, 2007); yellow, red and pink onions contain higher amounts of quercetin than white varieties that contain only very low levels of flavonols (Patil *et al.* 1995a). Yellow onions having the highest total flavonoid content, an amount 11 times higher than in white onions; red onions have considerable content of anthocyanin pigments with at least 25 different compounds identified, representing 10% of total flavonoid content (Slimestad *et al.*, 2007).

Some varieties appear to contain only the quercetin glycosides (Crozier *et al.*, 1997). Flesh colour is not the only determining factor for quercetin levels (Patil *et al.*, 1995a). In contrast, accessibility to light (*i.e.* skin colour) has been associated with flavonoid development (Patil and Pike, 1995). In addition, geographical location, growing conditions and storage factors also greatly affect the levels of quercetin found in onions (Patil *et al.*, 1995a,b). Not only polyphenols, but fructans and sulphur compounds too vary greatly with variety and growing conditions. Pungent genotypes may have more health benefits than mild varieties. The levels of fructans are usually higher in high dry matter onions though onions with low dry matter contain relatively little fructan and proportionately higher amounts of simple sugars (glucose, sucrose, fructose) (Griffiths *et al.*, 2002). Storage temperature and duration have significant effects on quercetin content but a relative patter n was not clarified (Patil *et al.*, 1995b). Differences in concentration due to growing location were also found but exact environmental factors were not determined. During storage, the fructan content drops with the release of free sugars (Jaime *et al.*, 2001a).

Processing

In one study no appreciable amounts of tannins and anthocyanins remained in the edible portion once the outer skin had been removed (Rhodes and Price, 1996) though this is not true for all the varieties. In peeled 'Tropea Red' onions the edible portion contained only 27% of the anthocyanins and 79% of the flavonols. Peeling may significantly reduce the flavonoid content (especially anthocyanins and to a lesser extent flavonols) and hence some of the health benefits of onions are reduced. Quercetin content is highest in the dry skin which decreases from the outer to inner rings (Patil and Pike, 1995). In contrast, fructans and sulphur compounds are richest in the fleshy layers (Jaime *et al.*, 2001b).

The organosulphur compounds in whole onions have little antiplatelet activity (AP), but upon cutting, sulphides are generated by enzymes that exhibit AP. Raw onions have significantly higher AP than cooked ones (Ali *et al.* 2000). As already mentioned above, several beneficial sulfur compounds are formed only when onion tissue is chopped, though these compounds may further get changed and loose their effect if left too long. However, when the onions are chopped and left for 30 minutes and then boiled these have similar AP to the raw onions (Hedges and Lister, 2007).

The quercetin 3,4'-diglucoside degrades rapidly in mashed tissues (about 50% decline after 5 hours) and gets converted to the quercetin monoglycoside and free quercetin though all these compounds too have antioxidant activity so this feature of onions might not be affected by chopping. Makris and Rossiter (2001) demonstrated that chopping had no significant effect on flavonol content or antioxidant activity. Ewald *et al.* (1999) reported greatest loss of flavonoids in onion during pre-processing step when the onion is peeled, trimmed and chopped before blanching.

Microwave cooking without water retains both flavonoids and ascorbic acid and frying does not affect flavonoid intake, however, since flavonoids are water-soluble, boiling onions can result to about 20-30% loss of flavonol (particularly quercetin glycosides) which gets transferred to the boiling water, *vis-a-vis* antioxidant activity also gets decreased (Ioku *et al.,* 2001; Makris and Rossiter, 2001). Crozier *et al.* (1997) reported that boiling reduces flavonoid content significantly while microwaving slightly less of an effect while frying results in the lowest loss.

Adam *et al.* (2000) reported that drying temperatures above 65°C exert a pronounced influence on colour ~ the pyruvate content decreases with increasing temperature and slice thickness; the sugar content is also significantly influenced by the drying temperature; *vis-a-vis* rate of ascorbic acid degradation gets decreased with increase in temperature and slice thickness. The carotenoids in spring onions are fat-soluble and are best absorbed in the body if accompanied by some form of oil or fat. Chopping and cooking help in releasing carotenoids from the food matrix thus increasing their bioavailability (Hedges and Lister, 2007).

Dehydrated onion

Onions can be dehydrated and stored for later use. Dehydrated onions are sold in many forms (slices, chopped ones, minced ones, granulated and powdered ones) and colours (yellow, red and white). Commercial processing plants prescribe a shrinkage ratio of 7:1 to 17:1 with a moisture content of 4% in final product. The dehydrated onions have longer shelf life than fresh onions, though they have milder taste and odour as compared to fresh raw onions. The preferred bulb qualities for preparing dehydrated onions are ~ pungent varieties of bulb onions having strong odour and high TSS value (15-20% Brix) because dehydrated product is primarily used as flavouring agent. Large bulbs due to economy in harvests; the bulbs with small neck and root zone and those with tall globe shape are preferred than flat types to permit greater efficiency in topping. White onions are preferred to red or yellow onions. Bulbs should have good stability with minimum shrinkage loss and rotting and there should be some 80% moisture content. Varieties like 'Pusa White Red', 'Pusa White Flat', 'S-48', 'N-257-9-1' and 'Udaipur-102' are suitable for dehydration.

Uses

Culinary

Onions can be used, usually chopped or sliced, in almost every type of food, including cooked foods and fresh salads, and as a spicy garnish. They are commonly used as a base for curries or made into a paste and eaten as a main course or as a side dish. It can be eaten raw along with meals, with lemon juice or vinegar, in chutneys, as salad alone or mixed with other salad items, as raw vegetable when cut-pieces are mixed with salt and mustard oil, as pickles or cooked with potato as vegetable and for cooking vegetables or meats and with various other edible preparations raw or cooked. They are rarely eaten on their own but usually act as accompaniment to the main course. Depending on the variety, an onion can be sharp, spicy and pungent or mild and sweet. Onions are a staple food in India and are therefore fundamental to Indian cooking.

The immature young plants may be harvested before bulbing or after bulbing has begun and used whole as green onions or spring onions. These greens are sometimes referred to as 'summer onions'. Scallions or green onions can be used in soups, stews, omelets, stir-fries or as a finishing herb.

Onions are available in fresh, frozen, canned, caramelised, pickled and chopped forms. The dehydrated product is available as kibbled, sliced, ringed, minced, chopped, granulated and powder forms. Onions pickled in vinegar are eaten as a snack. Onion powder is widely used for seasoning when the fresh ingredient is not available. Pearl and boiler onions may be cooked as a vegetable rather than as an ingredient and pickler onions are often preserved in vinegar as a long-lasting relish (Ministry of Agriculture, 1968).

Shallots: Though there is a weak distinction between onions and shallots in northern part of India and the larger varieties of shallot are sometimes confused with small red onions, and used interchangeably; the southern regions of India distinguish shallots from onions in recipes quite clearly, especially the tiny varieties which are widely used in curries and different types of *sambar*. Shallots pickled in red vinegar are common in many Indian restaurants, served along with sauces and *papad* on the condiments tray. In Kashmir region of India, shallots are widely used in preparation of *Wazwan* Kashmiri cuisine, as they add distinct flavour and prevent curry from becoming black which is common with onions. Indians also use shallots mixed with jaggery or sugar (en.wikipedia.org/wiki/Shallot) as a home remedy for sore throats.

Medicinal

The bulb is anthelmintic, anti-inflammatory, antiseptic, antispasmodic, carminative, diuretic, expectorant, febrifuge, hypoglycaemic, hypotensive, lithontripic, stomachic and tonic so when used regularly in the diet it offsets tendencies towards angina, arteriosclerosis and heart attack (naturalmedicinalherbs.net). Dhaliwal (2017) states that onion is the rich source of calories, vitamin C *cum* minerals, especially iron, phosphorus and calcium, *vis-a-vis* is used medicinally to treat various ailments ~ protecting against various infections, relaxing spasms *cum* blood clotting and reducing blood sugar levels. He further states that its consumption helps in prevention and treatment of arthritis, coronary heart diseases, diabetes, cancer and asthma. Liquid extracts of onion bulbs is used against bronchial and gastric infections. Raw onion helps in boosting good type HDL cholesterol, reducing total cholesterol levels and increasing blood-clot-dissolving activity. However, the sulphur compounds get destroyed through heating, hence the cooked onions do not possess blood thinning properties (Dhaliwal, 2017). Onions have been shown to possess antibacterial and antifungal properties (Hughes and Lawson, 1991; Augusti, 1996), though believed to be less active than garlic. Onion oil has been shown to be highly effective against gram positive bacteria and some fungi, and inhibits the growth and aflatoxin production of fungi genera (Zohri *et al.*, 1995). Organosulphur compounds are used as protective agents, *vis-a-vis* its antibacterial extract is effective against oral pathogenic bacteria and tooth decay (Kim, 1997; naturalmedicinalherbs.net).

Saponins have been found to have anti-inflammatory activity (Sparg *et al.*, 2004). Suresh Babu and Srinivasan (1997) found that a 3% onion-powder diet also reduces hyperglycemia, circulating lipid peroxides and blood cholesterol (LDL-VLDL exclusively). Onion consumption has been found to be inversely proportional to glucose levels in oral or intravenous glucose tolerance tests

~ which is suggested in tracing the allyl propyl disulphide in onions which lowers blood sugar levels by increasing availability of free insulin (GMF, 2007). It further elaborates that chromium is also considered to help cells responding appropriately to insulin with clinical trials showing that chromium can improve glucose tolerance, lowering insulin levels, decreasing fasting blood glucose levels and decreasing total cholesterol levels. Further it states that onion consumption has been found to lower high cholesterol levels and high blood pressure ~ due to its chromium, sulphur compounds and vitamin B6.

Exter nally onion is also used against acne and boils. An ointment or paste made from onion is said to prevent infection in wounds and burns. Fresh onion juice is very useful 'first aid' treatment for bee and wasp stings, bites, grazes or fungal skin complaints, when warmed the juice can be dropped into the ear to treat earache and the baked onions can be used as a poultice to remove pus from sores (naturalmedicinalherbs.net). Those taking raw onions with their meals or having white onion in their pockets when going out of home, it is said that they are protected from loo (heat current). It also aids the formation of scar tissue on wounds, thus speeding up the healing process, and has been used as a cosmetic to remove freckles (naturalmedicinalherbs.net).

Flavonoids

The flavonoids present in onions are potent antioxidants (Hedges and Lister, 2007) and have also shown to activate the detoxifying enzymes (Myhristad *et al.*, 2002; Munday and Munday, 2001). Flavonoids have a wide range of biochemical functions. They are involved in immune function, gene expression, capillary and cerebral blood flow, liver function, enzyme activity, platelet aggregation, and collagen, phospholipid, cholesterol and histamine metabolism (Hedges and Lister, 2007). Beneficial health effects associated with these compounds, such as reduced risk of coronary heart disease and different types of cancer, are thought to be primarily from antioxidative activity, including metal ion chelation and inhibition of lipid peroxidation (Formica and Regelson, 1995).

Flavonoids also reduce the susceptibility of LDL cholesterol to oxidation (O'Reilly *et al.*, 2000; Hertog *et al.*, 1993) which is an important step in the development of atherosclerosis so prevention has significant health benefits. Analysis of the effects of quercetin on human diabetic lymphocytes showed a significant increase in protection against DNA damage from hydrogen peroxide at the tissue level (Lean *et al.*, 1999). Studies have shown that quercetin is a powerful antioxidant (Dhaliwal, 2017) which can decrease cancer tumour initiation; promotes healing of stomach ulcers and inhibits the proliferation of cultured ovarian, breast and colon cancer cells (Hedges and Lister, 2007); and has been shown to inhibit platelet aggregation both *in vitro* (Hubbard *et al.*, 2003) and *ex vivo* (Janssen *et al.*, 1998). Tannins and anthocyanins from the skin of red onion also have antioxidant activity (Augusti, 1996).

Fructans

Fructans, the most abundant soluble carbohydrate in onions are a type of dietary fibre that helps reducing the chances of colorectal cancers (Dhaliwal, 2017) by promoting the growth of beneficial bifidobacteria and reduction of detrimental bacteria in the colon (Hedges and Lister, 2007). Fructans ingestion also reduces toxic metabolites and detrimental enzymes, prevents constipation, protects liver function, and reduces serum cholesterol and blood pressure (Hedges and Lister, 2007). After getting fermented in the bowel, fructans produce short chain fatty acids (SFAs) that have several beneficial effects, including providing energy for colonic mucosa, protection against various diseases of the colon and lowering colonic pH, so preventing the transformation of primary bile acids to co-carcinogenic secondary bile acids (Ekvall *et al.*, 2006).

Fructans have also been reported to reduce lipids and insulin levels in humans thus potentially have a cardioprotective effect (Jackson *et al.*, 1998). This is also the case for sulphur-containing compounds but these experiments have only been performed with cells and not with humans so being inconclusive (Hedges and Lister, 2007). Fructans may also promote resorption of calcium and reducing risk of osteoporosis (Ritsema and Smeekens 2003).

Sulphur compounds

Sulphides in chopped onions have antiplatelet activity (AP) and help in reducing platelet aggregation (Ali *et al.*, 2000; Bayer *et al.*, 1989). Cepaenes are strong antiflammatory chemicals with potency almost same as aspirin to inhibit platelet aggregation (Block and Zhao, 1992). The organosulphur compounds in onions exhibited antioxidant activity *in vitro* (Higuchi *et al.*, 2003) so proving to be strong anticarcinogenic (Fukushima *et al.*, 1997; Munday and Munday, 2001; Hatono *et al.*, 1996; Chu *et al.*, 2002).

This is believed to be partially due to its role in activation of detoxifying enzymes which remove potentially the cancer-causing substances. Two organosulphur compounds were linked to significant amelioration of weight loss, hyperglycemia, low liver protein and glycogen, and other characteristics of diabetes mellitus in rats (Chu *et al.*, 2002).

Carotenoids

Carotenoids in spring onions may protect against macular degeneration. Carotenoids are best known for their antioxidant activity though those predominant in spring onions (lutein and zeaxanthin) have been mostly researched in relation to eye diseases because these carotenoids absorb blue light, thus protecting retina from photochemical damage that could occur from light at these wavelengths. Some epidemiologic evidences suggest that lutein and zeaxanthin protect against age-related eye diseases (Sies and Stahl, 2003).

Other uses

Plant juice is used as a moth repellent and can also be rubbed onto the skin to repel insects. Plant juice can be used as a rust preventive on metals and as a polish for copper and glass. A yellow-brown dye is obtained from the skins of the bulbs. Onion juice rubbed into the skin is said to promote the growth of hair and to be a remedy for baldness. It is also used as a cosmetic to get rid of freckles. The growing plant is said to repel insects and moles. A spray made by pouring enough boiling water to cover 1kg of chopped unpeeled onions is said to increase the resistance of other plants to diseases and parasites (naturalmedicinalherbs.net).

Allergies

Some people suffer from allergic reactions after handling onions (Cantisani *et al.,* 2014). Symptoms can include contact dermatitis, intense itching, rhinoconjunctivitis, blurred vision, bronchial asthma, sweating and anaphylaxis. Allergic reactions may not occur when eating cooked onions possibly due to the denaturing of proteins from cooking (Arochena *et al.,* 2012).

Eye irritation

Cut onions emit volatile liquid, *syn*-propanethial-S-oxide and its aerosol which stimulates nerves in the eye by causing the lacrimal glands in the eyes to become irritated, releasing tears. This gas is produced by a chain of reactions which serve as a defence mechanism. Chopping an onion damages the cells which releases enzymes called alliinases. These break down amino acid sulphoxides and generate sulphenic acids. A specific sulphenic acid, 1-propenesulfenic acid is rapidly acted on by a second enzyme ~ the lacrimatory factor synthase (LFS) producing the *syn*-propanethial-S-oxide. This gas gets diffused through the air and after reaching the eyes it activates sensory neurons. Lacrimal glands produce tears to dilute and flush out the irritant. The amount of sulfenic acids, lacrimal factor released and the irritation effect differs among *Allium* species. Guinea hen weed (*Petiveria alliacea*) in the family Petiveriaceae and honey garlic [*Allium siculum* (syn. *Nectaroscordum siculum*)] contain a similar lachrymatory factor. Synthetic onion lachrymatory factor has been used in a study related to tear production (Higashihara *et al.*, 2010) and can be used as a deterrent against thieves and intruders.

Eye irritation can be avoided by cutting onions under running water or submerged in a basin of water. Leaving the root end intact also reduces irritation as the onion base has a higher concentration of sulphur compounds than the rest of the bulb.

Refrigerating the onions before use reduces the enzyme reaction rate, and using a fan can blow the gas away from the eyes. Eady *et al* (2008) developed 'tear-free onion' by suppressing the lachrymatory factor synthase gene, using RNA interference silencing.

Seed production

Onion is a cross-pollinated crop and isolation distance of 1000-1600m and 500m from other varieties is recommended for production of foundation and certified seeds, respectively. Onion seeds are produced by two methods, *viz.* seed to seed and bulb to seed method. Although seed yield is more under seed to seed method, bulb to seed method is preferred to produce true to type seeds because bulb selection cannot be exercised in seed to seed method. 1000kg bulb is required to plant one hectare. Seed production of varieties not storing well in storage is done by seed to seed method.

The bulbs for seed production are produced the same way as for the table crop. Bulbs are stored for replanting in next crop season. Usually medium-sized bulbs of 2.5-3cm diameter are planted on the side of ridges or in beds at 45×30-45 cm spacing. Bulbs are planted at a spacing of 45×30cm to raise a seed crop. Less than 12^{0}C low temperature coupled with low humidity are desirable for initiation of flowering, which is however, little affected by day length. In onions, primary umbels mature first, followed by secondary umbels.

Flower stalks will be produced during third month after planting of bulbs and seeds usually ripen within six weeks after formation of flower clusters. Both bulb and seed crops are inspected regularly to maintain genetic purity of the variety. Since seed maturity is not uniform, heads are harvested in 2 to 3 shifts when seeds start turning black but before they start shedding.

Average seed yield is 8-12 quintals per hectare. Before storing, seed is threshed, cleaned and dried to 6-8 per cent moisture level under shade in well ventilated rooms. Onion seed has a short storage life. Under ambient conditions seed maintains acceptable level of germination for one season only.

Cultivation of common onion

Propagation

Common onions may be grown from seeds, seedlings and bulbsets. Onion seeds are short-lived as compared with most of the other vegetables, therefore only fresh seeds should be used to obtain optimum germination. Fresh seeds germinate more effectively when sown in shallow rows or drills, each drill being 30 to 45cm apart (en.wikipedia.org/wiki/Onion). Onion sets are produced by thick sowing of seeds which result in stunted plants that produce very small bulbs in the first season. These bulbs are very easy to set out and grow into mature bulbs the following year but these are reputed of producing a less durable bulb than usually produced onions grown directly from seeds and thinned.

Raising nursery

This is the most common method practiced for irrigated crop as it results in high yield and large size bulbs.To raise a nursery for one hectare planting, Dhaliwal (2017) recommends about 200m^2 area and 10-12kg of seeds otherwise 500m^2 area and 5-7kg of seeds are required (vikaspedia.in). However, the nursery area and seed quantity depend on the varieties and growing conditions. Generally the ratio between nursery area and main field is about 1:20.

The field should be ploughed 2-3 times to break clods, and then well pulverized to hold water. The debris of previous crops, weeds and stones should be removed before bed preparation. It is recommended to mix well about 5-10 quintals of decomposed FYM at the time of last ploughing into the soil of nursery area. For raising nursery, raised beds of 10-20cm height is prepared having 1-1.2m width though length as per convenience. The inter-bed distance should be 30-60cm so that water movement is uniform with perfect drainage of excess water (vikaspedia.in; Dhaliwal, 2017).

Application of pre-emergence herbicide pendimethalin @ 0.2% is recommended to control weeds in nursery (vikaspedia.in). Seeds before sowing should be treated with Captan, Captaf or thiram @ 2-3g per kg of seed (Dhaliwal, 2017; vikaspedia.in) before sowing to avoid damage from damping off disease. Sowing is done in lines approximately 5-7.5cm apart to facilitate easy removal of seedlings for transplanting, quick weeding, spray of pesticides, *etc.* at a depth of 1-2 cm depending upon the soil type. Deep sowing is recommended in light soils. After sowing, the seeds should be covered with fine powdered farmyard manure or compost followed by light watering (vikaspedia.in). Application of water through drip or micro-sprinkler irrigation system helps in saving irrigation water (dogr.icar.gov.in; vikaspedia.in). Foliar spray of benomyl @ 0.2% is recommended to control soil-borne diseases in the nursery (vikaspedia.in). When severity of thrips infestation is high, foliar application of Fipronil or profenophos @ 0.1% is recommended (vikaspedia.in).

Rabi onion nursery is sown from mid-October to mid-November in the north and north-east India, in September- October in South and hilly regions and in November-December in Maharashtra and in parts of Gujarat.

Kharif onion nursery is sown in mid-June, although due to hot weather especially in north Indian plains the beds require protection from the scorching heat by providing shade which can immediately be removed once the seedlings get established. Moreover, the beds need to be irrigated twice a day (in morning and evening). Transplanting is done in first fortnight of August when seedlings are 6-8 weeks old. Since the *kharif* onion nursery often gets failed due to hot weather conditions, the 'bulbset technique' method can be used where bulb sets are produced by sowing in mid-March the same way as for the main (*rabi*) season crop. When the bulbsets attain a pencil-size thickness by June-end, they are uprooted and stored in baskets in a ventilated room to be transplanted in the field in second fortnight of August. The crop is harvested in November-end or during early December, however, the tops remain green and do not fall over due to low temperature (Dhaliwal, 2017). Varieties like '**Agrifound Dark Red**', '**Baswant 780**', '**N-53**' and '**Arka Kalyan**' are well suited for bulbsets method (vikaspedia.in).

In suitable climate, certain cultivars can be sown in late summer and autumn to overwinter in the ground and produce early crops the following year (Brickell, 1992). Where the season is sufficiently long to provide early pre-bulbing growth, onion can also be raised by direct seeding as is being practiced in some parts of Maharashtra (Dhaliwal, 2017).

Land preparation and transplanting

Seedlings are available for transplanting in 35-40 days after sowing for *kharif* and in 45-50 days for late *kharif* and *rabi* seasons (vikaspedia.in). Delayed planting may result in pre-mature bolting and reduced bulb size. Prior to transplanting, field should be ploughed and disked properly to eliminate debris and soil clods. At the time of last ploughing, approximately 15t/ha FYM or 7.5t/ha vermicompost or poultry manure should be incorporated and beds with appropriate sizes should be prepared after levelling (vikaspedia.in).

Seedlings are transplanted usually in the flat beds in the field when soil has good moisture level, maintaining bed spacing at 60cm where each bed may accommodate 3 rows (Dhaliwal, 2017). However, planting on raised beds improves bulb size and quality, *vis-a-vis* prevents waterlogging, especially during *kharif* or rainy season. Waterlogging favours anthracnose disease which is most devastating during *kharif* season (vikaspedia.in). Broad bed furrows (BBF) of 15cm height and 1.2m top width with 45cm furrow are formed to achieve proper spacing and population density. BBF is also suitable for drip and sprinkler irrigation and is also the best method for *kharif* onion production because the excess water can be drained out through the furrow, thus improving aeration and reducing the incidence of anthracnose disease. Optimum planting distance is 15cm between the rows and 10cm between the plants (vikaspedia.in).

Only healthy and proper-sized seedlings should be selected for transplanting and over- or under-aged seedlings are discarded, and planted after cutting about $1/3^{rd}$ of the seedling top for better establishment, and after dipping the roots in 0.1% carbendazim solution to reduce the incidence of fungal diseases during establishment (vikaspedia.in). First irrigation is applied 1-2 days after transplanting.

Factors influencing growth and development

Climate

Though onion is a cool season crop but can be grown under a wide range of climatic conditions such as temperate, subtropical and tropical, though bulbing onions are day-length sensitive and their bulbs begin growing only after the number of daylight hours has surpassed some minimal quantity.

There are, therefore, specific short-day varieties requiring 11-13h day length and long-day varieties requiring day length of 13-15h, however, short-day varieties will develop small bulbs if grown under long-day conditions and long-day varieties will not form bulbs if grown under short day conditions. Intermediate-day varieties are also available. All the varieties adapted to north Indian plains are short-day varieties whereas long-day onion is grown in hills. Lower temperature combined with short photoperiod is required for vegetative growth, whereas relatively higher temperature *cum* longer photoperiod is required for bulb development and maturity (vikaspedia.in).

Though young seedlings can tolerate freezing temperature, onion grows well under mild climate without extreme cold or hot *cum* excessive rainfall (vikaspedia.in; Dhaliwal, 2017). Optimum soil temperature for germination is 15-25^{0}C, for vegetative growth is 13-24^{0}C and for bulb development it is 16-25^{0}C (vikaspedia.in; Dhaliwal, 2017). The temperature below normal at initial stages of growth may result in bolting, whereas at later stages of crop growth the temperature above normal results in early maturity *cum* reduced bulb size (Dhaliwal, 2017). Bolting is driven by temperatures between 7.2° to 10°C. Variety, seedling size, planting date, temperature, *vis-a-vis* temperature duration ~ all collectively account towards onion bolting.

Onion requires about 70% relative humidity for good growth and can grow well in places where average annual rainfall is 650-750mm with good distribution during monsoon period though areas with low (< 650mm) or heavy rainfall (>750mm) are not generally suitable for rainfed crop (vikaspedia.in).

Soil

Onions can be grown in various types of soils such as sandy loam, clay loam, silt loam and heavy soils though the best soil for successful onion cultivation is deep, friable loam and alluvial soils having sufficient organic matter, good moisture-holding capacity and good drainage (Dhaliwal, 2017; vikaspedia.in). In general, sandy loam to clay loam soils are preferred. Sandy loams are good as they are low in sulphur while clayey soils usually have high sulphur content and produce pungent bulbs (en.wikipedia.org/wiki/Onion).

Sandy soils favour early maturity and require more and frequent irrigation whereas heavy soils produce misshapen bulbs. Onion crop is more sensitive to highly acidic, alkali and saline soils, *vis-a-vis* waterlogging conditions (vikaspedia.in). Optimum pH range in the soil for onion cultivation is 6-6.8, however, highly alkaline and saline soils are not suitable at all (Dhaliwal, 2017), though can somehow be grown in mild-alkaline soils (vikaspedia.in).

Onions do not thrive in soils having pH below 6.0 because of trace element deficiencies or occasionally Al or Mn toxicity. The threshold electrical conductivity of a saturation extract (ECe) for onion crop is 4.0dS/m. When the ECe level exceeds this, crop yield starts declining (vikaspedia.in).

Season

Though onion is mainly *rabi* season crop, it can also be cultivated in *kharif* and late *kharif* seasons. Based on seasonal requirement, varieties have been developed andrecommended areawise, *e.g.* for *kharif* season the varieties are 'Baswant–780', 'N-53', 'Agrifound Dark Red', 'Arka Kalyan' and 'Bhima Super'; for late *kharif* these are 'Baswant–780', 'Phule Samarth',Bhima Red' and 'Agrifound Light Red'; and for *rabi* season the varieties developed are 'N-2-4-1','Arka Niketan', 'Agrifound Light Red', 'Pusa Red', 'Pusa Madhavi', 'Bhima Raj', 'Bhima Red'. 'Bhima Super', 'Bhima Red' and 'Bhima Raj' have the potential to grow in all the three seasons, *viz. kharif*, late *kharif* and *rabi* season in Maharashtra (Lawande *et al.*, 2009).

Sowing, transplanting and harvesting timings of onion in different regions of India:

Season	*Time of seed sowing*	*Time of transplanting*	*Time of harvesting*
Maharashtra and some parts of Gujarat			
1. Early Kharif 2. Kharif 3. Late Kharif 4. Rabi	Feb.-Mar. May-June Aug.-Sept. Oct.-Nov.	April-May July-Aug. Oct.-Nov. Dec.-Jan	Aug.-Sept. Oct.-Dec. Jan.-Mar. Apr.-May
Tamil Nadu, Karnataka and Andhra Pradesh			
1. Early Kharif 2. Kharif 3. Rabi	Feb.-April May-June Sept.-Oct.	April-June July-Aug Nov.-Dec.	July-Sept. Oct.-Nov. Mar.-Apr.
Rajasthan, Haryana, Punjab, U.P. and Bihar			
1. Kharif 2. Rabi	June-July Oct.-Nov.	July-Aug. Dec.-Jan.	Oct.-Nov. May-June
West Bengal and Odisha			
1. Kharif 2. Late Kharif 3. Rabi	June-July Aug.-Sept. Sept.-Oct.	Aug.-Sept. Oct.-Nov. Nov.-Dec.	Nov.-Dec. Feb.-Mar. Mar.-Apr.
Hilly areas			
1. Rabi 2. Summer (Long day type)	Sept.-Oct. Nov.-Dec.	Oct.-Nov. Feb.-Mar.	June-July Aug.-Oct.
Source: *(vikaspedia.in/agriculture/crop-production/package-of-practices/vegetables-1/onion-allium-cepa)*			

Multiplier onions are propagated through offsets. In the long-season regions, especially the northern hemisphere, the offsets are usually planted during September or October (autumn) though in certain other parts of northern hemisphere the suggested planting time for the principal crop is either in February or beginning of March (early spring) where these mature in summer, whereas, fresh shallots can now be found year-round in supermarkets.

Manures and Fertilizers

Onion responds well to organic manures and removes 85:42:130kg of NPK to produce 30t/ha bulb yield. PAU (Ludhiana) recommends 20t of decomposed FYM, 100kg nitrogen and 50kg each of phosphorus and potash per hectare, whereas in Maharashtra 20-30 tonnes of FYM per hectare is recommended along with 90:60:60kg of NPK. The recommended NPK in Gujarat is 75:25:75kg though in Karnataka it is 125:50:125kg (Dhaliwal, 2017).

Phosphorus is often present in sufficient quantity in our soils but even can be applied before planting due to its low availability in the cold soils (en.wikipedia.org/wiki/Onion). Generally whole of phosphorus and potash and half of nitrogen is applied before transplanting and remainder of nitrogen is applied as top-dressing one month after transplanting. Fertilizers are applied either as broadcast or in bands 5-10cm directly below the transplants. The later method is practical when transplanting is done mechanically (Dhaliwal, 2017). Some growers apply whole amount of phosphorus before transplanting, whereas nitrogen and phosphorus at regular intervals during growing season, the last application of nitrogen being at least four weeks before harvesting (en.wikipedia.org/wiki/Onion).

Nitrogen deficiency enhances maturity but reduces bulb size which ultimately affects over all bulb yield. Higher doses of nitrogen result in excessive vegetative growth, increased bulb size with large thick neck and poor storage quality. Therefore, fertilizer application, needs to be judicious (Dhaliwal, 2017).

Sulphur is also an essential nutrient for improving yield and pungency of onion bulbs, therefore at transplantiung sulphur is recommended as basal dose at the rate of 15kg/ha where sulphur level is above 25 kg/ha; 30kg/ha where soil has below 25 kg/ha of sulphur, and 50kg/ha in long day onion crops (vikaspedia.in), however, in the soils manured recently the shallots there should not be planted immediately.

Fertilizer schedule for onion (per hectare)

Source: (vikaspedia.in/agriculture/crop-production/package-of-practices/vegetables-1/onion-allium-cepa)

Schedule	*N*	*P_2O_5*	*K_2O*	*Organic manures*
Kharif **onion (yield potential: 25-30 t/ha)**				
Basal	25kg	40kg	40kg	Organic manures equivalent to 75kg N (FYM: approx. 15t/ha or poultry manure: approx. 7.5t/ha or vermicompost: approx. 7.5t/ha)
30 DAT	25kg	-	-	-
45 DAT	25kg	-	-	-
Total	75kg	40kg	40kg	-
Late *Kharif* and *Rabi* onion (Yield potential: 40-50t/ha)				
Basal	40kg	40kg	60kg	Organic manures equivalent to 75kg N (FYM: approx. 15t/ha or poultry manure: approx. 7.5t/ha or vermicompost: approx. 7.5t/ha)
30 DAT	35kg	-	-	
45 DAT	35kg	-	-	
Total	110kg	40kg	60kg	

Schedule	N	P_2O_5	K_2O	Organic manures
Long day onion (Yield potential-100 t/ha)				
Basal	60kg	60kg	70kg	Organic manures equivalent to 75kg N (FYM: approx. 15t/ha or poultry manure: approx. 7.5t/ha or vermicompost: approx. 7.5t/ha)
30 DAT	60kg	-	-	
60 DAT	60kg	-	-	
Total	180kg	60kg	70kg	
*__Basal:__ Before or at the time of planting *DAT: Days After transplanting				

One third of recommended N and full dose of P_2O_5 and K_2O are applied at the time of planting while remaining two third N is applied in two equal splits at 30 and 45 days after planting (vikaspedia.in).

Manures and fertilizers as recommended by DOGR

Based on the field experiments carried out through All India Network Research Project on Onion and Garlic at different locations, DOGR recommends application of 75:40:40:40kg NPKS/ha along with a combination of two or three organic manures (FYM, poultry manure and vermicompost) equivalent to 75kg N/ha (7.5t poultry manure or 7.5t vermicompost or 15t FYM/ha) for Bihar, Chhattisgarh, Gujarat, Karnataka, Madhya Pradesh, Maharashtra, Odisha, Rajasthan and Tamil Nadu (Ooty) (dogr129).

For Haryana, Uttarkhand and Uttar Pradesh, DOGR recommends application of 100:50:50:50kg NPKS + 20t FYM/ha. Recommended organic manures should be applied before last ploughing and mixed well in the soil.

One third of nitrogen and full recommended dose of phosphorus, potassium and sulphur should be applied as basal at the time of planting. Remaining two third of nitrogen should be applied in two equal splits at 30 and 45 days after planting (dogr129).

If the soil test shows a deficiency of any micronutrients besides NPKS, the deficient micronutrient should also be applied to correct the deficiency. If the plant is diagnosed with micronutrient deficiency during growth stages, the deficiency should be corrected by either foliar or soil application of concerned nutrients immediately (dogr129).

- $ZnSO_4$ @ 10kg/ha as basal is recommended in areas having Zn-deficiency (dogr129).
- Borax @ 10kg/ha is recommended for areas having boron deficiency (dogr129).

- FYM @ 15t/ha is recommended in areas having multi-micronutrient deficiency along with foliar application of micronutrient mixture (Fe: 2.5%, Zn: 0.3%, Mn: 1%, Cu: 1.0%, B: 0.2%) at 45 and 60 days after transplanting for increasing onion productivity (dogr129).

Irrigation

Onion plants are shallow-rooted and require frequent light irrigations throughout the crop season to maintain optimum soil moisture for proper growth and bulb development. Excessive irrigation may produce bulbs with thick necks in which water deficiency results in cracking and doubling (splitting) of bulbs (Dhaliwal, 2017). Irrigation requirement of onion depends upon the season, soil type, age of the crop and method of irrigation. A dry spell followed by irrigation will result in splitting of outer scales, *vis-a-vis* formation of bolters (dogr.icar.gov.in; vikaspedia.in). First irrigation is given usually 1-3 days after transplanting, followed by subsequent irrigations at 7-10 days intervals depending upon the soil moisture and weather conditions (dogr.icar.gov.in; Dhaliwal, 2017; vikaspedia.in). Storage life of bulbs is improved by stopping irrigation one fortnight earlier before harvesting, *i.e.* when crop attains maturity and the top starts falling (dogr.icar.gov.in; Dhaliwal, 2017). Depending upon soil and weather conditions, total number of irrigations would generally be 12-18 in *rabi* crops, 5-10 in *kharif* crops and 10-12 in late *kharif* crops (dogr.icar.gov.in; Dhaliwal, 2017; vikaspedia.in). Irrigationat bulb formation is a must as moisture stress at this stage results in poor bulb development and consequently low yield (Dhaliwal, 2017).

Mordern irrigation techniques

Since water loss with flood irrigation is too high due to conveyance, seepage and percolation losses, drip- and micro-sprinkler irrigations saves the water to maximum extent and improves the marketable bulb yield significantly (vikaspedia.in). If drip irrigation is being practiced, seedlings should be planted at a spacing of 10×15cm in a 15cm high broad bed furrow (BBF) having 120cm top width with 45cm furrow (dogr.icar.gov.in; vikaspedia.in). Each BBF should have two drip laterals (16mm size) at 60cm distance with inbuilt emitters having the discharge flow rate of 4 l/h (dogr.icar.gov.in; vikaspedia.in). The distance between two inbuilt emitters should be around 30-50cm (dogr.icar.gov.in; vikaspedia.in). If micro-sprinkler irrigation technique is being used, the distance between two laterals (20mm) of micro-sprinkler should be 6m with a discharge rate of 135 l/h.

Fertigation

Fertigation is an efficient method of using drip irrigation to apply fertilizers in onion/garlic farming. In fertigation both irrigation water as well as crop nutrients are applied through drip irrigation only in the root zones, which not only helps in saving water but also reduces nitrogen losses by leaching into ground water. Application of fertilizers NPK @ 40:40:60kg/ha as basal and the remaining 70 kg N in seven splits through drip irrigation is recommended for achieving higher marketable bulb yield and cost-benefit-ratio (dogr.icar.gov.in; vikaspedia.in).

Weeds and their management

Onion is a poor competitor of weeds and requires routine care during the growing season to keep the rows free of competing weeds, especially when the plants are young (dogr.icar.gov.in).

Major weeds

Kharif	*Rabi*
Broad leaf	
• Pigweed: *Amaranthus viridis* Hook. F. (Amaranthaceae) • Swine cress: *Coronopus didymus* (L.) Sm. (Brassicaceae) • Black nightshade: *Solanum nigrum* L. (Solanaceae) • Common purselane: *Portulaca oleracea* L. (Portulacaceae) • False amaranth: *Digera arvensis* Forssk. (Amaranthaceae)	• Lamb's **quarter:** *Chenopodium album* L. (Chenopodiaceae) • Scarlet **pimpernel:** *Anagallis arvensis* L. (Primulaceae) • **Sweet** clover: *Melilotus indica* (L.) All. (Fabaceae) • **Fine leaf fumitory:** *Fumaria parviflora* Lam. (Fumariaceae) • **Corn spurry:** *Spergula arvensis* L. (Caryophyllaceae) • **Carrot grass:** *Parthenium hysterophorus* L. (Asteraceae)
Grassy	
• Rabbit/crow foot grass: *Dactyloctenium aegyptium* (L.) Beauv. (*Poaceae*) • Crab grass: *Digiteria sanguinalis* (L.) Willd. (Poaceae) • Barnyard grass: *Echinochloa crusgalli* (L.) Scop. (Poaceae)	• **Blue grass:** *Poa annua* L. (Poaceae) • **Canary grass:** *Phalaris minor* Retz. (Poaceae) • **Chinese love grass:** *Eragrostis unioloides* (Retz.) Nees. Ex Steud. (Poaceae) • **Goose grass:** *Eleusine indica* (L.) Gaertner (Poaceae)
Sedge	
• Purple nutsedge: *Cyperus rotundus* L. (Cyperaceae) • Flat sedge: *Cyperus iria* L. (Cyperaceae)	• **Purple nutsedge:** *Cyperus rotundus* L. (Cyperaceae)
Source: *[vikaspedia.in (a)]*	

Much of the weed problems can be managed while preparing the field for transplanting. Pre-sowing or pre-planting methods for weed control includes: i) nursery beds should be kept weed-free throughout, ii) seed bed technique or field solarization during summer, iii) deep ploughing during summer, iv) keeping the boundary *cum* bunds of the field free from weeds, v) keeping the onion planted in beds free of any weed throughout the crop season [vikaspedia.in (a)].

The crop should be maintained weed free initially for 30-45 days by following timely intercultural and hand weeding if required. In a highly weed infested fields, it is recommended to apply approved herbicides such as spray of oxyflourfen 23.5% EC @ 425-850ml in 500-1700 l of water/ha on soil as pre-plant application or quizalofop-ethyl 5% EC @ 750-1000ml 375-450 l of water/ha 20 days after planting of onion seedlings followed by one hand weeding if required [vikaspedia.in (b)].

Dhaliwal (2017) suggested spraying Stomp 30EC (pendimethalin) @ 2.5 l/ ha or 1.87 l/ha + one manual weeding or Goal 23.5 EC (oxyfluorfen) @ 1 litre + one manual weeding 90-100 days after herbicide spray which controls the weeds effectively in onion (Dhaliwal, 2017). Spraying should generally be done after first irrigation. Alter natively, spraying of Basalin @ 2.5 litres per hectare or 1.87 litre per hectare as pre-plant application + one manual weeding 45 days after transplanting can prove effective (Dhaliwal, 2017).

Crop rotation and mixed cropping

Rotation and mixed cropping onion crop should follow the vegetable crops which require large quantities of organic manures such as cauliflower, tomato and potato (vikaspedia.in). As a winter crop, it follows cereals, groundnut and cowpea while as a summer crop it follows paddy (vikaspedia.in). When cultivated as monsoon or late monsoon crop, coriander, French bean, sorghum, pearl millet, chillies and groundnut are planted after harvesting onion (vikaspedia. in). During the initial five months after planting sugarcane, onion is grown as an inter-crop (vikaspedia.in).

Harvest

At any stage the crop can be harvested for greens for fresh cooking or raw eating. Shallots and young onions can be harvested before bulbing to be used whole as spring onions or scallions. The immature onion is harvested after bulbing has begun to be used fresh or as vegetable.

Depending upon variety and season, bulbing in onion usually takes place after 84-120 days (en.wikipedia.org/wiki/Onion) and requires 95-150 days to reach maturity when the neck tissues begin to soften and the tops are about to decolourize and fall over; development of pigment and characteristic pungency of the variety are other indices of maturity (Dhaliwal, 2017). Bulbs can be gathered when needed to eat fresh though for storage they are harvested at maturity when the leaves have died back naturally (en.wikipedia.org/wiki/Onion). Early harvest may cause sprouting of the bulbs while late harvest results in formation of secondary roots during storage (vikaspedia.in). Onions for sale as dried bulbs or for storage should be harvested progressively after tops have started falling over (vikaspedia.in). For immediate marketing, onion is harvested when atleast 25 per cent of tops dry up and fall down though for long-term storage, harvesting should be done when atleast 50 per cent of tops dry up and fall down (vikaspedia.in; Dhaliwal, 2017).

In *kharif* crop, since the tops do not dry completely or fall over due to low temperature, bulbs are harvested soon after the colour of leaves changes to slightly yellow and red pigmentation on bulbs develop (vikaspedia.in; Dhaliwal, 2017). In *kharif* season, late harvesting results in bulb splitting and bolting. Onion bulbs are normally formed at the soil surface and in sandy soils it is sometimes possible to pull the mature bulbs even by hand. If hand pulling is not possible, a fork or hoe can be used. The clump of multiplying onions and shallots can be pulled out and a stalk can be separated out from it for replanting which will re-establish and begin to divide again if the weather is not too hot.

Yield

Under appropriate water conditions, yield of good bulbs varies between 300-450 quintals per hectare, depending on the variety.

Post harvest

Curing

After harvesting, the bulbs are properly cured and dried in shade for 7-10 days to remove field heat and excess moisture, *vis-à-vis* for proper skin colour development (Dhaliwal, 2017). Before storage the tops are cut retaining 2-2.5cm of neck but in case where the tops are cut too close, the neck does not close too well so providing entry for decay organisms (vikaspedia.in). This practice helps to increase the dry matter content. Bulbs are either cured in field or in open shade before storage. Generally, they are cured in field for 3-5 days in wind row method and then placed in shade and cured for 7-10 days to remove field heat. This shade curing improves bulb colour and reduces losses during storage. Onions are considered cured when neck is tight and the outer scales are dried until they rustle.

Grading

Thick-necked, bolted, doubles, injured and decayed bulbs are carefully sorted out for immediate consumption and the rests are graded as large, medium and small sizes as per market demand.

Storage

Firm, mature, cured and dried onions without cuts or wounds store much longer than soft, moist and wounded ones (Dhaliwal, 2017). Irrespective of type, any cut pieces of onion are best tightly wrapped, stored away from other produce and used within two to three days (en.wikipedia.org/wiki/Onion). Sweet onions can be stored for only 1-2 weeks at room temperature or can be refrigerated to increase the shelf life by 1-2 more weeks.

Bulbs cannot be stored in refrigerators as they develop mold due to high humidity inside (Dhaliwal, 2017). Bulbs can be stored in a single layer or in large mesh bags in a dry, cool, dark and well-ventilated room at 24-30^{0}C temperature for several weeks without sprouting and excessive water loss (Brickell, 1992; Dhaliwal, 2017; en.wikipedia.org/wiki/Onion). The shelf life of pungent red varieties is better as compared to the less pungent white varieties (Dhaliwal, 2017; en.wikipedia.org/wiki/Onion). However, pungent varieties will absorb odours and moisture from apples, pears and from other fruits and vegetables with which they are stored which may cause them to decay (en.wikipedia.org/wiki/Onion).

The bulbs should be regularly inspected and the rotten ones should immediately be removed to prevent infection to healthy ones. Thick necked, bolted, doubles, injured and diseased bulbs should be sorted out before marketing (Dhaliwal, 2017). In cold storage, the temperature of stored onions is lowered gradually to the optimum temperature of 0^{0}C. Low humidity of 60-70 per cent inhibits sprouting of bulbs in storage, however, the bulbs sprout faster after removal from the cold stores than from the conventional stores. Bulbs need to be conditioned for several days at 20-50 per cent relative humidity before removing from cold stores (Dhaliwal, 2017). The bulbs are packed in open mesh jute bags or plastic netted bags, usually of 8-25kg size, for export and 40kg size for domestic market (Dhaliwal, 2017).

Physiological Disorders

Rubberification and premature sprouting of bulbs

Rubberification (when bulbs become rubber-like, *i.e.* tough *cum* flexible) and premature sprouting of onion bulbs are the main physiological disorders noticed mostly in fields which are located in low-lying areas of watershed where there is heavy deposition of nutrients along with silt during heavy rains. These problems are severe in onion fields which are more frequently irrigated than the normal ones. With application of higher levels of nitrogen, there is an increased level of pre-mature sprouting of bulbs which results in bulb splitting and its rubberification {vikaspedia.in (e)}.

Premature sprouting of bulbs varies from variety-to-variety and can be seen in the field when there are winter rains or excessive soil moisture and nitrogen supply. Early-planting also causes sprouting. Delayed harvesting during rainy season increases premature sprouting and splitting of bulbs. Planting bulbs at wider spacing may increase uptake of nitrogen and water by the individual plants which also increases premature sprouting and rubberification of onion bulbs {vikaspedia.in (e)}.

Application of micronutrients *i.e.* zinc sulphate and ammonium molybdate controls rubberification completely in onion. Also it is controlled through use of growth regulator such as gibberellic acid and neem cake insecticides.

Splitting of bulbs

Sometimes the bulb may split into two or more, especially in large-sized varieties due to water stress during bulb development stage or due to other factors such as early planting, delayed harvesting, heavy soil texture, excessive fertilizer application (especially of nitrogenous fertilizers) and poor seed stock (en.wikipedia.org/wiki/Onion).

Formation of aerial bulbil

The aerial bulbil formation is also quite common in onion where lower temperature prevails for long period or if there is more temperature variation. It may be due to varietal variation though significantly it does not affect yield or quality of bulbs [vikaspedia.in (e)].

Pre-mature bolting

Early *cum* delayed planting of seedlings, poor seed stock and unstable weather conditions may result in the development of seed stalks before forming proper bulb size (en.wikipedia.org/wiki/Onion). Indian varieties pose problems of bolting and doubling of bulbs especially in September–October planting.

Nutritional Deficiencies

Being a tropical country, most of the Indian soils have low organic carbon and are deficient in available nitrogen (researchgate.net). Out of 133 onion and garlic districts, 43 districts are deficient in available phosphorus and 40% of districts deficient in sulphur. Almost all the soils of onion growing areas are having medium to high potassium, calcium and magnesium content. Common micronutrient deficiency problems observed in onion and garlic growing areas are Fe, Zn and B (researchgate.net).

Nutrient deficiency	*Symptoms*	*Correction measure*
Nitrogen {vikaspedia.in (e)} 	Leaves become yellowish green erect and upright curled, wilted and dwarf. At maturity tissue above bulbs become soft {vikaspedia.in (e)}	Foliar spray of urea 1% or DAP 2% twice at weekly intervals {vikaspedia.in (e)}
Phosphorus {vikaspedia.in (e)} 	Growth is slow, leaf colour becomes light green and bulbs have few dried outer peels. Tip burn in older leaves {vikaspedia.in (e)}	Recommended dose of phosphorous should be applied in the soil at the time of sowing or planting. Foliar spray of DAP 2% twice at fortnightly intervals {vikaspedia.in (e)}
Potassium {vikaspedia.in (e)} 	Since potassium moves in the plant at quite faster rate, symptoms develop only on young leaves in case of extreme deficiency. Leaves become dark green and erect with tip-burn symptoms. Bolting is promoted. Older leaves become yellow and necrotic {vikaspedia.in (e)}	Foliar application of K_2SO_4 @1% twice at weekly intervals {vikaspedia.in (e)}
Sulphur {vikaspedia.in (e)} 	Over all the leaves show a general chlorosis. Yellowing is much more uniform over the entire plant including young leaves {vikaspedia.in (e)}	Foliar spray of K_2SO_4 or C_aSO_4 @ 1% twice at fortnightly intervals {vikaspedia.in (e)}
Manganese {vikaspedia.in (e)} 	Deficiency occurs in well drained calcareous soil with free calcium carbonate nodules which release free calcium due to rain and thereby the soil pH around the root zone shoots up above 8.5 (researchgate.net). Leaves become light coloured, show tip burn and curling. Growth is restricted. Bulbing is delayed with thick necks {vikaspedia.in (e)}.	When visual deficiency symptoms appear, foliar application of 0.25-0.5% M_nSO_4 should be given. If required the spray may be repeated after 10-15 days {researchgate.net; vikaspedia.in (e)}.

Zinc {vikaspedia.in (e)} 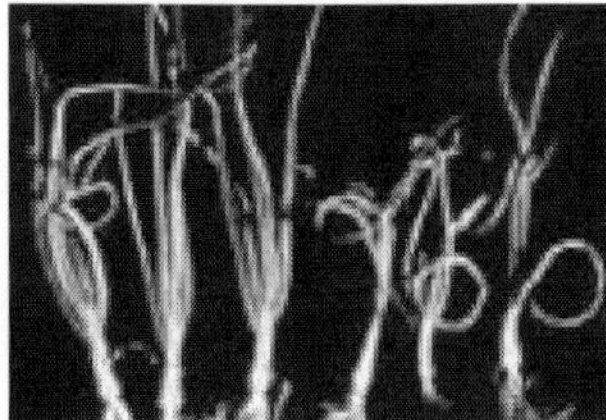	Zn deficiency may occur in soils having high pH coupled with high P content as under these conditions Zn becomes unavailable to the plants (researchgate.net) so growth is restricted with interveinal necrosis in leaves. In the early stages of deficiency the younger leaves become yellow and pitting develops in the interveinal upper surfaces of the mature leaves. As the deficiency progresses these symptoms develop into an intense interveinal necrosis but the main veins remain green {vikaspedia.in (e)}.	Soil application of zinc sulphate @ 10kg/ha is recommended for soils having low zinc content. In such cases, twice fortnightly foliar spraying of 0.25-0.5% zinc sulphate is recommended (researchgate.net; vikaspedia.in (e)).
Iron {vikaspedia.in (e)} 	Iron deficiency usually occurs in alkaline soils having more calcium carbonate (researchgate.net) where complete yellowing of young leaves with interveinal chlorosis, followed by an overall chlorosis which end up into totally bleached leaf. Since iron has a low mobility, iron deficiency symptoms appear first on the youngest leaves. Iron deficiency is strongly associated with calcareous soils, anaerobic conditions, and is often induced by an excess of heavy metals {vikaspedia.in (e)}.	When the deficiency of iron is diagnosed in field, iron chelates @ 0.2-0.5% Fe is recommended to correct the deficiency (researchgate.net). Foliar spray of F_eSO_4 @ 0.5% {vikaspedia.in (e)| is quite useful.
Boron {(researchgate.net)} 	Boron deficiency severely affects onion and garlic growth. Leaves become misshapen, wrinkled, thicker and darker. Emergence of young leaves will be restricted with 3-4 small leaves in B-deficient plants instead of usually 12-13. Leaves and stems are brittle due to poor cell wall formation or due to accumulation of phenols. With increased severity, the leaves spiral or twist (researchgate.net).	Foliar application of borax @ 0.1-0.2% through boric acid is recommended to overcome the boron deficiency (researchgate.net).

Available micronutrient content and critical limit of micronutrients in Indian soils

Source (researchgate.net).

Element	*Available nutrient content (mg/kg)*	*Critical limit content (mg/kg)*
Zinc	0.1-24.6	0.6
Boron	0.08-2.6	0.5
Iron	0.36-174	4.5
Manganese	0.6-164	2.5
Copper	0.1-32	0.2

Insects–pests and diseases

Of the several problems with which onions suffer the most serious for the home gardeners are likely to be the onion fly, stem and bulb eelworm, white rot and neck rot (en.wikipedia.org/wiki/Onion). Diseases such as rust and smut, downy mildew and white tip affect the foliage. The bulbs may get affected by splitting, white rot and neck rot. Most of these disorders are best treated by removing and bur n ing affected plants (Hessayon, 1978). The larvae of the onion leaf miner or leek moth (*Acrolepiopsis assectella*) sometimes attack the foliage and may burrow down into the bulb (Landry, 2007).

Insects–pests

Onion thrips (*Thrips tabaci* Lindeman) and Western flower thrips (*Frankliniella occidentalis* Pergande, 1895)

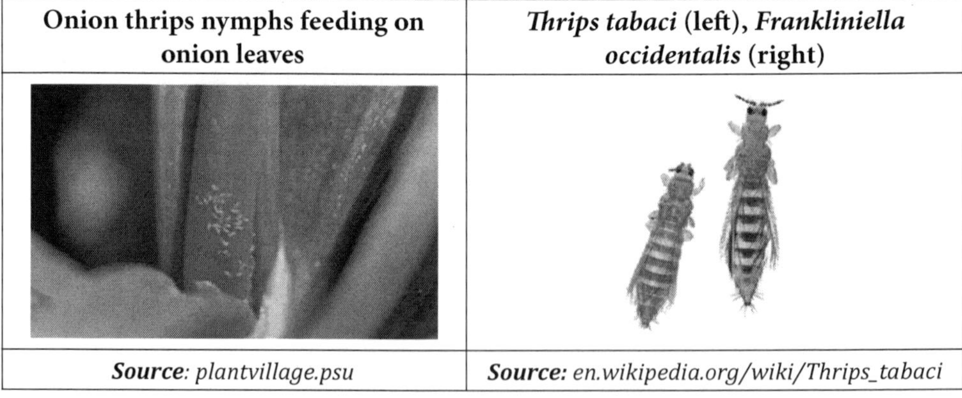

Onion thrips nymphs feeding on onion leaves	***Thrips tabaci* (left), *Frankliniella occidentalis* (right)**
Source: *plantvillage.psu*	***Source:*** *en.wikipedia.org/wiki/Thrips_tabaci*

Thrips are most damaging at the early bulbing stage of the plant. Both these thrips have an extensive host range so from other plants these migrate on onion crops (plantvillage.psu). Both adult and nymphs feed within the mesophyll layer using a punch-and suck motion. A significant portion of onion and garlic yield is lost due to onion thrips, particularly *Thrips tabaci* that remains the major pest species worldwide though other species, *i.e. Frankliniella occidentalis* and *F. fusca* have been recorded only in some areas though their damage is not so extensive (Lawande *et al.*, 2009).

In addition to direct damage, thrips infestation aggravates the infection of fungal pathogen, especially *Alternaria porri* that causes purple blotch through mechanical transmission (Bhangale and Joi, 1983). Similarly, severity of *Stemphylium*blight also increases in presence of large number of thrips. *Thrips tabaci* also acts as a vector of the 'iris yellow spot virus' which has a detrimental effect on bulb and seed yield (Gent *et al*, 2004; Hoepting *et al*, 2007).

Feeding area appears white to silvery and the leaf injury can be seen as patches and streaks. In severe cases, the leaves take on a silvery cast and can wither. Tiny black 'tar' spots of thrips excrement are evident on leaves with heavy feeding injury [vikaspedia.in (d)].

Field sanitation; use of resistant/tolerant varieties and avoiding the successive planting of garlic or other preferred/alternate host such as cabbage, cotton, tomato, cucumber, melons, pumpkins, strawberries, *etc.* [vikaspedia.in (b)] should be taken into account. Growers should avoid planting onion in close proximity to grain fields as thrips population build up on these plants in the spring. Overhead irrigation of plants may help reduce their population (plantvillage.psu). Reflective silvery plastic, *vis-a-vis* painted black aluminium mulches repel the thrips. Thrips are weak fliers and can be carried by wind, therefore, planting live-barriers such as maize can effectively block adult thrips from reaching onion plants. Planting two rows of maize or inner row of wheat and outer row of maize surrounding the onion plots will act as barrier crop [vikaspedia.in (b)], *vis-a-vis* yellow and bright blue sticky traps should widely be used to control these (Lawande *et al.*, 2009). Some species of predatory mite, pirate bugs and natural lacewing enemies of thrips (plantvillage.psu) feed on these. Conserving parasitoids such as *Ceranisus menes* (nymph) *etc.*; predators such as syrphid flies, minute pirate bug/anthocorid bug (*Blaptosthethus* sp., *Buchananiella whitei, Orius tantilus*), praying mantis, predatory thrips (*Aeolothrips fasciatum*), damselbug, lacewings, coccinellids (*Cheilomenes sexmaculata*), spiders, *etc.* minimize thrips population [vikaspedia.in (b)].

Application of appropriate insecticides at first sign of thrips damage; spray of dimethoate 30% EC @ 660ml in 500-1000 l of water/hectare, oxydemeton methyl 25% EC @ 1200ml in 500-1000 l of water/hectare, quinalphos 25% EC @ 1200ml in 500-1000 l of water/hectare or lambda cyhalothrin 5% EC @ 300ml in 300-400 l of water/hectare is quite useful. Spraying should be under high pressure and volume for good coverage of the crop by directing the nozzle at base of leaves. There should be addition of a spreader or sticker (0.05-1.0%) for retention and spread of spray fluid on erect leaf of onion. In seed crop, thrips should be controlled well before opening of flowers [vikaspedia.in (b)].

Onion maggot (*Delia antiqua* Meigen)

Onion maggot and bulb damage	Onion maggot and bulb damage
Source: *plantvillage.psu*	***Source:*** *en.wikipedia.org/wiki/Delia_antiqua*

Greyish adult fly lays white elongated eggs on the leaves and stems and around the base of the onion, shallot, leek and garlic plants. Females can lay several hundred eggs during their 2-4 week lifespan. These insects overwinter as pupae in the soil (plantvillage.psu). The fly is attracted to the crop by the smell of damaged tissue which is liable to occur after thinning (en.wikipedia.org/wiki/Onion). Plants grown from sets are less prone to attack. White tiny larvae emerging from the eggs bore into the onion plant. Mature larvae are about 1cm long with feeding hooks (plantvillage.psu). Only the larva causes damage by using its hooked mouth parts to enter the base of the plant (plantvillage.psu).

The larvae tunnel into the bulbs so foliage turns yellow and wilts; bulbs disfigure and rot especially in wet weather and the seedlings stunt or wilt (en.wikipedia.org/wiki/Onion). In its infestation the plant usually break at soil level if infestation occurs when plants are in bulbing stage. In its infestation, the bulbs are deformed and become susceptible to storage rot after harvest (plantvillage.psu). Frequently attacked seedlings die before the maggots are fully grown and then these maggots move to adjacent plants; second generation maggot feeding on developing bulbs usually results in distorted growth accompanied by rotting tissues; and feeding by third generation maggots on late season onion bulbs results in an unmarketable product [vikaspedia.in (d)].

Wider spacing and floating row covers may provide protection by preventing females from laying eggs around the plants; crop rotation; field sanitation; removal of debris and bulbs at the end of the season to starve the maggots without a food source; use of seed dressings; early sowing or planting; and removal of infested plants (en.wikipedia.org/wiki/Onion) are some of the precautionary measures.

Conservation of predators such as ground beetle, rove beetles, spiders, *etc.* by providing grassy refuge strips [vikaspedia.in (b)] will also control this pest. Commercial onion growers often apply appropriate granular insecticides, *vis-a-vis* insecticidal sprays when required (plantvillage.psu).

Leafminers [*Lyriomyza* spp. (*Lyriomyza* Mik, 1894)]

Adult leafminer is a small black *cum* yellow fly which lays its eggs inside the leaves which after hatching into larvae start feeding on leaf interior (plantvillage.psu). Mature larvae drop from leaves to pupate into the soil and entire lifecycle can take as little as 2 weeks in warm weather. Insect may go through 7 to 10 generations per year. Thin, white, winding mining trails on leaves appear in its attack. Heavy mining results leaves to fall from the plant prematurely. Early infestation causes reduced bulb yield.

Management: Transplants for signs of leafminer damage should be checked prior to planting. Insecticides should be used only when leafminer damage has been identified as unnecessary spraying will also reduce populations of their natural enemies (plantvillage.psu).

Leek moth [*Acrolepiopsis assectella* (Zeller, 1839)]

Leek moth with mottled brown and white wings that span approximately up to 12mm across (Landry, 2007) is a pest of leek crops as the larvae feed on several species of *Allium* by mining into the leaves or bulbs (Mason *et al.*, 2010). The leaf mining can occur in the tubular leaves or in the stem. In case of onions and shallots, the larvae mine into the bulb. Larvae feed on plants of the genus *Allium*, including onion (*Allium cepa*), garlic (*Allium sativum*), leek (*Allium porrum*), elephant garlic (*Allium ampeloprasum*), shallot (*Allium cepa* var. *aggregatum*), chive (*Allium schoenoprasum*), nodding onion (*Allium cernuum*) and Prairie onion (*Allium stellatum*) (Landry, 2007; Mason *et al.*, 2013). The preference of leek moth is for garlic, leek and onion compared to other *Allium* species (Mason *et al.*, 2013).

Plants infested by leek moth may rot and die. It makes white patches on leaves and tunnels or mines in the stems.

It is managed through trap-cropping and intercropping (Asman *et al.*, 2001; Asman, 2002). Its menace can be checked by spreading an insect-proof mesh as a means to protect the crops. Regular inspection and picking off pupae will help controlling the future generations.

Bulb mite (*Rhizoglyphus robini* Claparede)

Although not considered a primary pest of bulbs, bulb mites are often responsible for serious losses. They enter and establish themselves even through the slightest injuries to a bulb [vikaspedia.in (d)]. Damage to plants by bulb mites allows secondary invasion by other pathogens which cause bulb rot (plantvillage.psu). Apparently bulb mites can attack healthy new roots and bulbs, especially in greenhouses.

The pest is a cream-white bulbous mite <1mm in length which resembles a pearl with legs (plantvillage.psu). Its invasion causes reduced stand, stunted growth, bulbs failing to produce new growth or the new growth may be off-coloured, distorted and stunted [vikaspedia.in (d)], *vis-a-vis* bulb rotting may occur in the ground as well as in storage.

Growers should avoid planting successive onion or garlic crops. Its planting should also be avoided after the cole crops because decaying debris of cole crops, especially the cauliflower may harbour a very high population of bulb mite in the field. Flood irrigation reduces mite levels in the soil [vikaspedia.in (b)].

Conservation of predators such as anthocorid bugs (*Orius* spp.), mired bugs, syrphid/hover flies, green lacewings (*Mallada basalis* & *Chrysoperla zastrowi*), predatory mites (*Amblyseius alstoniae, A. womersleyi, A. fallacies* & *Phytoseiulus persimilis*), predatory coccinellids (*Stethorus punctillum*), staphylinid beetle (*Oligota* spp.), predatory cecidomyiid fly (*Anthrocnodax occidentalis*), predatory gall midge (*Feltiella minuta*), spiders, *etc.* [vikaspedia.in (b)] reduce the mite population drastically. This pest has developed resistance to many insecticides though metam sodium fumigation is very effective, *vis-a-vis* various carbamates and organophosphates also control this pest effectively. Onions may be dipped in an organophosphate (like chlorpyrifos) before planting.

Eriophyid mite [*Aceria tulipae* (Keifer)]

Adults and nymphs both feed on young leaves and in between the bulb scales of onion [vikaspedia.in (d)]. Its infestation causes plant stunting *cum* twisting, scarification, curling and discoloration of foliage and drying of bulb tissues. This damage has been attributed to various viruses thought to be transmitted by the mites [vikaspedia.in (d)]. Avoiding planting of successive onion or garlic crop in the same field will minimize its incidence. Flood irrigation is also quite effective to minimize its menace. Its biological control is the same as for bulb mites [vikaspedia.in (b)].

Red spider mite [*Tetranychus cinnabarinus* (Boisduval)]

Adults and nymphs feed primarily on the undersides of the leaves [vikaspedia.in (d)]. Upper leaf-surface becomes stippled with little dots that are the feeding punctures. The mites tend to feed in pockets, often near the midrib and veins of leaves producing visible silky-webbing which causes the leaves eventually becoming bleached and discoloured and may fall off [vikaspedia.in (d)].

A thorough water spray washes off the mites from the plant [vikaspedia.in (b)]. Biologically it is controlled as in case of 'bulb mites' [vikaspedia.in (b)].

Onion eelworm/stem nematode/stem and bulb eelworm/onion bloat (*Ditylenchus dipsaci*).

These tiny parasitic nematodes, with an extensive host range, penetrate into plants either from soil or infested planting material and occasionally from seeds. They live between the cells of onion or garlic leaves and/or bulb scales where these feed on cell sap and multiply (en.wikipedia.org/wiki/Ditylenchus_dipsaci). The parasite enters the plants through wounds or stomata creating galls or malformations in plant growth and allowing for the entrance of secondary pathogens such as fungi and bacteria. Some other plant hosts of this nematode are peas, beetroot, vegetable marrow, pumpkin, rhubarb and or n amental bulbs. Some weeds also act as hosts including *Stellaria media*, *Linaria vulgaris*, *Polygonum aviculare*, *Fallopia convolvulus* and *Galium aparine*.

In alliums, it primarily infests onions, garlic (Aftalion and Cohn, 1990) and leeks causing stunted growth, leaf curl, yellow spots and foliage lesions. Stems often have swollen regions called spikkles. Young plants are killed and older ones produce soft bulbs. As adult nematodes migrate into the scales of the bulb, the scales become soft, grey and loosely packed.

Highly infested bulbs can also split apart or show malformed bloating. The leaves of the plant become flaccid and may collapse. This can lead to defoliated plants, *vis-à-vis* missing portions of the root system when harvested (Agrios, 1997). Garlic shows similar symptoms of leaf yellowing and stunted bulbs (Mollov *et al.*, 2012). It includes seed *cum* field sanitation, heat treatment, crop rotation and field fumigation. Only clean seeds and bulbs should be used. Heat treatment can be given by soaking them in 43.4-53°C water with formalin for 2-3h which will kill the nematodes successfully (Qui *et al.*, 1993). Affected plants should be uprooted and burnt.

The site should not be used for growing onions again for several years and should also be avoided for growing carrots, parsnips and beans as these crops are also susceptible to eelworm. Cooler temperatures and lower humidity can suppress the infestation rates. Field fumigation during fall can control nematodes on a susceptible crop in the spring.

Fungal diseases

Alliums are infected by several soil-bor n e fungal pathogens. Early infection can result in collapse of seedlings or failure of emergence which may be aggravated by drought. The major root-infecting fungal diseases of onion and garlic are white rot (*Sclerotium cepivorum*), pink root (*Phoma terrestris*, syn *Pyrenchaeta terrestris*), basal rot (*Fusarium oxysporum* f. sp. *cepae*, *F. oxysposum* f. sp. *allii*), southern blight (*Sclerotium rolfsii*), onion smut (*Urocystis cepulae*) and damping-off (*Pythium* spp., *Fusarium* spp., *Rhizoctonia solani*) (Lawande *et al.*, 2009). Major foliar diseases of onion and garlic in India are purple blotch (*Alternaria porri*), blight (*Stemphylium vericariuno*), anthracnose/Twister disease, downy mildew and black mold which usually defoliate the crop prematurely (Lawande *et al.*, 2009). Black mold is the most important post-harvest disease under hot climates, followed by onion smudge first reported by Singh (2007).

Black mold (*Aspergillus niger* van Tiegham)

The fungus infects onion bulbs preferably through injured portion of the outer scales and colonizes bulbs, roots, neck, flowers, peduncle and leaves of onion plants in the field in that order of preponderance (Rajasab and Rao, 1992). Spores of this fungus are very common in air and soil. Fungus survives in air, soil and infected plant debris [vikaspedia.in (c)]. Free moisture for six hours or longer on the onion surface is necessary for infection to occur. Black mold is most common when temperatures are higher than 30°C in the field or 24°C in storage.

Post-harvest black neck discolouration occurs in the form of lesions on outer scales; black streaks under outer dry scales; and entire surface of bulb turning black and shrivelling (plantvillage.psu). Infection may advance from the neck into the central fleshy scales [vikaspedia.in (c)].

Symptoms of black mold

Source: *plantvillage.psu*

Seeds should be treated with appropriate fungicide prior to planting to reduce rot in mature bulbs (plantvillage.psu); bruising bulbs during and after harvest should be avoided [plantvillage.psu; vikaspedia.in (b)]; onions after harvesting should be left for drying in the field for two days and then dried further in shade for 10-15 days before storage [vikaspedia.in (b)]; properly dried seeds and bulbs should be stored at temperatures below 15°C to prevent mold from spreading but it may resume once temperature increases [plantvillage.psu; vikaspedia.in (b)]; and proper aeration in the storage is must. Ahir and Maharishi (2008) reported 0.2% pre-harvest spray (20 days before harvesting) of 12% carbendazim + 63% mancozeb *cum* Iprodion as quite effective.

Botrytis leaf blight (*Botrytis squamosa* J.C.Walker)

This pathogen affects only *Allium* spp. such as onion, garlic and leek (en.wikipedia.org/wiki/Botrytis_squamosa). Disease emergence is favoured by high humidity and warm temperatures. Fungus survives either in soil or on piles of crop debris. On plants older leaves are more susceptible to blighting than younger leaves (plantvillage.psu).

Botrytis leaf blight

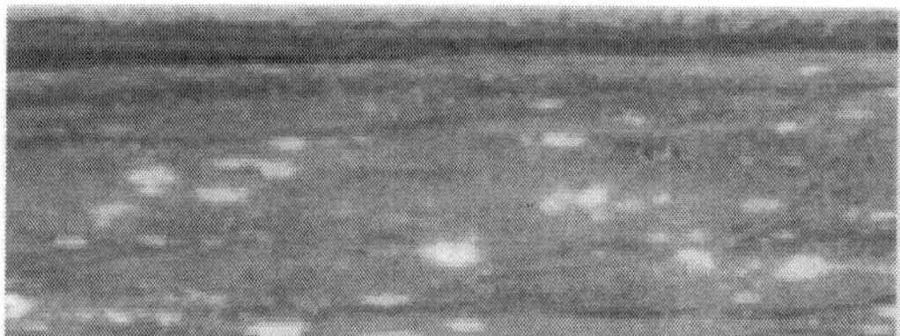

By Rasbak - Own work, CC BY-SA 3.0, https://commons.wikimedia.org/w/index.php?curid=11169508

First symptom occurs on the leaves 24–48 hours after initial exposure to the pathogen, appearing as long white lesions 1–5mm in length, surrounded by a greenish-white halo (en.wikipedia.org/wiki/Botrytis_squamosa). These small white lesions with light green halos may expand slightly as they age, and in prolonged periods of moisture fungus may develop rapidly and cause leaf blighting (plantvillage.psu).

Cultural controls include avoiding cull piles and 2–3 years of rotation with non-host (species other than *Allium*) plants; discarding damaged and diseased onions at harvest; planting in single rows and allowing at least 30cm between plants to promote good air circulation and quick drying of foliage after rain; and allowing plants to dry out sufficiently.

Under chemical control, when plants have attained at least five true leaves and disease symptoms start appearing, 0.2% dithiocarbamates have been identified as the most effective protectant fungicides (en.wikipedia.org/wiki/Botrytis_squamosa).

Botrytis neck and bulb rot [*Botrytis allii* Munn (1917) and *Botrytis aclada* Fresenius]

It is a disease of onion, garlic, leek and shallot. *Botrytis allii* and *Botrytis aclada* cause this disease in onion and *Botrytis porri* in garlic. The fungus spreads through airborne conidia. It may also be seedborne. Sclerotia are formed in the soil which persist until the next season. In both onion and garlic, initial infections may not be noticeable, and symptoms may develop only when leaves senesce and become necrotic. The greatest incidence of infection occurs in persistent cool (10-24°C) and moist weather. Infections are favoured by excessive use of nitrogenous fertilizers, applications of nitrogen after bulb initiation, and poor drying of the tops *cum* curing of the necks. Healthy onions that are stored properly and harvested at the right time are seldom affected.

In onions, *Botrytis* bulb rot generally appears during storage although infection originates in the field. In the stored crop, the rot typically starts in the neck region and upper parts of the bulb but can occur in other parts if there is physical injury (Chilvers and du Toit, 2006). The scales inside the bulb become progressively translucent and watery and a mycelium develops between them. A mass of grey conidiophores and conidia develop on the mycelium and blackish sclerotia form at the site of initial infection (Chilvers and du Toit, 2006). In time, a dry rot sets in and the bulb becomes a dry and mummified structure. In a humid atmosphere, a gray and feltlike growth (where spores are produced) appears on rotting scales, and mycelia may develop between scales. Dark-brown-to-black sclerotia (the resting bodies of the pathogen) may eventually develop in the neck or between scales.

Cultural controls include harvesting of bulbs only when crop is perfectly mature and the necks are well cured. During growing season, there should not be bulb damage, *vis-a-vis* excessive or late application of nitrogenous fertilizer should be avoided. Late-season watering to facilitate harvesting should be avoided so that foliage and neck dry properly before harvest. Before storage of bulbs, the neck tissue should be well cured. Bruised and wounded bulbs should be sorted out so that in storage pathogens may not enter the bulbs. Bulb storage temperature should be 5^{0}C or below with minimum relative humidity and good air circulation. Wide fluctuations in temperature can cause condensation on the bulbs which favours the development of the disease. Seed and bulb treatment with fungicides prior to planting can reduce incidence of *Botrytis* neck rot. However, late fungicide applications during last month of harvest are most effective at preventing infection of the necks since late-season infection poses the greatest risk for bulb rot in storage. Not chemical but only cultural practices may be the most reliable way to avoid infection once the symptoms develop at the time of lifting.

Botrytis allii grows on *Allium* species including onion (*A. cepa*), aggregating onion (*A. cepa* var. *aggregatum*), garlic (*A. sativum*) and leek (*A. porrum*). It may also infect wild *Allium* species and can grow saprophytically on decaying crop residues such as cereal, pea and bean straw. It is capable of colonizing and producing spores on sterilized poppy straw (*Papaver somniferum*) (Chilvers and du Toit, 2006).

Cercospora leaf spot/blight (*Cercospora duddiae* Welles)

This disease appears on leaves as ash-coloured, small and irregularly shaped spots scattered on leaf lamina which after gradual coalescing result in blightening of foliage (eagri.org2).

Spraying of ziram or captan @ 2g/litre of water or copper oxychloride @ 3g/litre of water at fortnightly intervals gives good control (eagri.org2).

Colletotrichum blight/Anthracnose/Twister disease {*Colletotrichum gloeosporioides* [(Penz.) Penz. and Sacc]}

Ebeneb (1980) reported that onion twister disease and onion anthracnose are caused by *Colletotrichum gloeosporioides* whose perfect stage is *Glomerella cingulata*. The fungus can survive for many years as sclerotia in the soil or for shorter periods on infected plant debris [vikaspedia.in (c)]. It is most severe in warm [25-30°C] and moist soils that are high in organic matter, though rapidly decreases below 15°C resulting in little disease development [vikaspedia.in (c)].

Initially the symptoms appear on the leaves as water-soaked pale yellow spots which spread lengthwise covering entire leaf blade so the affected leaves shrivel and droop down [vikaspedia.in (c)]. Characteristic field-symptoms are curling, twisting and chlorosis of leaves and abnormal elongation of the neck (false stem) and the bulbs becoming smaller in size where some even rot before the harvest while others rot in store (Lawande *et al.*, 2009).

Curling and abnormal elongation due to *Colletotrichum* blight

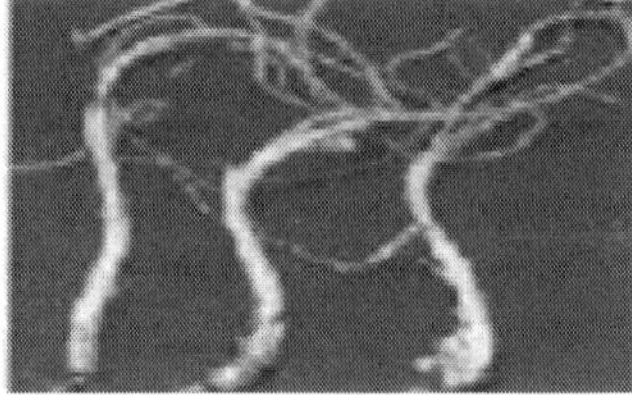

Source: *vikaspedia.in (c)*

This disease can be managed by destroying infected plant debris and through use of resistant varieties such as 'IPA 3', 'Belem', 'IPA 9', 'Franciscana IPA 10', 'Vale Ouro IPA II' and 'Roxinha de Belem' (Assuncao *et al.*, 1999). Application of Benomyl @ 0.2% as soil-treatment [vikaspedia.in (b); Remiro and Kirmati, 1975] or 0.25% Mancozeb spraying gives good control (Lawande *et al.*, 2009).

Damping off [*Pythium* spp., *Rhizoctonia solani* (Kuhn), *Fusarium oxysporum* (Schlecht), *F. oxysporum* f. sp. *cepae* (Schlecht. emend. Snyder & Hansen); *Sclerotium rolfsii* (Sacc.), *S. cepivorum* (Berk) *and Colletotrichum* sp.]

Pythium is a genus of parasitic oomycetes, a distinct phylogenetic lineage of fungus-like eukaryotic microorganisms. Pathogen(s) survives on infected crop debris and soil which are source of primary inoculums. The disease is more prevalent during *Kharif*/rainy season and causes about 60-75% damage. High soil moisture and moderate temperature along with high humidity lead to the development of the disease [vikaspedia.in (c)].

Wilt due to *Fusarium* damping-off

Source: *plantvillage.psu*

Rotting seeds are covered in mold with discoloured root tips which may be pink, tan, yellow, red or black, and thus slowly growing seedlings wilt and die (plantvillage.psu). Pre-emergence damping-off results in seed and seedling rot. In post-emergence damping-off, the pathogen attacks the collar region of seedlings on the surface of soil where the collar portion rots and ultimately the seedlings collapse and die [vikaspedia.in (c)].

Apart from using well-decomposed FYM in the nursery, solarization of nursery beds with transparent polythylene sheet for 30 days before sowing in summer provides good control of all the diseases and pests present in the soil [vikaspedia.in (b)]. Raised beds with at least 15cm height will have ample access to drainage and so there won't be any waterlogging problem [vikaspedia.in (b)]. Excessive watering should be avoided.

Watering should be light and that too in the morning or by the evening, if possible but not under full sun. Only disease-free seeds or planting material should be used and that too after seed/bulb treatment. To reduce the level of pathogens in the soil, crop rotation with cereals or grasses should be followed. Before sowing/planting the field should be given steam treatment or fumigated to reduce levels of *Fusarium* in the soil (plantvillage.psu). Dense sowing/planting of onion seeds/plants should also be avoided.

Downy mildew [*Perenospora destructor* (Berkeley) Caspary]

Downy mildew is a serious problem throughout the onion growing areas world over where these are grown in cool and humid conditions (Lawande *et al.*, 2009). Fungus survives in soil and infected plant debris and its development is favoured by warm and humid climate [vikaspedia.in (c)]. Cultural practices such as late planting, poor drainage, higher dose of fertilizer and frequent irrigation encourage high disease-incidence (Ahmad and Karimullah, 1998).

Downy mildew symptoms

Source: *plantvillage.psu*

Cottony white mycelial growth develops on leaves which gradually turn pale-yellow to dark brown and so the leaves dry up [vikaspedia.in (c)].

If possible only resistant cultivars should be selected for sowing/planting. Crop rotation can be followed for 3-4 years with non-host crop and only healthy bulbs should be selected for seed production (Lawande *et al.,* 2009). Infected crop debris and refuse heaps of onion culled from storage should be disposed off properly. Late planting, poor drainage, higher doses of fertilizers and frequent irrigation should be avoided (Ahmad and Karimullah, 1998). Onion bulbs meant for seed crop should be exposed to sun for 12 days to destroy the fungus [vikaspedia.in (b)].

Allium roylei possesses resistance to downy mildew (Kofoet and Zinkernagel, 1990). Onion lines 'IC-48045', 'IC-32149', 'IC-49371' and 'DOP-2' have been reported to be resistant to downy mildew (Sharma, 1997). Spraying zineb 75% WP @ 1.5-2kg in 750-1000 litres of water/hectare is quite effective against this problem [vikaspedia.in (b)]. SprayingMancozeb@ 0.25% and Ziram @ 0.1% at

10-12 day intervals is quite useful (Marikhur et al, 1977). Bulb and seedling-dip in Ridomil MZ @ 0.25% for 12h followed by 2 foliar sprays give effective disease control. Metalaxyl and Cyomaxanil have been found most effective in reducing disease severity upto 88% (Palti, 1989). The low degree of fertility exhibited by hybrids between *A. cepa* and other *Allium* species restrict successful introgression of disease resistance.

Basal *Fusarium* rot (*Fusarium oxysporum* f. sp. *cepae* Schlecht)

Fusarium basal rot of onion and garlic occurs worldwide and is a common problem in onion seed crop. The chlamydospore (resting spore) of this fungus survives in soil and onion bulb for many years as a primary source of inoculum. The development of disease is favoured by moist soil and 27°C temperature [vikaspedia.in (c)].

Discoloration around basal plate due to Basal rot

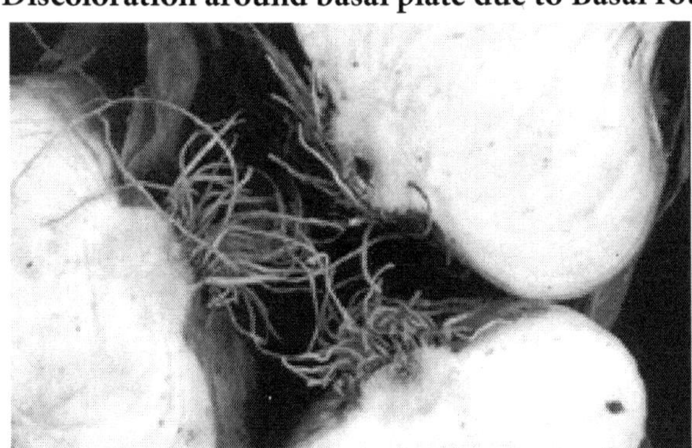

Source: *plantvillage.psu*

Curved, yellowed or necrotic leaves appear, the necrosis starting from the leaf tips to downward with ultimate plant wilting. Infected bulbs may be brown and watery with rot spreading as browning from stem plate to basal leaves (plantvillage.psu).

Rotation with non-susceptible crops for at least 4-5 years; mixed cropping with tobacco and sorghum; growing resistant onion varieties [plantvillage.psu; vikaspedia.in (b)]; avoiding injuries during cultural practices; maintaining proper drainage; deep summer ploughing and bed solarization by using polythene sheet of 25 guage in summer season for 30 days reduces the pathogen inoculum [vikaspedia.in (b)].

Pre-harvest spray of Carbendazim (0.1%) results in least decay of stored onion after 5 months (Srivastava *et al,* 1996). Girija *et al* (1998) found three lines, *viz.* 'IIHR 141', 'IIHR 506' and 'Selection 13-1-1' to be consistently resistant to *Fusarium oxysporum* in the field under different growing seasons.

Green mold [*Aspergillus fumigatus* Fresenius, *A. alliaceus* Thom & Church]

Spores of these fungi are very common in air and soil which are the sources of infection. Fungus survives in air, soil and infected plant debris [vikaspedia.in (c)]. Free moisture for six hours or longer on the onion surface is necessary for infection to occur. Black mold is most common when temperatures are higher than 30°C in the field or 24°C in storage.

Infection usually occurs through neck tissues when foliage dies down at maturity. In this case infected bulbs are discoloured green around the neck, and affected scales show shriveling along with masses of powdery green spores generally arranged as streaks along veins on and in between outer dry scales. Later on, infection may advance from the neck into the central fleshy scales. Its management is similar to black mould.

Pink root rot [*Phoma terrestris* (Hansen) Gorenz, Walker & Larson, *Fusarium solani* (Mart.) Sacc.]

Fungus can survive in the soil, in diseased roots and on the debris of susceptible crops for several years and can be spread through soil movement and with irrigation water [vikaspedia.in (c)]. Fungus colonizes plant through root tips and can survive in soil down to a depth of 45cm (plantvillage.psu).Reduced bulb size *cum* water-soaked and/or transparent roots which turn pink to maroon or purple and the plants looking as if having nutrient deficiency [vikaspedia.in (c); plantvillage.psu]. In severe cases the roots may die and the plants become weak, stunted and undersized, and the bulbs shrivel especially in drier areas of the field [vikaspedia.in (c); plantvillage.psu] whereas infected seedlings may die [plantvillage.psu].

Pink root symptoms

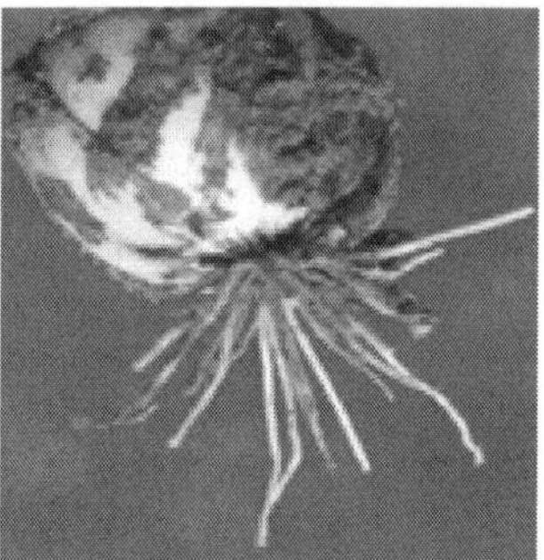

Source: *plantvillage.psu*

A rotation of 3-6 years with cereal and green manure crops is preferred. Resistant varieties or Welsh onion (*Allium fistulosum*) being resistant to pink root should be used [vikaspedia.in (b); plantvillage.psu]. Solarization through polythene sheet of 25 guage in summer for 30 days will reduce the pathogen inoculum and in turn will reduce the disease [vikaspedia.in (b)] in the field. Fumigation will help reducing the levels of pathogen in the soil (plantvillage. psu). Late planting will reduce disease incidence [vikaspedia.in (b)].

Powdery mildew [*Leveillula taurica* (Lev.) G. Arnaud (1921)]

This disease prefers warm *cum* dry environment. It is rare in United States, and is currently restricted to western states. Globally, it is a minor problem with limited occurrences in the Middle East, Europe and South America. *Leveillula taurica* causes powdery mildew of onions, but is also known to infect other alliums, solanaceous crops and cucurbits (en.wikipedia.org/wiki/Leveillula_taurica).

Distinct pale-yellow circular or oblong lesions of variable sizes on abaxial surface of leaf associated with powdery mass are its main characteristics (eagri. org2). These have chlorotic or necrotic appearance. The lesions having white mycelium with conidiophores bearing either lanceolate or rounded conidia appear on older leaves before the bulb of the onion begins to form, though also occur on the younger leaves towards the end of the season (en.wikipedia.org/wiki/Leveillula_taurica).

The crop residues from previous onion crop should be removed and field should be deeply tilled. Same field should not be used for growing *Allium* crop again, however, a non-host crop should be grown. Onions with adequate moisture will be more resistant to the pathogen, and onion crops with overhead irrigation rarely see powdery mildew attack in the field (en.wikipedia.org/wiki/Leveillula_taurica). After appearance of the disease, regular fortnightly sprayings of sulphur fungicides @ 2g/l gives good control (eagri.org2).

Purple blotch [*Alternaria porri* (Ellis)]

This soil-borne fungus survives in the soil as well as in the infected bulbs, and may persist in plant debris or on roots of weeds. The development of the disease is favoured by hot and humid climate with temperature ranging from 21 to 30°C and high relative humidity (80-90%) [vikaspedia.in (c)]. Disease emergence is favoured by wet foliage, with sporulation occurring in the night during periods of high humidity (plantvillage.psu). Small water-soaked lesions on leaves or stalk with white centres appear which enlarge to become zonate and brown to purple with red or purple margin surrounded by yellow zone. Large lesions may coalesce and girdle the leaf killing any tissue between the lesions and the leaf tip. Severely infected foliage may die (plantvillage.psu) and infected plants fail to develop bulbs [vikaspedia.in (c)].

Disease can be managed through use of healthy seeds for planting; 2-3 years crop rotation with non-host crops; creating proper drainage facilities; use of recommended doses of N and P fertilizers; hot water soaking of onion seed (50°C for 20 minutes) and use of resistant varieties [vikaspedia.in (b)]. Chemical control includes spray of difenaconazole 25% EC @ 100ml in 500 litres of water/hectare or kitazin 48% EC @ 200ml in 200 litres of water (or as required depending upon crop stage and plant protection equipment used [vikaspedia.in (b)].

Badly Damaged leaves	**Infection on stem**
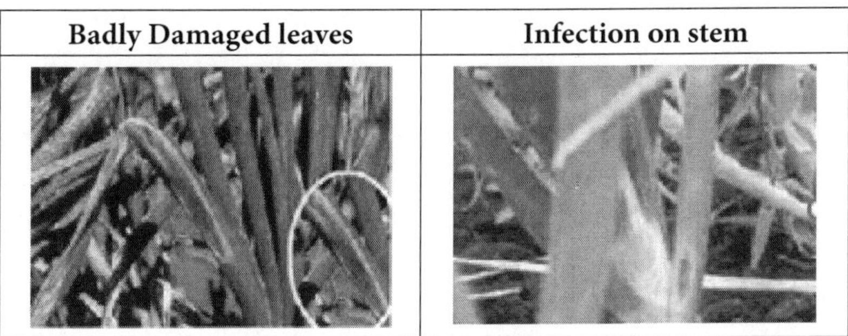	

Onion varieties reported to be moderately resistant are 'Agrifound Light Red' (Sharma, 1997), '53-3' (Pandotra, 1965), 'Agrifound Dark Red', 'Red Globe' (Sugha *et al.*, 1992), 'VL Piyaz 3' (Mani *et al.*, 1999) and 'RO 59' (Mathur *et al.*, 2006). Application of Mancozeb (0.25%) and Captafol@ 0.2% (Gupta *et al*, 1986), Iprodione @ 0.25% (Gupta *et al*, 1996a,b) or Metalaxyl and Dinocap (Upadhyay and Tripathi, 1995; Srivastava *et al.*, 1996) were found effective in controlling the disease.

Rust/ Leek rust [*Puccinia porri* (Sowerby) G.Winter (1881)]

The fungus causes leek rust, but it also affects garlic, onion and chives (en.wikipedia.org/wiki/Puccinia_porri). It develops more quickly under high humidity but low rainfall and the spores can be transported over long distances by wind (plantvillage.psu).

Rust symptoms on onion leaves	**Orange rust pustules on onion leaves**

By Rasbak- Own work, CC BY-SA 3.0, https://commons.wikimedia.org/w/index.php?curid=358054	***Source:*** *plantvillage.psu*

Small white flecks, later on turning circular to elongated orange pustules on leaves and stems appear. Severe infections can cause leaves to yellow and die. Since no resistant varieties are yet known against onion rust hence only disease-free seed should be used and planting should be done only in well-draining soil. Weeds around the crop should be controlled regularly. Appropriate protective fungicide can be used.

Smut [*Urocystis colchici* (Schltdl.) Rabenh. or *U. cepulae*]

Smut found in almost all onion-growing states, is caused by soil-bor n e fungi, *Urocystis colchici* or *U. cepulae*, which primarily affects seedlings. Onions are susceptible to smut infection only during germination and emergence, after which plants become resistant. Susceptibility decreases as tissue ages, and transplant seedlings are rarely attacked. The pathogens attack the members of the onion family including onion, leek and green bunching onion but is a serious problem on seeded onions grown in muck soils. The smut pathogens persist in soil for many years and are mainly introduced through infected sets, transplants, and by the movement of soil particles or infected crop debris. Its spores spread through wind, equipments or irrigation water (wiki.bugwood.org/HPIPM:Onion_Smut).

Symptoms appear on cotyledons (seed leaves) and young leaves as thickened longitudinal blackish blisters with a silver sheen. Blisters may become large causing leaves to bend downwards. Lesions mature and become covered with black powdery fungal masses. Due to infection the plant growth stunts and seedlings often die before or within six weeks after emergence (plantvillage.psu; wiki.bugwood.org/HPIPM:Onion_Smut).

Symptoms of smut

Source: *plantvillage.psu*

Three-year or longer crop rotation to non-hosts such as small grains and corn should be practiced. In infested soils the transplants should be used instead of seed. Movement of soil and plant debris among the fields should be avoided to prevent the smut pathogen from being introduced into other fields (wiki.bugwood.org/HPIPM:Onion_Smut). Healthy sets and transplants should be planted to promote rapid growth (plantvillage.psu). Chemical controls are most effective when combined with sound cultural practices. Only seed treated with a fungicide should be planted (wiki.bugwood.org/HPIPM:Onion_Smut). Chemical seed-treatment with Thiram or Captan (0.3%) and also soil application of these fungicides in the nursery reduces seedling infection.

Stemphylium leaf blight {*Stemphylium cesicarium/vesicarium* ((Wallr.) E.G. Simmons}

The fungus survives in the soil or plant debris and is favoured by warm (18-25°C) humid conditions and long periods of leaf wetness (16 hours or more) [vikaspedia.in (c)]. This onion disease is found throughout the country though causes severe losses only in northern India (Gupta and Pandey, 1986). It infects plants after long warm periods when leaves remain wet (Lawande *et al.*, 2009). Infection occurs on radial leaves of transplanted seedlings at 3-4 leaf stage during late March and early April. The disease on the inflorescence stalk causes severe damage to the seed crop [vikaspedia.in (c)]. Due to its attack about 90% loss in seed yield is recorded (Lawande *et al.*, 2009).

This disease has been reported on garlic too (Sinha *et al*, 1995, 1998). The disease is more severe in *rabi* than in *kharif* season (Lawande *et al.*, 2009). Another species of the same fungus, S. *botryosum* causes black stalk-rot (Singh and Sharma, 1977a). The fungus can survive on alternate hosts in absence of onion crop (Lawande *et al.*, 2009).

The symptoms appear as small yellowish to orange flecks or streaks in the middle of the leaves, soon developing into elongated spindle-shaped spots surrounded by pinkish margin [vikaspedia.in (c)].

Cultural control includes field sanitation by burning the crop residues; long rotations with non-host crops; proper field drainage; reduced plant density; soaking of onion seeds in hot water (50°C for 20 minutes); use of resistant varietieis and Welsh onion (*Allium fistulosum*) lines which are moderately resistant to stemphylium leaf blight though chemical control includes spray of mancozeb or zineb 75% WP @ 1.5-2 litres or kilogrammes as the case may be, dissolved in 750 litres of water/hectare [vikaspedia.in (b)]. Mancozeb @ 0.3% spraying should start from 45 days after transplanting (Lawande *et al.*, 2009). Among the newer fungicides, two sprays of Hexaconazole(0.1%) is very cost-effective (Barnwal *et al.*, 2006). Foliar spraying of leaf extracts (20%) of *Azadirachta indica* and *Datura metel* was also found quite effective though *Pseudomonas fluorescens* was found comparatively less effective (Barnwal *et al.*, 2003).

Stemphylium leaf blight

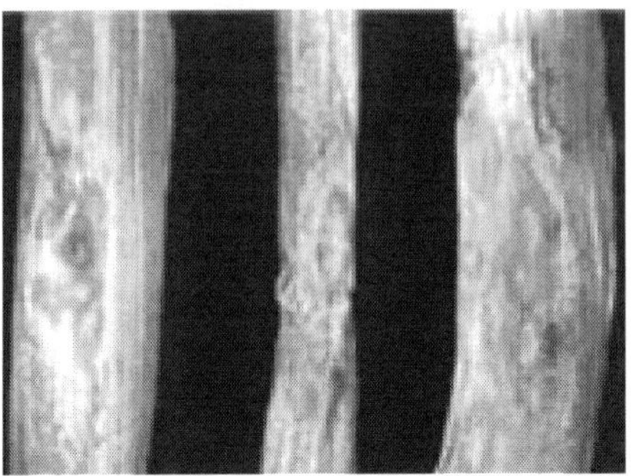

Source: *vikaspedia.in (c)*

White or Sclerotial rot of onions, leeks & garlic (*Sclerotium cepivorum* Berk.)

Fungus survives for many years in soil as sclerotia and it can over-winter in infected onion debris and in diseased onion sets. It is most severe in cool and dry soils. The initial symptoms include yellowing and dieback of leaf tips though later stage infection causes scale's stem plate and roots getting destroyed, bulbs becoming soft and water-soaked. Later, white fluffy or cottony growth of mycelium with abundant small and black to globular black structures called sclerotia, resembling mustard grain, develop on the infected bulbs and these resting structures remain in the soil to reinfect its future crop [vikaspedia.in (c); en.wikipedia.org/wiki/Onion].

Affected plants should be removed and destroyed; disease-free areas should be used for its cultivation; crop rotation should be followed with cereal crops; hot water treatment of bulbs at 49°C is quite effective; and solarization of soil at high temperature, *i.e.* 35°C for 18 hours or 45°C for 6 hours reduces the disease incidence by 50-70% [vikaspedia.in (b)].

Bacterial diseases

Bacterial decay of onion may severely damage onion crops, especially in warm climates. Seedling blight is caused by *Pseudomonas siccata* (Moniz and Patel, 1958) and *P. gladioli* pv. *alliicola* (Moniz and Bhinde, 1964) in field crops resulting in streaking of leaves and premature death. The same pathogen is also known to cause stalk (Swarup *et al*, 1973) and bulb rot (Kumar *et al*, 2001). Brown rot is caused by *P. aeruginosa* (Gupta *et al*, 1986), while soft rot is induced by many bacteria, *i.e.*, *Pectobacterium carotovorum* (Patel, 1972), *Pseudomonas marginalis* pv. *marginalis* and *Erwinia carotovora* pv. *carotovora* (Raju and Raj, 1980).

Usually, all the onion cultivars are susceptible to bacterial infection and bulb decay so the best way to control bacterial diseases of onion is to grow the crop under best possible condition of tilth, fertilization, drainage, crop rotation and freedom from weeds; drying the bulbs quickly after harvest, however, during rainy season artificial curing should be adopted (Lawande *et al.*, 2009).

Bacterial brown/slippery skin rot [*Pseudomonas aeruginosa* (Schroter) Migula, *P. allicola* Starr and Burk]

This bacterium is soil-borne requiring moisture for infection, especially at temperature range of 5-41°C. Physically its spores can readily be water-splashed from the foliage and necks where it can enter through wounds. Heavy irrigation and persistent dews are also the reason to cause this disease [vikaspedia.in (c)].

During early stages of infection, the bulbs may appear healthy except for a softening of the neck tissue. In a longitudinal section, one or more inner scales become watery or cooked. It progresses from the top of the infected scale to the base; eventually, all the internal tissues rot and finally the internal scales dry and bulbs shrivel. Squeezing the base of infected plants causes the rotted inner portion of the bulbs to slide out through the neck hence the name slippery skin [vikaspedia.in (c)].

Bacterial soft rot

Source: *vikaspedia.in (c)*

Growing of crop following proper crop rotation and under optimum conditions of weed-free environments; proper tilth, fertilizer and drainage; *vis-à-vis* sorting out of healthy bulbs and discarding the diseased ones immediately after harvesting; their proper curing *cum* quick drying after harvesting, as well as rapid artificial curing during rainy season [vikaspedia.in (b)] are some of the tips for getting rid of this disease.

Bacterial soft rot (*Erwinia carotovora* pv. *carotovora* Jones)

This pathogen is soil-borne with the main sources of inoculum being contaminated soil and crop residues. Its infection is noticed usually after heavy rains when leaves start drying though it also spreads through irrigation water and insects. This disease is favoured by warm *cum* humid conditions with an optimum temperature range of 20-30°C. However, during storage or transit soft rot can develop when temperatures are above 3°C [vikaspedia.in (c)].

It is mainly a problem on mature bulbs where affected scales first appear water-soaked and pale-yellowing to light browning, followed by invaded fleshy scales becoming soft and sticky with the interior of the bulb breaking-down and emitting a watery foul-smelling thick liquid on squeezing from the neck of diseased bulbs [vikaspedia.in (c)]. Its control measures are the same as in case of 'bacterial brown rot', leaf streak and bulb rot [*Pseudomonas viridiflava* (Burkholder, 1930); Dowson 1939].

The bacterium efficiently colonizes the leaf surfaces of onions, several weeds and various other crops which during cool *cum* wet weather conditions initially infect the leaves and then progresses downward into the bulb where this causes a decay of inner scales (wiki.bugwood.org/HPIPM:Onion_Bacterial_Streak_Bulb_Rot). Rapid spread of disease on infected plants is promoted by rainfall (plantvillage.psu).

Bacterial leaf streak and bulb rot	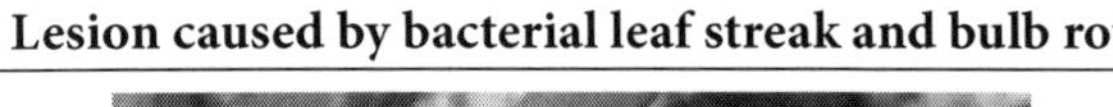 **Lesion caused by bacterial leaf streak and bulb rot**
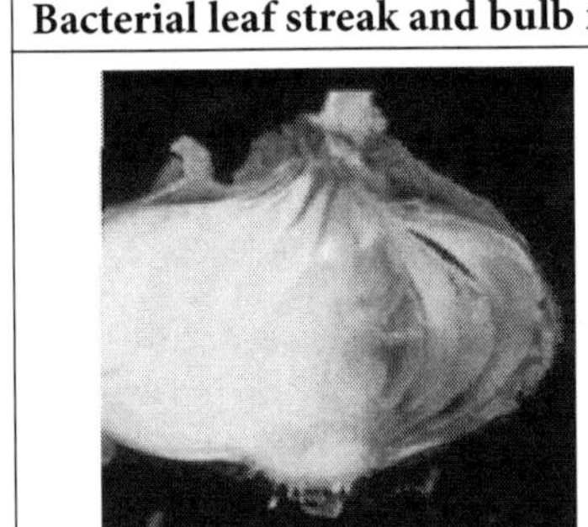	
Source: *plantvillage.psu*	

Water-soaked dark green oval lesions or streaks on leaves, leaf tip-burn, dark spots on wrapper scales of bulbs and reddish-brown discoloration of inner scales in ring-like pattern (plantvillage.psu) are observed in its infection. Rotted scales are firmer than those caused by other bacteria but infected bulbs can be colonized by certain other secondary pathogens that can rot the bulb entirely (wiki.bugwood.org/HPIPM:Onion_Bacterial_Streak_Bulb_Rot).

Fertilization during winter should be avoided. Though nitrogenous fertilizer application is important but not excessive and that too not especially post-bulbing or after storm-damage. Crop rotation for three years with non-host crops such as small grains but not the pulses such as alfalfa, soybean and dry bean which may serve as an alternate host. Crop residues after harvest should deeply be buried to reduce pathogen survival.

Weeds and volunteer onion in and around the field should be eliminated. Overhead irrigation and reuse of irrigation tail water should strictly be avoided. Drip irrigation may reduce disease severity compared to furrow irrigation. Bulbs after curing should be stored at 0-2.2°C with less than 70% RH (wiki.bugwood.org/HPIPM:Onion_Bacterial_Streak_Bulb_Rot). In severe cases, bactericidal sprays should be applied (plantvillage.psu).

Viral diseases

Some of the viruses infecting the crop are members of the potyvirus group such as 'garlic mosaic virus' (Ahlawat, 1974), 'onion yellow dwarf virus', *etc.* while others are Carla virus such as 'garlic latent virus' (Majumdar *et al*, 2007), 'shallot latent virus' and 'carnation latent virus', however, garlic is more vulnerable to viral infection than onion (Lawande *et al.,* 2009).

Iris yellow spot virus

This is a relatively new disease of onion which has become widespread in western counties, especially in the US. In India, it was first reported from Jalna and Nashik region of Maharashtra by Ravi *et al.* (2006) and now has been reported from some other onion growing states of the country.

Transmission of virus is usually through onion thrips (*Thrips tabaci*) in a persistent manner. Disease severity is positively correlated with thrips population in field. This virus is not seed transmitted. Over-wintering onions, volunteers from prior productions, infected transplants and alternate hosts can all serve as sources of both vector and virus [vikaspedia.in (c)].

Often seen as a clear to less obvious creamy elliptical spots on the leaves, scapes or floral stalks. Through its infection the leaves and scapes may collapse at the site of the spots. Initially the spots may be insignificant though have potential of devastating the whole crop [vikaspedia.in (c)].

Diamond shaped viral lesions

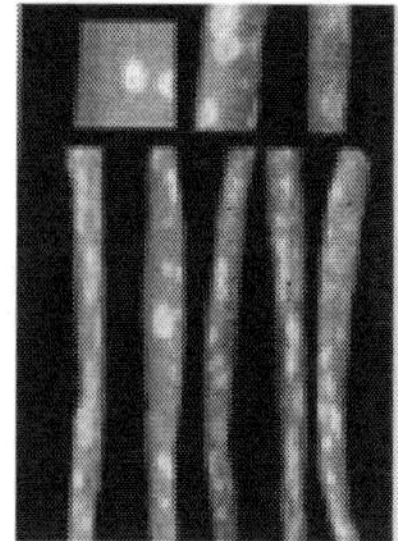

Source: *vikaspedia.in (c)*

Cultural control includes use of healthy seedlings free from thrips and 'iris yellow spot virus'; at least three year's rotation between onion crops; eliminating alternate host plants, culls, and weeds in and around onion field; use of sprinkler irrigation for suppression of thrips and disease; and by avoiding thin, patchy plant stands and crop stress [vikaspedia.in (b)]. Biological control is similar to 'onion yellow dwarf' disease.

Onion yellow dwarf virus

This virus in India was first reported by Dhingra and Nariani (1963) which also infects garlic and leek, and is a common problem in seed production. The virus is carried by infected seed bulbs, onion sets and volunteer onions. Many aphid species can transmit this virus from infected to healthy plants [vikaspedia.in (c)]. Initially the symptoms on young onions appear as yellow streaks at the bases of the first true leaves though all the later leaves show symptoms ranging from yellow streaks to complete yellowing. Sometimes the leaves are crinkled and flattened and tend to fall over and the bulbs produced are undersized [vikaspedia.in (c)].

Onion yellow dwarf disease infection

Source: *vikaspedia.in (c)*

Cultural control includes the management of the vector population and use of virus-free transplants; crop rotation of at least three years; destroying of infected plants; rogueing of diseased plants and isolation from other susceptible crops or volunteer onions; and use of blue sticky trap. Biological control includes the predators such as coccinellids, lacewings, spiders, wasps, *etc.* for controlling thrips [vikaspedia.in (b)].

Stagewise IPM

Management	***Activity***
Pre-sowing *	
Nutrients	• Soil testing before sowing should be done to find out the soil fertility status. Nutrient should be provided as per recommendations of soil test. • Add well rotten FYM @ 25t/ha or vermicompost @ 10t/ha. Incorporate FYM at the time of field preparation at 2 to 3 weeks before transplanting. Vermicompost should be incorporated one week before transplanting.
Weeds	• Deep ploughing during summer. • At the time of field preparation, adopt stale seed bed technique or soil solarization to minimize menace of weeds in field. • The boundary and bunds of the field should be free from weeds. • Keep the nursery weeds free by manual weeding.
Soil-borne pathogens, nematodes, and resting stages of insects	**Cultural control** • Deep field ploughing during summer to control juveniles and adults of nematodes, vis-a-vis overwintering stages. • Solarize the beds with polythene sheet of 45 gauge (0.45mm) thickness for three weeks before sowing which will help in reducing the soil-borne pests.
Damping off	**Cultural control** • Avoid excessive watering • Use 15cm high raised beds for better water drainage
Sowing / planting*	
Nutrients	• **Biofertilizers:** Seed/seedling treatment with Azotobacter and PSB cultures @ 600g each/ha seed/seedling. • Apply NPK fertilizers @ 100:50:100kg/ha. Apply 1/3 of N, entire P and 60% K as basal dose at the time of planting. • Based on soil test for micronutrient, the deficient micronutrient should be applied in soil at sowing/transplanting.
Weeds	• In high weed infested fields, apply approved herbicides as mentioned below; • Spray oxyflourfen 23.5% EC @ 425-850ml in 500-750 litres of water/ha on soil as pre-plant application or quizalofop-ethyl 5% EC @ 750-1000ml in 375-450 litres of water/ha 20 days after planting of onion seedlings followed by one hand weeding if required.

Management	Activity
Vegetative stage	
Nutrients	• Apply 1/3 of N at 21days and remaining 1/3 at 45 days after transplanting. • Apply remaining 40% potash as top dressing in standing crop at 21 days after transplanting. • Micronutrient application may be made after soil test.
Weeds	• Crop should be maintained weed-free initially for 30-45 days by following timely intercultural operations and hand-weeding.
Onion thrips	**Cultural control** • Use resistant/tolerant varieties. • Practice field sanitation. • Avoid successive planting of garlic or other preferred/alternate host such as cabbage, cotton, tomato, cucumber, melons, pumpkins, strawberries, etc. • Plant the new crop in upwind direction of already planted crop which helps in escaping infestation from old planting to some extent at initial stages. • Adjust the transplanting dates. • Use of reflective plastic silver colour and aluminium painted black mulches to repel the thrips, especially in seed crop. • Use sprinkler irrigation to reduce thrips population considerably compared to drip and surface irrigation • Plant two rows of maize or inner row of wheat and outer row of maize surrounding the onion plots as barrier crop (NHM 2012). **Biological control** • Conserve parasitoids such as *Ceranisus menes* (nymph), etc. • Conserve predators such as syrphid flies, minute pirate bug/ anthocorid bug (*Blaptosthethus* sp., *Buchananiella whitei*, *Orius tantilus*), praying mantis, predatory thrips (*Aeolothrips fasciatum*), damselbug, lacewings, coccinellids (*Cheilomenes sexmaculata*), spiders, etc. **Chemical control** • Spray dimethoate 30EC @ 660ml in 500-1000 litres of water, oxydemeton methyl 25EC @ 1200 ml in 500-1000 litres of water, quinalphos 25EC @ 1200ml in 500-1000 litres of water or lambda cyhalothrin 5EC @ 300 ml in 300-400 litres of water, per hectare. • Use high pressure and volume of spray for good coverage of the crop and direct the nozzle at base of leaves. • Add a spreader or sticker (0.05-1.0%) for retention and spread of spray fluid on erect leaves of onion. • In seed crop, thrips should be controlled well before flowers open.

Management	*Activity*
Onion maggot	**Cultural control** • Avoid close spacing while planting. • Follow crop rotation. • Maintain field sanitation. **Biological control** • Conserve predators such as ground beetle, rove beetles, spiders, etc. by providing grassy refuge strips.
Bulb mite	**Cultural control** • Avoid planting onion after cole crops as decaying cole crops, especially the cauliflowers may harbour very high bulb mite populations in the field. • Avoid planting successive onion or garlic crops. • Use flood irrigation to reduce mite levels in the soil. **Biological control** • Conserve predators such as anthocorid bugs (Orius spp.), mired bugs, syrphid/hover flies, green lacewings (*Mallada basalis* & *Chrysoperla zastrowi* sillemi), predatory mites (*Amblyseius alstoniae, A. womersleyi, A. fallacies* & *Phytoseiulus persimilis*), predatory coccinellids (*Stethorus punctillum*), staphylinid beetle (*Oligota* spp.), predatory cecidomyiid fly (*Anthrocnodax occidentalis*), predatory gall midge (*Feltiella minuta*), spiders, etc.
Eriophyid mite	**Cultural control** • Use flood irrigation to reduce mite level in the soil. • Avoid planting successive onion or garlic crops. **Biological control** • Conserve predators such as anthocorid bugs (*Orius* spp.), mired bugs, syrphid/hover flies, green lacewings (*Mallada basalis* and *Chrysoperla zastrowi sillemi*), predatory mites (*Amblyseius alstoniae, A. womersleyi, A. fallacies* and *Phytoseiulus persimilis*), predatory coccinellids (*Stethorus punctillum*), staphylinid beetle (*Oligota* spp.), predatory cecidomyiid fly (*Anthrocnodax occidentalis*), predatory gall midge (*Feltiella minuta*), spiders, etc.

Management	*Activity*
Red spider mite	**Cultural control** • A thorough Jet water spraying to wash off the mites from the plant. **Biological control** • Conserve and augment the natural enemies. Conserve predators such as anthocorid bugs (*Orius* spp.), mirid bugs, syrphid/hover flies, green lacewings (*Mallada basalis* & *Chrysoperla zastrowi sillemi*), predatory mites (*Amblyseius alstoniae, A. womersleyi, A. fallacies* & *Phytoseiulus persimilis*), predatory coccinellids (*Stethorus punctillum*), staphylinid beetle (*Oligota* spp.), predatory cecidomyiid fly (*Anthrocnodax occidentalis*), predatory gall midge (*Feltiella minuta*), spiders, etc.
Gram pod borer, Tobacco caterpillar**	**Cultural control** • Field sanitation. • Adopt ecological engineering of onion/garlic with growing intercrops such as cowpea, maize, coriander, urdbean, etc. and with growing sorghum or maize in 4 rows all around onion/garlic crop as guard crop. • Rotate the onion crop with a non-host cereal crop, cucurbit or cruciferous vegetables. • Adopt ecological engineering of garlic with repellant plants: Ocimum and marigold as ovipositional trap crop. • Erecting of bird perches @ 50/ha for encouraging predatory birds such as king crow, mynah, etc. • Install pheromone traps @ 10-12/ha for monitoring adult moth activity. Replace the lures after every 2-3 weeks. • Set up the light trap @ 2-3/ha. **Biological control** • Inundatively release *Tricogramma* spp., *T. pretiosum* @ 1,00,000/hectare 4-5 times from flower initiation stage at weekly intervals. • Conserve parasitoids such as *Tetrastichus* spp. (egg), *Telenomus* spp. (egg), *Campoletis chlorideae* (larval), etc. • Conserve predators such as *Chrysoperla zastrowi sillemi*, coccinellids, King crow, common mynah, wasp, dragonfly, spider, robber fly, reduviid bug, preying mantis, fire ants, big-eyed bugs (*Geocoris* sp.), pentatomid bug (*Eocanthecona furcellata*), earwigs, ground beetles, rove beetles, etc. • Apply entomopathogenic nematodes (EPNs) @ 50-300 crore infective juveniles of *Steinernema feltiae*/hectare.

Management	*Activity*
Damping off	**Cultural control** • Solarize the nursery beds with transparent polythylene sheet for 30 days before sowing. • Avoid continuous nursery raining in the same plot. • Avoid excessive soil moisture. • Follow crop rotation. • Maintain proper drainage through land levelling, and installing drains. • Sow clean and healthy seeds on raised beds. • Avoid overcrowding or dense sowing of onion seeds. • Apply well-decomposed FYM in the nursery.
Purple blotch	**Cultural control** • Use healthy seeds for planting. • Follow 2-3 years crop rotation with non-host crops. • Provide proper drainage facilities at regular intervals. • Use recommended doses of N and P fertilizers. • Soak onion seeds in hot water at 500C for 20 minutes. • Use resistant varieties. **Chemical control** • Spray difenaconazole 25 EC @ 100 ml in 500 litres of water/ha or kitazin 48% EC @ 200ml in 200 litres of water (or as required depending upon crop stage and plant protection equipment used)/ha
Stemphylium leaf blight	**Cultural control** • Keep proper field sanitation by collecting and burning crop residues regularly, vis-à-vis reduce plant density. • Follow long rotations with non-host crops. • Follow proper field drainage. • Treat onion seeds in hot water at 500C for 20 minutes. • Plant resistant Welsh onion (*Allium fistulosum*) lines which are moderately resistant to *Stemphylium* leaf blight. **Chemical control** • Spray mancozeb 75% WP @ 1.5-2kg in 750 litres of water/hectare, zineb 75% WP @ 1.5-2kg in 750-1000 l of water/ha or spray zineb 75% WP @ 1.5-2kg in 750-1000 litres of water/hectare.

Management	*Activity*
Colletotrichum blight/ anthracnose/ twister disease	**Cultural control** • Keep the field clean by destroying the infected plant debris. • Use resistant varieties
Downy mildew**	**Cultural control** • Select healthy bulbs for seed production. • Follow crop rotation for 3-4 years with non-host crops. • Dispose off properly the infected crop debris and refuge heaps of onion culled from storage. • Avoid late planting, poor drainage, higher doses of fertilizers and untimely irrigation. • Use resistant varieties. • Sun-expose the seed-crop onion bulbs for 12 days to destroy the fungus. Chemical control • Spray zineb 75% WP @ 1.5-2kg in 750-1000 litres of water/ha.
Fusarium basal rot/ basal plate rot**	**Cultural control** • Follow crop rotation for 4-5 years with non-host crop. • Follow mixed cropping with tobacco and sorghum as it is effective in reducing pathogen inoculum and the disease. • Solarize the soil by using polythene sheet of 25 guage in summer for 30 days to reduce the pathogen inoculum, vis-à-vis disease. • Follow proper drainage practices. • Plough the field deeply in summer. • Avoid injury during cultural practices. • Flood the soil in non-growing season for basal rot control. • Use resistant varieties.
White rot**	**Cultural control** • Use only disease-free areas. • Follow crop rotation with cereal crops. • HWT of bulbs at 49°C. • Solarize the soil at high temperature either at 35°C for 18 hours or 45°C for 6 hours to reduce disease incidence by 50-70%.

Management	*Activity*
Pink root rot**	**Cultural control** • Follow crop rotation with cereal and green manuring. • Solarize the soil by using polythene sheet of 25 guage in summer for 30 days to reduce pathogen inoculum, vis-a-vis diseases and pink root incidence. • Use resistant onions such as lines of Welsh onion (Allium fistulosum) which are resistant to pink root. • Delay planting to reduce disease incidence. • Follow long rotations with non-host crop.
Black mold/green mold**	**Cultural control** • Leave onions for drying in the field for two days after harvesting and then for 10-15 days shade-drying before storage at 1-150C. • Store seed and bulb after proper drying. • Maintain proper aeration in the storage • Avoid bruising of onion bulbs during harvest, storage or transport.
Bacterial rots (brown rot/ soft rot/slippery skin)**	**Cultural control** • Grow the crop under optimum conditions of tilth, crop rotation, weed-free environment, drainage and fertilizer. • Follow proper curing and rapid bulb drying after harvesting. • Follow artificial curing during rainy season. • Discard the infected and injured bulbs while storing.
Onion yellow dwarf disease	**Cultural control** • Manage the vector population. • Plant virus-free transplants. • Follow crop rotation of at least three years. • Remove and destroy infected plants to check spread of disease. • Rogueing of diseased plants and isolation from other susceptible crops or volunteer onions. • Use blue sticky trap. **Biological control** • Conserve the predators such as coccinellids, lacewings, spiders, wasps, etc. for controlling thrips.

Management	*Activity*
Iris yellow spot disease	**Cultural control** • Plant 'iris yellow spot virus' free and insect-free healthy seedlings. • Follow at least three year's rotation between onion crops. • Eliminate alternate hosts, culls and weeds in vicinity of onion field • Use sprinkler irrigation for suppression of thrips and disease. • Avoid thin, patchy plant stands and crop stress. **Biological control** • As in onion yellow dwarf disease.
Reproductive stage	
Nutrients	**Cultural control** • Apply third dose (second top-dressing) of N, i.e. 33.33/ha at 60 days after transplanting.
Weeds	**Cultural control** • Left over weeds before shedding of seeds should be removed to avoid further spread.
Insects and diseases	**Cultural control** • Remove damaged leaves and shoots. • Follow crop rotation with cereal and green manuring crops. • Other practices same as in vegetative stage. • Insect and disease management same as in vegetative stage.

Note: The pesticide dosages and spray fluid volumes are based on high volume sprayer.
* Apply *Trichoderma* spp. and *Pseudomonas fluorescens* as seed/seedling/planting material, nursery treatment and soil application (if commercial products are used), check for label claim. However, biopesticides produced by farmers for own consumption in their fields, registration is not required.
** Pests of regional significance

Cultivation of multiplier onions

Propagation

Multiplier onions are almost exclusively propagated by division using offsets or from daughter bulbs, although reproduction from seed is possible (Fritsch and Friesen, 2002). Nesting bulbs should not be confused with sets that are used to start common biennial onions. Sets are actually small onions that somebody else started from seed. Multiplier onions directly produce bulbs that can be used to propagate the plant (cultivariable.com). Varieties being hybrids do not grow true from seed.

When they are started from seed and transplanted, the plants produce just 3-4 bulbs per plant but when a bulb set is planted it will develop a cluster of about 8-12 bulbs around the original (gardenerspath.com). To grow shallots from seed, the seeds should be sown roughly 8-10 weeks before growing area's last projected frost date in the spring. Seeds should be sown around 0.5-1cm deep in a seed-starting mix in a shallow tray placed in a bright location. Germination may take about a week. Soil should not be allowed to dry completely. Seedlings must be hardened off before planting. After the danger of frost has passed, the seedlings can be planted 10-15cm apart in the garden (Iannotti, 2022).

Dormancy

The dormancy period of potato onions is 8-10 weeks although this can be extended substantially with cool *cum* dry storage conditions. Shallots require a cool dormant period of at least one month with temperatures between 0 to 10 ^{0}C before they start growing (Iannotti, 2022).

Climate and planting season

Shallots or other multiplier onions should be planted from the falls to early spring (greyduckgarlic.com). Planting time depends on whether shallots are being grown from seeds or shallot sets (immature bulbs), in case from seeds it should be sown outdoors 4 weeks before expected frost in spring or indoors 8-10 weeks before average last frost date, however, the sets can be roughly planted 2-4 weeks before projected last frost date in the spring (Iannotti, 2022).

Bulb sets planted in early spring can be harvested in the fall, whereas shallot sets planted in late fall can be harvested by early summer, however, fall plantings don't always work out in the colder parts of this plant's growing zones (Iannotti, 2022). Under Tamil Nadu condition in India, bulbs are planted during April-May (eagri.org1).

Soil

Shallots like sunny situation (at least 6 hours of direct sunlight; Iannotti, 2022) with a rich and well-drained soil having a pH of 6-7 (greyduckgarlic.com).

Field preparation and planting

As in common onion, field is ploughed to a fine tilth. In Tamil Nadu, planting is generally done by preparing ridges and furrows at 45cm apart. Medium-sized bulbs are planted at a distance of 10-15cm on either side of ridges. For planting one hectare of land some 15 quintals of bulbs are required.

As in common onions, seedlings raised from seeds are also used for planting (eagri.org1). Individual bulbs should be planted root-end down with tops of the bulbs just be under the soil (greyduckgarlic.com) or just above the soil line (Iannotti, 2022). The soil surrounding the bulbs is often drawn away when the roots have taken hold. These should not be planted on ground recently manured (eagri.org1).

Cultural practices

Cultural practices and climatic conditions are same as in common onion. Some shallot growers believe that mulching may rot the bulbs so this should be avoided. Moreover, the shoots of multiplier onions are quite weak and may not be able to push through the mulch. Watering should be done only when the soil is dry as excess moisture will rot the bulbs.

It should be lightly fertilized in the spring before bulbs begin to develop. As the plant grows it will form a cluster of 4-12 bulbs around the mother bulb (greyduckgarlic.com). The flower stalks should be cut off so that the plant can concentrate its full energy to develop the bulbs (Iannotti, 2022).

Harvest

Harvesting can be done when tops become yellow and die. The bulb clusters are pulled up and dried in a well-ventilated shaded area. Like garlics, shallots too get sunburnt and rot when exposed to too much sun (greyduckgarlic.com).

Storage

Bulbs can be stored for up to 12 months at their optimal temperature of 2-7^{0}C. French red shallots (*Allium cepa* var. *aggregatum*) keep very well in a cool, dry and well-ventilated dark area compared to French grey shallots (*A. oschaninii*) which is a different species (greyduckgarlic.com). Shallot bulbs as become dormant during winter can overwinter in the ground but once the temperatures begin to warm, the plant awakens and then spurring the growing process (Iannotti, 2022).

Yield

Yield varies from 10-15t/ha with crop duration of 2-3 months (eagri.org1).

Pests and diseases

Shallots are subject to many of the same problems as onions.

Cultivation of tree onion and triploid onion

Propagation

Tree onions reproduce themselves either by sprouting the cluster of underground bulbs attached to the mother bulb or through the bulblets formed on the stalk that may bend down under the weight of the new growth and fall over, creating new plants where the bulblets hit the ground.

Bulbils are harvested in late summer and can be replanted immediately or stored in a cool dry frost-free place to be planted in late winter or early spring. Division of the underground bulbs can be done after the leaves die down in late summer. These bulbs are harvested in autumn and planted individually at new locations. They also multiply from the base, best being during spring, so established clumps can be divided as well (hort.extension.wisc.edu).

Since it does not set seeds together with underground bulbs having poor development so it is multiplied vegetatively through bulbils formed on the inflorescences (usually 4–8 bulbils per inflorescence) (Jasna, 2013). Propagation of *A.* × *cornutum* is also similar to tree onion.

Factors influencing growth and development

Most of the cultural practices such as weeding, mulching, irrigation, manure and fertilization, *etc.* are almost the same as in common onion.

Climate and soil

This vigorous and trouble-free plant is hardy in zones 3-10 and its aboveground parts die back to look scruffy over the winter, but very early in the spring new green shoots emerge from the brownish bases (hort.extension.wisc.edu). There is a rare variety that can tolerate cold winters, grow prolifically and is one of the first varieties to sprout in the spring (specialtyproduce.com). If the winter is mild, the plant will even remain green and provide the spring onion-like stems (plantura.garden). The triploid hybrid *A.* × *cornutum* being quite drought-tolerant can tolerate even poor soils and is able to persist even in wild habitats.

Though tree onion can successfully be grown in most soils that are in good condition, it should be grown in fertile and well-drained moisture retentive soil in full sun. Liming can be done if the soil is acidic, *i.e.* pH below 6.5 though it can tolerate the pH range of 4.5 to 8.3.

Land preparation and planting

As in common onion, field is ploughed to a fine tilth and ridges and furrows are prepared at 45cm apart. Medium-sized bulbs or bulbils are planted at a distance of 10cm on either side of ridges. The planting depth can be 2cm and the tops of the bulbs are kept a little above the ground. Planting distance is usually 25-30cm with row spacing of 35-45cm.

Crop rotation and mixed cropping

These hybrids grow well with several plants, especially roses, carrots, beet and chamomile but it inhibits the growth of legumes.

Pruning

Bulbils can be harvested from the top of the flower stem for culinary purposes and to reduce spread around the garden. Shearing should be done once the foliage has died down (rhs.org.uk).

Harvest

Stems are used as spring onions or chives and can be harvested all year round once they have started sprouting, although excessive hardening of green stems should be avoided especially when the plant is still to form the aboveground daughter bulbs. These small bulbs (bulblets) are formed between June and September and can be harvested when they have reached the desired size. The small onions are ready for harvesting from July and can be harvested for storage from the end of September (plantura.garden).

It is a good idea to dig up the plant at least after 3 years and then replanted in composted and freshly dug ground. This way it will continue to do well. The clumps can be divided at this time to make more plants.

Storage

Only the mature and undamaged bulbs of tree onion that have not yet sprouted are suitable for storing. They should be stored in cool, dark and very dry area. In suitable conditions these can be stored until the next spring (plantura. garden).

The bulbs of *A.* × *cornutum* are not good for long storage because they sprout and produce roots quickly after harvest. They are usually kept by being preserved in vinegar (Jasna, 2013).

Disorders

These plants too are subject to many of the same problems as onions. Bolting, *i.e.* initiating quite earlier flowering may be a problem as it inhibits the formation of bulblets (rhs.org.uk). Slug infestation is a great problem as here full-grown and established plants are completely gone due to its attack. The plants are also susceptible to onion fly, leek moth and narcissus eelworm or to diseases such as onion white rot, onion downy mildew, leek rust and onion neck rot (rhs.org.uk).

References

Adam, E., W. Muhlbauer, A. Esper, W. Wolf and W. Spiess, 2000. Quality changes of onion (*Allium cepa* L.) as affected by the drying process. *Die Nahrung*, **44**(1): 32-7.

Aftalion, B. and E. Cohn, 1990. Characterization of two races of stem and bulb nematode (*Ditylenchus dispaci*) in Israel, **18**: 229-32.

Aggie-horticulture.tamu.edu/plantanswers/publications/vegetabletravelers/onions.html.

Agrios, G.N. 1997. Fitopatología. In: Plant Pathology (4t h ed.), pp. 858-61.

Agropedia.iitk.ac.in/content/onion-varieties.

Ahir, R.P. and R.P. Maharishi, 2008. Effect of preharvest application of fungicides and biocontrol agent on black mold (*Aspergillus niger*) of onion in storage. *Ind. Phytopath.*, **61**:130–31.

Ahlawat, Y.S. 1974. A mosaic disease of garlic in Darjeeling Hills. *Sci. & Cult.*, **40**:466-7.

Ahmad, S. and F. Karimullah, 1998. Relevance of management practices of downy mildew in onion. *Sarhad J. Agri.*, **14**:161-2.

Alamgir, A.N.M., 2017. Origin, definition, scope and area, subject matter, importance, and history of development of pharmacognosy. doi:10.1007/978-3-319-63862-1_2. https://www.sciencedirect.com/science/article/pii/S2667142522000811.

Ali, M., M. Thomson and M. Afzal, 2000. Garlic and onions, their effect on eicosanoid metabolism and its clinical relevance. *Prostaglandins, Leukotrienes & Essential Fatty Acids*, **62**: 55-73.

Amagase, H. 2006. Clarifying the real bioactive constituents of garlic. *J. Nutrit.*, **136**(3): 16S-72.

Anon. *Allium × proliferum* (Moench) Schrader. *Acta Botanica Croatica,* **58**, 65–77.

Anon., 2023. *Allium cepa* Linnaeus. Flora of North America (FNA), **26**: 225, 228, 244. http://www.efloras.org/florataxon.aspx?flora_id=1&taxon_id=200027457.

Anon. 2023a. *Allium asarense* R.M.Fritsch & Matin. Plants of the World Online (POWO). *Royal Botanic Gardens, Kew*. Retrieved 22 January, 2023. https://powo.science.kew.org/taxon/20002422-1

Anon. 2023b. *Allium vavilovii* Popov & Vved. Plants of the World Online (POWO). Royal Botanic Gardens, Kew. Retrieved 22 January, 2023. https://powo.science.kew.org/taxon/529098-1

Arochena, L., C. Gamez, V. del Pozo and M. Fernandez-Nieto, 2012. Cutaneous allergy at the supermarket. *J. Investigational Allergology & Clin. Immunol.*. **22** (6): 441–442.

Arshad, M.S., M. Sohaib, M. Nadeem, F. Saeed, A. Imran, A. Javed, Z. Amjad and S.M. Batool, 2017. Status and trends of nutraceuticals from onion and onion by-products: A critical review. *Cogent Food Agric.,* **3,** Article 1280254, 10.1080/23311932.2017.1280254. https://www.sciencedirect.com/science/article/pii/S2667142522000811.

Asman, K. 2002. Trap cropping effect on oviposition behaviour of the leek moth Acrolepiopsis assectella and the diamondback moth *Plutella xylostella. Entomologia Experimentalis et Applicata,* **105**(2):153–64. doi:10.1046/j.1570-7458.2002.01043.x. S2CID 85865566.

Asman, K., B. Ekbom and B. Ramert, 2001. Effect of intercropping on oviposition and emigration behavior of the leek moth (Lepidoptera: Acrolepiidae) and the diamondback moth (Lepidoptera: Plutellidae). *Environ. Entomol.*, 44 (6): 288–94. doi:10.1603/0046-225x-30.2.288.

Assuncao, I.P., R.S.B. Coclho, de A.G.S. Lima, J.A.S. Lima and S.C.C. Tavares, 1999. Reaction of onion cultivators to isolates of *Collectotrichum gloeosporioides* from sub medio San Francisco Region, Brazil. *Summa Phytopathologica,* **25**:205- 9.

Augusti, K. 1996. Therapeutic values of onion and garlic. *Ind. J. experim. Biol.*, **34**: 634-40.

Barnwal, M.K., S.M. Prasad. And D. Maiti, 2003. Efficacy of fungicides and bioagents against Stemphylium blight of onion. *Ind. Phytopath.*, **56**:291-2

Barnwal, M.K., S.M. Prasad and S. Kumar, 2006. Cost effective fungicidal management of *Stemphylium* blight of onion. *J. Res., Birsa Agricultural University*, **18**:153-5.

Bayer, T., W. Breu, O. Seligmann, V. Wray and H. Wagner, 1989. Biologically active thiosulphinates and alpha-sulphinyl disulphides from *Allium cepa. Phytochem.,* **28**(9): 2373-7.

Bhangale, G.T. and M.B. Joi, 1983. Role of thrips in development of purple blotch of onion. *J. Maharashtra Agril. Univ.*, **8**:299-300

Bilyk, A., P.L. Cooper and G.M. Saper, 1984. Varietal difference in distribution of quercetin and kaempferol in onion (*Allium cepa* L.). J. agr. & *Food Chem.*, **32**: 274-6.

Block E, H. Gulati, D. Putman, D. Sha, N. You and Z. Shu-Hai, 1997. *Allium* chemistry, synthesis of 1-[alk(en)ylsulfinyl]propyl alk(en)yl disulfides (cepaenes), antithrombotic flavorants from homogenates of onion (*Allium cepa*). *J. agr. & Food Chem.*, **45**: 4414-22.

Block, E. and S.H. Zhao, 1992. *Allium* chemirombotic 'cepaenes' from onion and 'deoxycepaenes' from oil of shallot by reaction of 1-propenethiolate with sulfonyl halides. *J. Org. Chem.*, **57**: 5815-17.

Bozzini, A. 1964. On the karyotype of a viviparous onion known as *Allium* cepa var. viviparum (Metzg.) Alef. *Caryologia,* **17**: 459–64.

Brewster, J.L. 2008. O*nions and Other Vegetable Alliums.* CABI, pp. 152–3. ISBN 978-1-84593-622-8.

Brewster, J.L. 1994a. *Onions and Other Vegetable Alliums* (1st ed.), pp. 5.. CABI. Wallingford, UK. ISBN 978-0-85198-753-8.

Brewster, J.L. 1994b. *Onions and Other Vegetable Alliums* (1st ed.), pp. 16. CAB International, Wallingford, UK.. ISBN 978-0-85198-753-8.

Brewster, J.L. 1994c. Onions and other vegetable alliums (1st ed.). Wallingford, UK: CAB International. p. 3. ISBN 978-0-85198-753-8.

Brewster, J.L. 1994d. Onions and Other Vegetable Alliums (1st ed.), 212p. CABI, Wallingford, UK. ISBN 0-85198-753-2.

Brickell, C. (ed.), 1992. The Royal Horticultural Society Encyclopedia of Gardening. Dorling Kindersley. p. 345. ISBN 978-0-86318-979-1.

Britannica.com/plant/Allioideae

Cantisani, C., B. Visconti, G. Paolino, F. Frascani, S. Tofani, G. Fazia and S. Calvieri, 2014. Unusual food allergy: Alioidea allergic reactions overview. *Recent Patents on Inflammation & Allergy Drug Discovery,* **8** (3): 178–84.

Chilvers, M.I. and L.J. du Toit, 2006. Detection and identification of *Botrytis* species associated with neck rot, scape blight and umbel blight of onion (PDF). *Plant Health Progress,* **7**: 38. doi:10.1094/PHP-2006-1127-01-DG. Archived from the original (PDF) on 2016-03-04. Retrieved 2013-03-31.

Chu, Y.F., J. Sun, X. Wu and R.H. Liu, 2002. Antioxidant and antiproliferative activities of common vegetables. *J. agric. & Food Chem.*, **50**: 6910-16.

Corea, G., E. Fattorusso, V. Lanzotti, R. Capasso and A.A. Izzo, 2005. Antispasmodic saponins from bulbs of red onion, *Allium cepa* L. var. Tropea. *J. agric. & Food Chem.,* **53**(4), 935-40.

Corgan, J. and M. Holland, 1993. 'NuMex Starlite' onion. *HortSci.,* **28**: 66-7.

Crozier, A., M.E.J. Lean, M.S. McDonald and C. Black, 1997. Quantitative analysis of the flavonoid content of commercial tomatoes, onions, lettuce and celery. *J. agric & Food Chem.,* **45**(3): 590-95.

Cultivariable.com/instructions/root-crops/how-to-grow-potato-onion/

Cundell, D. 2014. Herbal phytochemicals as immunomodulators. *Curr. Immunol. Rev.*, **10**:64-81. 10.2174/1573395510666140915213156. https://www.sciencedirect.com/science/article/pii/S2667142522000811.

Curl, C.L., K.R. Price and G.R. Fenwick, 1985. The quantitative estimation of saponin in pea (*Pisum sativum-l*) and soya (*Glycine max*). *Food Chem.,* **18**(4):41-50.

Dhaliwal, M. 2017. Bulbous Vegetables. In:Handbook of Vegetable Crops (3rd ed.), pp. 177-92. Kalyani Publishers, New Delhi, India.

Dhingra, K.L. and T.K. Nariani, 1963. A virus disease of onion. *Ind. Phytopath.*, **16**:311-2

Dogr.icar.gov.in/index.php?option=com_ content&view=article&id=118&Itemid=189&lang=en#:~:text=Irrigation%20requirement%20of%20onion%20depends,depending%20upon%20the%20soil%20moisture.

Dogr129:- dogr.icar.gov.in/index.php?option=com_ content&view=article&id=129&Itemid=186&lang=en

Donner, H., L. Gao and G. Mazza, 1997. Separation and characterization of simple and malonylated anthocyanins in red onions, *Allium cepa* L. *Food Res. Intern.,* **30**(8): 637-43.

Dorsch, W. and H. Wagner, 1991. New antiasthmatic drugs from traditional medicine? *Intern Arch Allergy & Appl. Immun.*, **94**: 262-5.

Eady, C.C., T. Kamoi, M. Kato, N.G. Porter, S. Davis, M. Shaw, A. Kamoi and S. Imai, 2008. Silencing onion lachrymatory factor synthase causes a significant change in the sulfur secondary metabolite profile. *Pl. Physiol.*, **147**:2096-2106.

Eagri.org1: eagri.org/eagri50/HORT281/lec15.html

Eagri.org2: eagri.org/eagri50/HORT281/lec16.html

Ebadi, P., M.H. Karimi and Z. Amirghofran, 2014. Plant components for immune modulation targeting dendritic cells: implication for therapy. *Immunotherapy*, **6**: 1037-53, 10.2217/imt.14.77. https://www.sciencedirect.com/science/article/pii/S2667142522000811.

Ebaneb, A.C. 1980. Onion twister disease caused by *Glomerella cingulata* in North Nigeria. *Pl. Dis.,* **64**:1030-32.

Ecoursesonline.iasri.res.in/mod/page/view.php?id=12130

Ekvall, J., R. Stegmark and M. Nyman, 2006. Content of low molecular weight carbohydrates in vining peas (*Pisum sativum*) related to harvest time, size and brine grade. *Food Chem.,* **94**(4): 513-9.

en.wikipedia.org/wiki/Allium

en.wikipedia.org/wiki/Botrytis_allii

en.wikipedia.org/wiki/Botrytis_squamosa

en.wikipedia.org/wiki/Ditylenchus_dipsaci

en.wikipedia.org/wiki/Leek_moth

en.wikipedia.org/wiki/Leveillula_taurica

en.wikipedia.org/wiki/Onion

en.wikipedia.org/wiki/Potato_onion

en.wikipedia.org/wiki/Puccinia_porri

en.wikipedia.org/wiki/Shallot

en.wikipedia.org/wiki/Thrips_tabaci

en.wikipedia.org/wiki/Tree_onion

Etewa, S.E. and S.M. Abaza, 2011. Herbal medicine and parasitic diseases. *Herb. Med. Parasites*, **4**:3-14. https://www.sciencedirect.com/science/article/pii/S2667142522000811.

Ewald, C., S. Fjelkner-Modig, K. Johansson, I. Sjoholm and B. Akesson, 1999. Effect of processing on major flavonoids in processed onions, green beans and peas. *Food Chem.*, **64**(2): 231-5.

Farooqui, T. and A.A. Farooqui, 2017. Neuroprotective Effects of Garlic in Model Systems of Neurodegenerative Diseases. Elsevier Inc., the Netherlands. 10.1016/B978-0-12-811959-4.00016-X. https://www.sciencedirect.com/science/article/pii/S2667142522000811.

Finegardening.com/project-guides/fruits-and-vegetables/cultivating-the-alliums-from-mild-to-wild

Fiskesjo, G. 1975. Chromosomal relationships between three species of *Allium* as revealed by C-banding. *Hereditas*, **8**: 23–32.

Food.ndtv.com/ingredient/shallots-701108

Formica, J.V. and W. Regelson, 1995. Review of the biology of quercetin and related bioflavonoids. *Food Chem. & Toxicol.*, **33**(12): 1061-80.

Fossen, T., O.M. Andersen, D.O. Ovstedal, A.T. Pedersen and A. Raknes, 1996. Characteristic anthocyanin pattern from onions and other *Allium* spp. *J Food Sci.,* **61**(4): 703-6.

Friesen, N. and M. Klaas, 1998. Origin of some vegetatively propagated *Allium* crops studied with RAPD and GISH. *Genetic Res. & Crop Evol.,* **45** (6): 511–23. doi:10.1023/A:1008647700251. S2CID 26205471.

Fritsch, R.M. and N. Friesen, 2002. Evolution, Domestication, and Taxonomy. In: Allium Crop Science: Recent Advances (eds Rabinowitch, H.D. and Lesley Currah). CABI Publ., Wallingford, UK. doi:10.1079/9780851995106.0005. ISBN 0-85199-510-1. OCLC 228168061. S2CID 189956991.

Fritsch, R.M., Matin Farideh and Klaas Manfred, 2001. *Allium vavilovii* M. Popov et Vved. and a new Iranian species are the closest among the known relatives of the common onion A. cepa L. (Alliaceae). *Gen. Res. & Crop Evol.*; **48** (4): 401–8. doi:10.1023/A:1012034931024. S2CID 28813594.

Frutas-hortalizas.com: frutas-hortalizas.com/Vegetables/Types-varieties-Onion.html

Fukushima, S., N. Takada, T. Hori and H. Wanibuchi, 1997. Cancer prevention by organosulfur compounds from garlic and onion. *J. Cell Biochem.* (Suppl), **27**: 100-105.

Gardenerspath.com/plants/vegetables/grow-shallots/

Gent, D.H., H.R. Schwartz and R. Khosla, 2004. Distribution and incidence of IYSV in Colorado and its relation to onion plant population and yield. *Pl. Dis.*, **88**:446-52

GMF (George Mateljan Foundation). 2007. Onions. *World's Healthiest Foods.* Retrieved November 11, 2007.

Gibson, G. 1998. Dietary modulation of the human gut microflora using prebiotics. *British J. Nutr.*, **80**(2): S209-S212.

Girija, G., C.S. Pathak. And R.V. Gowda, 1998. Reaction of onion lines to basal rot disease caused by *Fusarium oxysporum* f. sp. *cepae.* PKV Res. J., **22**:53-4.

Goldman, I., M. Kopelberg, J. Devaene and B. Schwartz, 1996. Antiplatelet activity in onion is sulfur dependent. *Thrombosis and Haemostasis*, **76**(3): 450-52.

Green, Aliza, 2004. Field Guide to Produce: How to Identify, Select, and Prepare Virtually Every Fruit and Vegetable at the Market. *Quirk Books*, p. 256. ISBN 978-1-931686-80-8.

Gregrova, A., H. Cizkova, I. Bulantova, A. Rajchl and M. Voldrich, 2013. Characteristics of garlic of the *Czech origin, Czech J. Food Sci.*, 31:581-8. 10.17221/539/2012-cjfs. https://www.sciencedirect.com/science/article/pii/S2667142522000811.

Greyduckgarlic.com/How-to-Grow-Shallots.html

Griffiths, G., L. Trueman, T. Crowther, B. Thomas and B. Smith, 2002. Onions – a global benefit to health. *Phytoth. Res.*, **16**: 603-615.

Gupta, R.P. and U.B. Pandey, 1986. Stemphylium blight of onion. *Ind. Hort.*, **31**:13

Gupta, R.P., P.K. Srivastava.and U.B. Pandey, 1986. Control of purple blotch of onion seed crop. *Indian Phytopath.*, **39**:303-304

Gupta, R.P., P.K. Srivastava and R.C. Sharma, 1996a. Chemical control of purple blotch and *Stemphylium blight* diseases of onion. *NHRDF News letter*, **16**:14- 6.

Gupta, R.P., P.K. Srivastava and R.C. Sharma, 1996b. Effect of foliar spray of different fungicides on the control of Stemphylium blight disease and yield of onion bulb. *NHRDF News letter*, **16**:13-5

Haim, D.R. and Lesley, C. 2002. *Allium* Crop Science: Recent Advances. CABI intern., pp. 87–8, UK. ISBN 978-0-85199-510-6.

Hatono, S., A.Jimenez and M.J. Wargovich, 1996. Chemopreventive effect of S-allylcysteine and its relationship to the detoxification enzyme glutathione S-transferase. *Carcinogenesis,* **17**(5): 1041-4.

Hedges, L. and C. Lister, 2007. The nutritional attributes of *Allium* species. *Crop & Food Res. Confid. Rep.*, No. 1814. New Zealand Institute for Crop & Food Research Limited. 10.13140/2.1.4265.4402. Source: https://www.researchgate.net/publication/265192188_The_nutritional_attributes_of_Allium_species

Helen, O.N., A.O. Job and O. Oghenebrorhie, 2020. Antibiotic and antihelminthic effects of garlic bulbs (*Allium sativum*) and ginger rhizomes (*Zingiber officinale*) as growth promoters on broiler chickens in the tropics. *Adv. Anim. Verter. Sci.*, **8**: 1019-27. 10.23959/sfmrl-1000016. https://www.sciencedirect.com/science/article/pii/S2667142522000811.

Herbst, S.T. 2001. The New Food Lover's Companion: Comprehensive Definitions of Nearly 6,000 Food, Drink, and Culinary Terms. Barron's Cooking Guide. Barron's Educational Series. Hauppauge, NY (USA). ISBN 0764112589.

Hertog, M.G., E.J. Feskens, P.C. Hollman, M.B. Katan and D. Kromhout, 1993. Dietary antioxidant flavonoids and risk of coronary heart disease, the Zutphen Elderly Study. *Lancet,* No. 342, pp. 1007-11.

Hessayon, D.G. 1978. Be your own Vegetable Doctor. *Pan Britannica Industries,*. pp. 22–23. ISBN 978-0-903505-08-6.

Higashihara, H., N. Yokoi, M. Aoyagi, N. Tsuge, S. Imai and S. Kinoshita, 2010. Using synthesized onion lachrymatory factor to measure age-related decreases in reflex-tear secretion and ocular-surface sensation. *Jap. J. Ophthalm.*, **54** (3): 215–20.

Higuchi, O., K. Tateshita and H. Nishimura, 2003. Antioxidative activity of sulfur-containing compounds in *Allium* species for human low-density lipoprotein (LDL) oxidation in vitro. *J. agr. & Food Chem.*, **51**: 7208- 14.

History of onions. US National Onion Association, Greeley, CO. 2011. Archived from the original on 24 June 2017. Retrieved 23 January 2017.

Hoepting, C.A., H.R. Schwartz and H.R. Pappu, 2007. First report of iris yellow spot virus on onion in New York. *Pl. Dis.*, **91**:327

Hort.extension.wisc.edu/articles/egyptian-walking-onions/

https://commons.wikimedia.org/w/index.php?curid=31181075

https://commons.wikimedia.org/w/index.php?curid=43420635

Hubbard, G.P., S. Wolffram, J.A. Lovegrove and J.M. Gibbins, 2003. The role of polyphenolic compounds in the diet as inhibitors of platelet function. *Proc Nutr Soc.*, **6B2**: 469-78.

Hughes, B. and L. Lawson, 1991. Antimicrobial effects of *Allium sativum* L. (garlic), *Allium ampeloprasum* L. (elephant garlic), and *Allium cepa* L. (onion), garlic compounds and commercial garlic supplement products. *Phytother. Res.*, **5**: 154-8.

Iannotti, M. 2022. How to Grow Shallots? Reviewed by Julie Thompson-Adolf. Fact checked by SARAH SCOTT. Updated on 09/05/22. The Spruce. https://www.thespruce.com/growing-shallots-in-the-vegetable-garden-1403464.

icar.org.in/node/323

ikisan.com/ka-onion-varieties.html

Ioku, K., Y. Aoyama, A. Tokuno, J. Terao, N. Nakatani and Y. Takei, 2001. Various cooking methods and the flavonoid content in onion. *J. Nutr. Sci. & Vitaminol.*, **47**(1): 78-83.

Jackson, K.G., G.R. Taylor, A.M. Clohessy and C.M. Williams, 1998. The effect of the daily intake of inulin on fasting lipid, insulin and glucose concentrations in middle-aged men and women. *Brit. J. Nutr.*, **82**: 23-30.

Jaime, L., F. Martinez, M.A. Martin-Cabrejas, E. Molla, F.J. Lopez-Andreu, K.W. Waldron and R.M. Esteban, 2001a. Study of total fructan and fructooligosaccharide content in different onion tissues. *J. Sci. Food & Agric.*, **81**(2): 177-82.

Jaime, L., M.A. Martin-Cabrejas, E. Molla, F.J. Lopez-Andreu and R.M. Esteban, 2001b. Effect of storage on fructan and fructooligosaccharide of onion (*Allium cepa* L.). J. *Agr. & Food Chem.*, **49**(2): 982-8.

Jaime, Prohens-Tomas and Fernando, N., 2007. Vegetables II: Fabaceae, Liliaceae, Solanaceae and Umbelliferae. Springer Science & Business Media (Holland), pp. 124, 152. ISBN 978-0-387-74110-9.

Janssen, K., R.P. Mensink, F.J. Cox, J.L. Harryvan, R. Hovenier, P.C. Hollman and M.B. Katan, 1998. Effects of the flavonoids quercetin and apigenin on hemostasis in healthy volunteers, results from an *in vitro* and a dietary supplement study. *Amer. J. Clin. Nutr.*, **67**: 255-62.

Jasna, P. 2013. Shallots in Croatia – genetics, morphology and nomenclature. *Acta Bot. Croatia*, **72** (2): 387–98 (Review paper). ISSN 0365–0588. eISSN 1847-8476. https://archive.sciendo.com/BOTCRO/botcro.2013.72.issue-2/botcro-2013-0016/botcro-2013-0016.pdf

Kim, J. 1997. Anti-bacterial action of onion extracts against oral pathogenic bacteria. *J. Nihon Univ. School Dentistry*, **39**(3): 136-41.

kitchengardenseeds.com/seed-index/tubers-sets-slips/french-gray-shallot-bulb-sets.html

Kofoet, A. and V. Zinkernagel, 1990. Resistance to downy mildew (*Peronospora destructor*) in *Allium* species. *Z. Pflanzenkr. Pflanzenschutz.*, **97**:13-23. krishijagran.com (a): krishijagran.com/agripedia/onion-varieties-for-different-seasons/

krishijagran.com: krishijagran.com/agripedia/best-varieties-of-onion-in-india-for-higher-yield-profits/

Krishna Kumar, N.K. Jai Gopal and V.A Parthasarthy, 2015. The Onion, pp. 114-118. ICAR, New Delhi. ISBN: 978-81-7164-159-8. https://ia802901.us.archive.org/28/items/onion0000unse/onion0000unse.pdf

Kumar, S., J..P. Sharma, A. Kumar, A..K. Sarkar and V.N. Sharma, 2001. Yield, quality attribute and *Pseudomonas* bulb rots influenced by nitrogen, boron, potassium and fly ash application in acid soil. In: Recent Trends in Horticultural Research, pp. 270-75 (eds Mathura Rai, S. Kumar, Ranvir Singh and V.S.R. Krishna Prasad). Horticultural Society of Chotanagpur, Ranchi.

Lancaster, J.E., M.L. Shaw and W.M. Randle, 1998. Differential hydrolysis of alk(en) yl cysteine sulphoxides by alliinase in onion macerates, flavour implications. *J. Sci. Food & Agr.*, **78**: 367-72.

Landry, J-F. 2007. Taxonomic review of the leek moth genus *Acrolepiopsis* (Lepidoptera: Acrolepiidae) in North America. *Canad. Entomol.*, **139** (3): 319–53. doi:10.4039/n06-098. S2CID: 86748199.

Lanzotti, V. 2006. The analysis of onion and garlic. *J. Chromatogr.*, **A1112**(1-2): 3-22.

Lawande, K.E., Anil Khar, V. Mahajan, P.S. Srinivas, V. Sankar and R.P. Singh, 2009. Onion and garlic research in India; *J. Hortl. Sci.*, **4**(2):91-119. https://www.researchgate.net/publication/332143333_Onion_and_garlic_research_in_India

Lean, M., M. Noroozi, I. Kelly, J. Burns, D. Talwar, N. Satter and A. Crozier, 1999. Dietary flavonoids protect diabetic human lymphocytes against oxidant damage to DNA. *Diabetes*, **48**: 176-81.

Linnaeus 1753, *Allium* I, pp. 294–301.

Majumdar, S., M. Arya, R.P. Pant and V.K. Baranwal, 2007. Shallot virus X in Indian shallot, new report for India. *New Dis. Rep.*, **15**:52.

Makris, D.P. and I.T. Rossiter, 2001. Domestic processing of onion bulbs (*Allium cepa*) and asparagus spears (*Asparagus officinalis*); effect on flavonol content and antioxidant status.*J. agr. & Food Chem.*, **49**(7): 3216-22.

Mani, V.P., V.S. Chauhan, H.C. Joshi and J.P. Tandon, 1999. Exploiting gene effects for improving bulb yields in onion. *Ind. J. Genet. Pl. Breed.*, **59**:511-4.

Marikhur, R.K., A.K. Dhar and M.R. Kaw, 1977. Downy mildew of *Allium cepa* and its control with fungicides in Kashmir Valley. *Ind. Phytopath.*, **30**:576-7.

Mason, P.G., M. Appleby, S. Juneja, J. Allen and J.F. Landry, 2010. Biology and development of *Acrolepiopsis assectella* (Lepidoptera: Acrolepiidae) in eastern Ontario. *The Canad. Entomologist,* **143**(4): 393–404. doi:10.4039/n10-026. S2CID 85817953.

Mason, P.G., W.H. Jenner, A. Brauner, U. Kuhlmann and N. Cappuccino, 2013. Biological control programmes in Canada 2001-2012, pp. 56–62. doi:10.1079/9781780642574.0056. ISBN 9781780642574.

Masterclass.com/articles/what-is-a-shallot. Retrieved on 29 November, 2022.

Mathur, K., S.N. Sharma. And R.S. Sain, 2006. Onion variety RO-59 has higher yield and resistance to purple blotch and *Stemphylicum* blight. *J. Mycol. Pl. Pathol.*, **36**:49-51.

McCollum, G.D. 1974. Hybrid origin of top onion, *Allium cepa var. viviparum. Zeitschrift für Pflanzenzüchtung*, **71**: 222–32.

McNeal, Dale W. Jr. and T.D. Jacobsen, 2002. *Allium cepa.* In: Flora of North America (Editorial Committee). *Flora of North America North of Mexico (FNA),* vol. 26. New York and Oxford – via eFloras.org, Missouri Botanical Garden, St. Louis, MO & Harvard University Herbaria, Cambridge, MA.

Ministry of Agriculture; Fisheries and Food (England), 1968. *Home Preservation of Fruits and Vegetables.* HMSO, 107p.

Mollov, D.S., S.A. Subbotin and C. Rosen, 2012. First report of *Ditylenchus dipsaci* on garlic in Minnesota. *Plant Dis.*, **96** (11): 1707. doi:10.1094/PDIS-06-12-0532-PDN. PMID 30727492.

Moniz, L. and V.P. Bhinde, 1964. Three new bacterial diseases of plants from Maharastra. *Ind. Phytopath.,* **17**:146-53.

Moniz, L. and M.K. Patel, 1958. Three new bacterial diseases of plant from Bombay State. *Curr. Sci.*, 27:494-5.

Moradi Younes, Hemen Moradi-Sardareh, Hasan Ghasemi, Nejad Mohamadi, Mohammad-Nabi Moradi and Seyed-Mostafa Hosseini-Zijoud, 2013. Medicinal properties of Persian shallot. *Europ. J. exp. Biol.*, **3**(1):311-2. https://www.primescholars.com/articles/medicinal-properties-of-persian-shallot-90731.html.

Morrison, R. 1993. Desktop Guide to Keynotes and Confirmatory Symptoms. Hahnemann Clinic Publishing. Grass Valley, CA: ISBN 096353680X.

Munday, R. and C.M. Munday, 2001. Relative activities of organosulfur compounds derived from onion and garlic in increasing tissue activities of quinone reductase and glutathione transferase in rat tissues. *Nutr. & Cancer*, **40**: 205-10.

Myhristad, M.C., H. Carlsen, O. Nordstrom, R. Blomhoff and J.O. Moskaug, 2002. Flavonoids increase the intracellular glutathione level by transactivation of the gamma-glutamylcysteine synthetase catalytical subunit promoter. *Free Radicals in Biology and Medicine*, **32**:386-93.

naturalmedicinalherbs.net/herbs/a/allium-cepa-proliferum=tree-onion.php

nhb.gov.in/pdf/vegetable/onion/oni013.pdf

O'Reilly, J.D., T.A. Sanders and H. Wiseman, 2000. Flavonoids protect against oxidative damage to LDL in *vitro*, use in selection of a flavonoid rich diet and relevance to LDL oxidation resistance *ex vivo*? *Free Radical Res.*, **33**:419-26.

Ohsumi, C., A. Kojima, K. Hinata, T. Etoh and T. Hayashi, 1993. Interspecific hybrid between *Allium cepa* and *Allium sativum*. *Theor Appl Genet.*, **85**(8):969-75. doi: 10.1007/BF00215036. PMID: 24196147.

Oyen, L.P.A. and C.M. Messiaen, 2004. *Allium fistulosum* L. [Internet] Record from PROTA4U (eds Grubben, G.J.H. and O.A. Denton). PROTA (Plant Resources of Tropical Africa/Ressources vegetales de l'Afrique tropicale), Wageningen, Netherlands. <http://www.prota4u.org/search.asp>.

Palti, J. 1989. Epidemiology, production and control of downy mildew of onion caused by *Perasospora destructor. Phytoparasitica*, **17**:1.

Pandey Anjula, Pavan Kumar Malav, K.M. Rai and S.P. Ahlawat, 2022. Genus *Allium* L. of the Indian Region: A Field Guide for Germplasm Collection and Identification. ICAR-National Bureau of Plant Genetic Resources, New Delhi, India, 109p + i-vi. http://www.nbpgr.ernet.in/Downloadfile.aspx?EntryId=9197.

Pandotra, V.R. 1965. Purple blotch disease of onion in Punjab. II: Studies on the life history, viability and infectivity of the causal organism *Alternaria porri. Proc. Ind. Acad. Sci. (Sec. B)*, **61**:326-30.

Patel, P.N. 1972. Bacterial diseases of plants in India. ICAR, New Delhi.

Patil, B.S. and L.M. Pike, 1995. Distribution of quercetin content in different rings of various coloured onion (*Allium cepa* L.) cultivars. *J. hort. Sci.*, **70**(4): 643-50.

Patil, B.S., L.M. Pike and K.S. Yoo, 1995a. Variation in the quercetin content in different colored onions (*Allium cepa* L.). *J. Amer. Soc. Hort. Sci.*, **120**(6): 909-13.

Patil, B.S., L.M. Pike and B.K. Hamilton, 1995b. Changes in quercetin concentration in onion (*Allium cepa* L.) owing to location, growth stage and soil type. *New Phytologist*, **130**: 349-55.

Pike, L.M., R.S. Horn, C.R. Andersen, P.W. Leeper and M.E. Miller, 1988. 'Texas Grano 1015Y' a mild pungency, sweet, short day onion. *HortSci.*, **23**:634-37.

plantura.garden/uk/vegetables/tree-onion/tree-onion-overview

plantvillage.psu.edu/topics/onion/infos

Puizina, J. and D. Papes, 1999. Classical and molecular cytogenetic studies of top onion,

Qui, J., B.B. Westerdahl, D. Giraud and C.A. Anderson, 1993. Evaluation of hot-water treatments for management of *Ditylenchus dipsaci* and fungi in daffodil bulbs. *J. Nematol.*, **25**(4): 686–94. PMC 2619428. PMID 19279827.

Rabinowitch, H.G. and R. Kamenetsky, 2002. Shallot (*Allium cepa* Aggregatum Group). *Allium Crop Science, Recent advances* (eds Rabinowitch, H.D.and L. Currah), pp. 5–30. CABI Publishing, CAB International, Wallingford Oxon OX108DE, UK.

Rajasab, A.H. and C.V. Rao, 1992. *Aspergillu niger* van Tiegh the causal organism of black mold of onion. *J. Aerobiol.*, pp.127-32.

Raju, K.S. and R.B. Raj, 1980. Soft rot of onion in storage. *Madras agril. J.*, **67**:194-5.

Ranjan, S., V.S. Jadon and S. Gupta, 2014. Studies on the anti-inflammatory and analgesic activities of *Allium stracheyi*, a medicinal plant of alpine Himalayas of Uttarakhand. *BDDD*, 1:27-34. https://www.sciencedirect.com/science/article/pii/S2667142522000811.

Ravi, K.S., A.S. Kitkaru and S. Winter, 2006. Iris yellow spot virus in onions: a new tospovirus record from India. *Pl. Path.*, **55**:288.

Remiro, D. and H. Kirmati, 1975. Control of seven curls disease of onion with benomyl. *Sum. Phytopath.*, pp. 51-54.

researchgate.net/publication/265192188_The_nutritional_attributes_of_Allium_species

researchgate.net/publication/342521336_Integrated_nutrient_management_in_onion_and_garlic

Rhodes, M. and K. Price, 1996. Analytical problems in the study of flavonoid compounds in onions. *Food Chem.*, **57**(1): 113-7.

rhs.org.uk/plants/95918/allium-cepa-proliferum-group/details

Ritsema, T. and S. Smeekens, 2003. Fructans, beneficial for plants and humans. *Current Opinion in Plant Biology*, **6**: 223-30.

Ruttle, Jack, 2011. Confessions of an onion addict. National Gardening Association. Retrieved 17 February 2011.http://www.garden.org/articles/articles.php?q=show&id=1727&page=4;.https://en.wikipedia.org/wiki/Tree_onion.

Satya, V.K., S. Gayathiri, R. Bhaskaran, V. Paranidharan and R. Velazhahan, 2007. Induction of systemic resistance to bacterial blight caused by *Xanthomonas campestris* pv. *malvacearum* in cotton by leaf extract from a medicinal plant zimmu *(Allium sativum L. × Allium cepa L.). Archives Phytopath & Pl. Prot.*, **40**(5): 309-22.

Schubert, I., H. Ohle and P. Hanelt, 1983. Phylogenetic conclusions from Giemsa banding and NOR staining in top onions (Liliaceae). *Pl. Syst. & Evol.*, **143:** 245–56.

Sharma, I.M. 1997. Screening of onion varieties/lines against purple blotch caused by *Alternaria porri* under field conditions. *Pl. Dis. Res.*, **12**:60-61.

Sheela, C., K. Kumud and K. Augusti, 1995. Anti-diabetic effects of onion and garlic sulfoxide amino acids in rats. *Planta Medica*, **61**: 356-7.

Shock, C.C., J.K. Ishida, E.P. Eldredge and M. Seddigh, 2000. Yield of yellow onion cultivars in eastern Oregon and southwestern Idaho. *HortTech.*, **10**:613-20.

Sies, H. and W. Stahl, 2003. Non-nutritive bioactive constituents of plants, lycopene, lutein and zeaxanthin. *Intern. J. Nutr. Res.*, **73**(2): 95-100.

Singh, B.M. and Y.R. Sharma, 1977a. Rust on garlic. *FAO Pl. Prot. Bull.*, **25**:41-2.

Singh, B.M. and Y.R Sharma, 1977b. Occurrence of leaf blight of garlic caused by *Stemphylium botryosum* in India. *Ind. Phytopath.*, **30**:272-3.

Singh, R.P. 2007. Some uncommon diseases of onion: smudge. *National Research Centre for Onion and Garlic News*, **11**:5.

Sinha, R.K.P., A.P. Singh, B..B...P. Sinha. And S.M.A. Asghar, 1995. *Stemphylium* blight: a new fungal disease of garlic. *J. Appl. Biol.*, **5**:93.

Sinha, B.B.P., R.K.P. Sinha, S.M.A. Asghar and A.P. Singh, 1998. Perfect stage of *Stemphylium vesicarium* on garlic. *J. Appl. Biol.*, **8**:82-3.

Sites.google.com/site/knowyourvegetables/know-your-onions/know-your-tree-onion

Slimestad, R., T. Fossen and I.M. Vågen, 2007. Onions: A source of unique dietary flavonoids. *J. agric. & Food Chem.*, **55** (25): 10067–80.

Sonomamg.ucanr.edu/Vegetable_of_the_Month/Shallots/

Sparg, S.G., M.E. Light, J. van Staden, 2004. Biological activities and distribution of plant saponins. *J. Ethnopharmac.*, **94**(2-3): 219-43.

Specialtyproduce.com/produce/Walking_Tree_Onions_12607.php

Srivastava, P..K., B.K. Tiwari, K.J. Srivastava and R.P. Gupta,.1996. Chemical control of purple blotch and basal rot diseases in onion bulb crop during *Kharif. NHRDFHortl. Found. News Letter*, **16**:7-9.

Sugha, S.K., R.K. Develash and R.D. Tyagi, 1992. Performance of onion genotypes against purple blotch pathogen. *South Ind. Hort.,* **40**:297.

Suresh Babu, P. and K. Srinivasan, 1997. Influence of dietary capaicin and onion on the metabolic abnormalities associated with streptozotocin induced diabetes mellitus. *Molec. & Cellul. Biochem.*, **175**: 49-57.

Swarup, J., M.S. Nagaroti and H.K. Saksena, 1973. A new bacterial stalk rot of onion caused by *Pseudomonas allicola i*n India. *Ind. J. Mycol. Pl. Path.*, **3**:187-9.

Terahara, N., M. Yamaguchi and T. Honda, 1994. Malonylated anthocyanins from bulbs of red onion, *Allium cepa* L. *Biosci. Biotechn. Biochem.*, **58** (7): 1324-5.

Terrepromise.ca/en/produit/tree-onion/

Thompson, Sylvia (1995). *The Kitchen Garden*. Bantam Books, 143p.

Upadhyay, J. and K.C. Tripathi, 1995. Field evaluation of fungicides against purple blotch of onion seed crop. *Rec. Hort.,* **2**:153-5.

USDA 2006. National Nutrient Database for Standard Reference Release 19. Agricultural Research Service.

Vikaspedia.in (a): vikaspedia.in/agriculture/crop-production/integrated-pest-managment/ipm-for-vegetables/ipm-strategies-for-onion/pests

Vikaspedia.in (b): vikaspedia.in/agriculture/crop-production/integrated-pest-managment/ipm-for-vegetables/ipm-strategies-for-onion/crop-stage-wise-ipm

Vikaspedia.in (c): vikaspedia.in/agriculture/crop-production/integrated-pest-managment/ipm-for-vegetables/ipm-strategies-for-onion/diseases-and-symptoms

Vikaspedia.in (d): vikaspedia.in/agriculture/crop-production/integrated-pest-managment/ipm-for-vegetables/ipm-strategies-for-onion/insect-and-mite-pests-management

Vikaspedia.in (e): vikaspedia.in/agriculture/crop-production/integrated-pest-managment/ipm-for-vegetables/ipm-strategies-for-onion/nutritional-deficiencies-disorders

Vikaspedia.in: vikaspedia.in/agriculture/crop-production/package-of-practices/vegetables-1/onion-allium-cepa

Vincent, E.R. and Mas, Y. 2012. World Vegetables: Principles, Production, and Nutritive Values. Springer Science & Business Media, pp. 307–8. ISBN 978-1-4615-6015-9.

Warid, W.A. and J.M. Loaiza, 1996. Attributes of onion yield in twelve cultivars of short day onions. *Onion Newslett. Tropics,* **7**:22-4.

wiki.bugwood.org/HPIPM:Onion_Bacterial_Streak_Bulb_Rot

wiki.bugwood.org/HPIPM:Onion_Smut

Yamashita, K. and Yosuke Tashiro, 2001. RFLP Analysis of Mitochondrial DNA in Wakegi Onion. *Engei Gakkai Zasshi,* **70** (2): 232–4. doi:10.2503/jjshs.70.232.

Zohary, D. and M. Hopf, 2000. Domestication of plants in the Old World (3rd ed.). Oxford University Press, Oxford, 198p. ISBN 978-0-19-850357-6. Zohary, D. and M. Hopf, 2000. Domestication of plants in the Old World (3rd ed.). Oxford University Press, Oxford, 198p. ISBN 978-0-19-850357-6.

Zohri, A., K. Abdel-Gawad and S. Saber, 1995. Antibacterial, antidermatophytic and antitioxigenic activities of onion (*Allium cepa* L.) oil. *Microbiol. Res.,* **150**: 167-72.

2

Garlic (*Allium sativum* L.)

Subgenus: *Allium*; **Section**: *Allium* (Pandey *et al.*, 2022).

Common name: Garlic, in Hindi: Lahsun.

Synonyms: *Allium arenarium* Sadler ex Rchb. 1830 not L. 1753, *A. controversum* Schrad. ex Wild., *A. longicuspis* Regel, *A. ophioscorodon* Link, *A. pekinense* Prokh., *Porrum ophioscorodon* (Link) Rchb., *Porrum sativum* (L.) Rchb. 1830 not (L.) Mill. 1768.

Introduction and origin

Allium sativum is an easy-growing bulbous plant native to Central Asia and NE Iran. The word *garlic* derives from Middle English, *garleak* (*gar* standing for 'spear'+ leek), as a 'spear-shaped leek' (en.wikipedia.org/wiki/Garlic). Garlic is a perennial herb which is now one of the important bulb crops grown throughout the world. It has a history of several thousand years of human consumption and use (Block, 2010) and grows in the wild in areas where it has become naturalized. Apart from having culinary uses for food flavouring, it has a long history of being used as a traditional medicine to treat various ailments. Closely related plants of the same genus include chives *(Allium schoenoprasum)*, leeks (*A. porrum*), Chinese onions (*Allium chinense*) and onions (*A. cepa*). Several other plants in the genus *Allium* also have the common name 'garlic', such as wild garlic (*Allium ursinum*), crow/wild garlic (*A. vineale*) and field garlic (*A. oleraceum*) of Britain. In North America, *Allium vineale* and *A. canadense* ('meadow garlic', 'wild garlic' or 'wild onion') are common weeds in fields. So-called elephant garlic is actually a variety of *Allium ampeloprasum* and not true garlic. Single clove garlic (also called 'pearl garlic' or 'solo garlic') originated in the Yunnan province of China.

Garlic has lost the ability to produce fertile seeds therefore it is propagated only vegetatively through bulbs and aerial bulbils. Some varieties don't even produce flower stalks or flowers. Numerous varieties available today are presumably the result of random mutations. Garlic varieties can be broadly categorized as hardneck ones (*A.s.* subsp. *ophioscorodon*) that produce flower stalks, and softneck ones (*A.s.* subsp. *sativum*) that don't produce flower stalks. Though hardneck varieties produce flower stalks but do not set seeds under standard growing conditions (hort.extension.wisc.edu).

Wild garlic probably came from areas with a harsh continental climate of hot, dry summers with intense sunlight and long cold winters. It would mature in late spring, then stay dormant until the cooler *cum* moister fall and then begin another cycle of root growth. In early spring, shoots would spring up rapidly and plants would form bulbs and complete their growth before the soil dries out and the scorching sun arrives. When it was moved to the Mediterranean and continental Europe, garlic had to adapt to year-round rain, cooler summers and/or milder *cum* wetter winters. This exposure to a variety of climatic conditions probably led to the diversification of varieties we have today (hort.extension.wisc.edu).

Garlic for India is a foreign exchange earner. As a spice or condiment, it is being grown throughout India. After China, India is the second largest onion growing country though garlic is second most widely cultivated *Allium* but after onion. It produces edible bulb composed of a number of cloves. The bulbs have a powerful pungent aroma which differs from onion-aroma *cum* pungency so are usually not eaten raw. It has higher nutritive value than other bulb crops and is rich in proteins, phosphorous, potassium, calcium, magnesium and carbohydrates. Ascorbic acid content is very high in green garlic.

Botany

It has diploid chromosome number 2n(2x)=16. Based on temperature, day-length response and formation of a flower stalk, garlic can be classified as hard-neck/long-day having varieties with a flower stalk, intermediate ones (weakly bolting hardnecks) producing floral stalks very rarely though soft-necked (short-day) varieties never produce a flower stalk. Garlic plants are quite similar to onion but here the bulb is compound and composed of many tightly packed cloves and the leaves are flat as compared to cylindrical ones in the onion. Though it is perennial plant but is grown as an annual vegetatively. Most varieties grow usually 30-60cm tall and rarely up to 1m in height at maturity. Depending on the variety, the long leaves typically arise from a short and flattened hard central stem (hardneck) above the bulb or emerge from a softer pseudostem made up of overlapping leaf sheaths (softneck). The plant contains up to 50cm long and to 1.25-2.5cm wide, 6-12 flat blade-like leaves with a crease down the middle, and are held erect in two opposite ranks. Shorter than scape, the leaves are broadly linear to linear-lanceolate with an acute apex.

Bulbs (Top-Setting,Grown In Lieu of Flower)

By h. zell- own work,cc by -sa3.0,http://commons.wikimedia.org/w/index.php?curid=8992840

The plant produces an underground edible bulb containing numerous fleshy bulblets (called cloves) but with exception of the single clove types. Depending on the bulb size, varieties and growing conditions, the number of cloves per bulb may vary from 4 to 24 or sometimes even more that are usually arranged in single layer around a hard central stalk in hardneck types and in 3-6 or more layers in various varieties of softneck types. Both the cloves and the entire bulb are encased in papery sheathes that can be white, off-white or have a pink/purple hue. The skin is inedible though in Korea whole immature heads are sometimes prepared with the tender skins intact. The root cluster attached to the basal plate of the bulb is the only part not typically considered edible in any form.

Hardneck varieties form a flower stalk known as a scape, which is followed by large umbellate flowers but these flowers are usually sterile. Scapes first grow straight out of the garlic bulb, and then coil. They are cylindrical, solid, up to 60cm long and have sheathing under middle part. When harvested, they look like long curly green beans.

If these scapes are not harvested, the umbels or flowerheads may open to reveal showy star-shaped blooms in shades of pink and white. The flowers are produced only under favourable conditions but usually these abort and form bulbils (tiny secondary bulbs that form in place of flowers) with tiny cloves identical to the parent bulb. It is a good idea to snip them at the base of their stalk so that all the energy can go back into growing the bulb underground. The spherical flower cluster is initially enclosed in a pair of papery tapered bracts which split open when the green-white or pinkish flowers open.

Garlic Flower

The flowers are hermaphrodite and are pollinated by bees and other insects though being sterile do not set seeds under standard growing conditions, but on those rare occasions when it does, garlic seed resembles the small, black seeds of onions. The flowers of garlic plants usually fade long before producing any seed. In modern garlic varieties, the presence of vegetative topsets (bulblets) which develop in garlic inflorescence is one of the major causes of the inability of this plant to develop normal flowers and true seeds.

Clonal lineages within this species show a remarkably high degree of phenotypic diversity. New genotypes have not been obtained through hybridization but through selection of spontaneous mutations.

Wild progenitor of common garlic

Identification of the wild progenitor of common garlic is difficult due to the sterility of its almost all the cultivars which all may have been descended from the species *Allium longicuspis*, which grows wild in C&SW Asia (Salunkhe and Kadam, 1998; Zohary and Hopf, 2000). Wild *A. longicuspis* is found in the Tien

Shan (Celestial or Heavenly) mountains, on the border between China and Kyrgyzstan (Hirst, 2021). Since *Allium longicuspis* is almost sterile, it is doubtful that it is the ancestor of *Allium sativum* (Zohary and Hopf, 2000). *A. longicuspis* is practically indistinguishable from garlic (*A. sativum*) and is currently considered a synonym for garlic (Shemesh-Mayer and Kamenetsky Goldstein, 2018; Maaß and Klaas, 1995; POWO, 2021). Other species that have been suggested as probable ancestor include *Allium tuncelianum*, *A. macrochaetum* and *A. truncatum*, all of which are native to the Middle East (Mathew *et al.*, 2011; Zohary and Hopf, 2000).

- ***Allium tuncelianum*** (Kollmann) Ozhatay, B.Mathew & Siraneci (Tunceli garlic and Ovacik garlic): It is endemic to Munzur Valley of Munzur Mountain Range within Tunceli Province in eastern Anatolia (East Turkiye). Botanists have suggested that this species may be a close relative or perhaps an ancestor of garlic, but genetic analysis shows that it is actually more closely related to leek (Ipek *et al.*, 2008). Though it has odour and flavour similar to garlic and similarly used locally; unlike garlics, it usually produces single bulb with no cloves (en.wikipedia.org/wiki/Allium_tuncelianum). *A. tuncelianum* shares some common characteristics with *A. sativum* such as odour, coiling of the flower stem before anthesis, pale colouration, small, glabrous & rather narrow perianth segments and glabrous filaments with very long lateral cusps (Ipek *et al.*, 2008).
- ***Allium macrochaetum*** Boiss. & Hausskn: *Allium macrochaetum* is native to Central Turkiye, Syria to NW Iran and Jordan (POWO, 2022a). It too is locally used similar to garlic.
- ***Allium truncatum*** (Feinbrun) F.Kollmann & D.Zohary: *Allium truncatum* is native to Israel, Palestine, Lebanon and Turkiye (POWO, 2022b).

Subspecies and varieties

Allium sativum has two subspecies, *viz. A. sativum* ssp. *Ophioscorodon* (Hardneck type or Long-day type) and *A. sativum* ssp. *sativum* (Softneck type or Short-day type); 10 major groups or classes of varieties within these two subspecies are named either for appearance of the bulb or for geographical origin, along with and hundreds of varieties or cultivars. Growers with freezing winter weather can grow either hardneck or softneck garlic but have the best success with hardnecks. Growers in warm winter areas may do better with softnecks. The optimal day length requirement for bulb formation is 13-14 hours for long-day garlic and 10-12 hours for short-day garlic (Farmingindia, 2017). There are at least 120 cultivars originating from Central Asia, thus making it the main centre of garlic biodiversity (Kamenetsky *et al.*, 2005).

Hard-necked garlic [*A. sativum* ssp. *ophioscorodon* (Link) Doll]

Hardneck garlic (*A.s. ophioscorodon*) is sometimes considered to be a separate species *Allium ophioscorodon* G.Don. These are quite similar to wild garlic and produce a woody flower stalk called scape that will remain upright and rigid even when these die back. The flowers are produced under favourable conditions but they usually abort and form bulbils (small aerial bulb) with tiny cloves identical to the parent bulb. These bulbils can be planted to produce clones but the bulbs they form will be very small and may require two or three years to reach usable size.

Some types may not produce large bulbs unless the flower stalks are removed. Scapes should be cut in the early summer otherwise the production of bulbils will share the energy that otherwise would have been utilized in the bulb development. Below the ground around this central flowering stem is a single row of 4-8 or up to 15 rounded cloves wrapped together in a papery sheath to form the 'head' or bulb of garlic. Though they have fewer cloves per head than softneck types, the cloves on hardneck garlic varieties are larger, easy to peel and have stronger flavour. These varieties are generally well adapted to cooler climates and have a slightly shorter shelf life, *i.e.* 4 to 7 months since cloves have loose, easily peeling clove skins. Because of being larger in size, their productivity is higher and these fetch a good price in local, national and international markets. Hard-neck varieties cannot be braided for storage whereas softneck varieties can be braided and stored.

Hardneck type includes eight sub-types ~ rocambole, porcelain, purple striped, glazed purple striped, marble purple striped, Asiatic, turban and Creole, each having several cultivars. These eight subtypes can be further classified as shown below:

Hardnecks

They produce a woody flower stalk called scape.

- Rocambole
- Porcelain
- Purple striped
- Glazed purple striped
- Marble purple striped garlic

Previously, the three varieties ~ purple striped, glazed purple striped and marble purple striped were classified under the same subtype – Purple Striped Garlic. But now they each form their own distinct subtype.

Intermediate garlic varieties (Weakly bolting hardnecks): These are varieties that can produce a scape in harsh conditions such as drought or excessive heat otherwise grow like a softneck and these are:

- Asiatic
- Turban
- Creole

There is some confusion with respect to the classification of the Asiatic, Creole and Turban garlic types. Through genetic analyses, Pooler and Simon (1993) classified these garlic types as softnecks despite the fact that they occasionally make scapes.

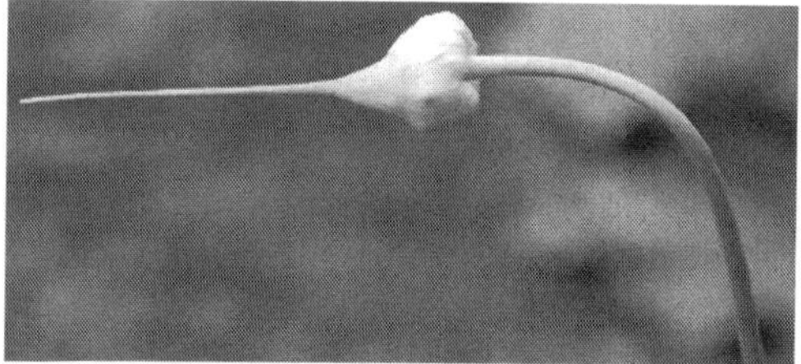

By Kickof - Own work, CC BY 3.0,

Rocambole garlics

Varieties in this class have intense savory flavour and are considered a top gourmet choice and sought-after by garlic lovers. These have dark to blue green leaves, a moderate spreading shape and a tightly coiled flower stalk with one to three loops or coils. Their off-white bulbs with various amounts of purple blush or streaking, usually have 6-12 cloves with very easy to peel thin papery skins. Their outer skin can range from striped to red and pink. Their storage life is generally 3-4 months. These garlics are very particular about overwatering and won't perform well in wet areas. These also need a bit hotter weather than other hardneck varieties. Generally these produce 10 to 40 brown, yellow or purple bulbils in the umbels.

Some popular rocambole-type garlic cultivars include 'Amish', 'Arcuri', 'Bavarian', 'Brown Saxon', 'Carpathian', 'Chamisal Wild', 'De vivo', 'Doukhobor', 'French rocambole', 'German Red', 'Hungarian Purple', 'Italian Purple', 'Killarney Red', 'Krasnodar Red', 'Legacy', 'Marino', 'Penasco Blue', 'Penn Wonder', 'Phillips', 'Pitarelli', 'Russian Red', 'Salt Spring Select', 'Sans Creek', 'Spanish Roja', 'Spicy Korean Red', 'Temptress', 'Turkish Red', 'Ukrainian Red', 'Yugoslavian', *etc.*

Porcelain garlics

Cv. 'Music' of Porcelain hardneck https://hoodrivergarlic.com/product

Varieties in this class are generally easy to grow and highly adaptive to different growing conditions. These are known for their attractive large, round and white bulbs having 4-6 elongated, large, satiny white cloves with classic garlic taste; closely related to Rocambole in flavour. Their flavour is often mild at harvest but intensifies during storage. Taller than Rocamboles, these garlics have pale to deep green leaves and looping or irregularly coiled scapes. These generally have many tiny white bulbils in the umbels. The peels are tighter than some other hardnecks and are only somewhat easy to peel. Although the outer skin is white, individual cloves are usually covered in red/brown skin. Their storage life is generally 4 to 8 months. They have the highest allicin content of all varieties.

Some popular porcelain-type garlic varieties include 'Armenian', 'Big Boy', 'Calabria', 'Dan's Russian', 'Floha', 'Georgia Fire', 'Georgian Crystal', 'German Extra

Hardy', 'German White', 'Iowa German White', 'Italian Red', 'Kyjev', 'Leningrad', 'Majestic', 'Montana Zemo', 'Music', 'Northern White', 'Polish Jenn', 'Polish', 'Reisig', 'Romanian Red', 'Rosewood', 'Zemo', *etc.*

Purple stripe or purple streaked garlics

Varieties in this class have bright purple streaks and blotches on both bulb wrappers and clove skins. The leaves on the plant grow at wider angles to the stem than Rocambole types, and the scapes often make perfect 270 degree curls. Most of these cultivars have 8-12 elongated crescent-shaped cloves with thin and easy to peel skins. These generally produce more than 50 purple bulbils in the umbels. Their storage life is slightly longer as compared to Rocambole cultivars. Roasted cloves have sweet flavour rather than spicy. Some of its popular cultivars include 'Chesnok Red', 'Oshala Purple'' Persian Star', 'Russian Red', 'Shatili', *etc.*

Marbled purple stripe garlics

Varieties in this class have easy to peel large bulbs with marbled purple cloves but white peels, comprising 4-7 rounded cloves. The scapes of these types too often make a 270^0 loop. They are known for their hot and spicy flavour. They can be stored for longer than other classes of hardneck garlic, generally for up to six months. These produce generally 20 to 60 purple bulbils in the umbels. Some popular cultivars of the group include 'Bogatyr', 'Brown Rose', 'Duganskij', 'Gaia's Joy', 'Gourmet Red', 'Khabar', 'Metechi', 'Nordic', 'Russian Giant', *etc.*

Glazed purple stripe garlics

Varieties in this class usually have smaller but more cloves (8-12) than marbled purple stripe, and have attractive glossy purple wrappers. The peels of the bulbs have an almost metallic sheen. These varieties are easy to peel and are perfect for roasting due to their balanced moderate flavour. Although these types have a longer storage life in comparison to other hardnecks, these are not commonly available for shipping long distances because of their delicate and very thin skins. Some of its popular cultivars include 'Oregon', 'Purple Glazer', 'Red Rezan', 'Vekan', *etc.*

Asiatic garlics

Varieties in this class are known for their diverse range of flavours. The bulbs usually have 6-9 plumpy cloves that are thick with glossy and tight peels. Some growers consider this a softneck garlic rather than a hardneck one as it rarely forms scapes. These types have drooping scapes with a few purple bulbils in an elongated, wrinkled bulbil capsule. These types usually have 6-9 plumpy and glossy skinned cloves per bulb. Some popular cultivars include 'Asian Tempest', 'Japanese', 'Keeper', 'Maiski', 'Pyong Vang Korean', *etc.*

Turban garlics

Varieties in this class are known for turban-shaped bulbil capsule. These types are very similar to Asiatic garlics but these mature a bit earlier. The bulbs usually have 7-11 cloves with course and thick skin. Some authorities consider this garlic to be softneck garlic, as it only occasionally makes scapes. Scapes have 30–100 small purple bulbils. Its some of the popular cultivars include 'Red Janice', 'Sonora', *etc.*

Creole garlics

Varieties in this class are known for their attractive peel and tall cloves with elongated tips. The bulbs usually have 4-8 cloves with tight skin. Some growers consider Creole garlic to be softneck garlic because it only occasionally makes scapes (and can vary widely by geographical region). Moreover, their scapes are generally short and weak and have numerous (usually more than 30) small to medium bulbils. Its some popular cultivars are 'Ajo Rojo', 'Burgundy', 'Creole Red', *etc.*

Soft-necked garlic (*A. sativum* ssp. *sativum*)

The softneck garlic cultivars are either nonbolting or produce only very weak flower stalks; however, climate can be a factor as to whether or not this actually happens. Often thought of as 'true garlic', softnecks account for most of what is found at the supermarket. This is because they are more productive, easier to grow, especially in warm climates or under mild winter conditions, and these store for long up to 9-12 months or even more under ideal storage conditions. These types are better adapted to areas with long seasons and mild winters. These large-bulbed, coarse-skinned garlics are highly productive in a wide range of climates and soils though lacking typical hardneck garlic flavour ~ having either a very hot or very mild taste.

Softnecks mature faster than hadrnecks and these produce more cloves per plant. Generally, most softneck types have 8 to 45 tightly packed cloves though mostly there are only 12 to 20 of varying sizes as compared to hardneck types as these have only a few single-row arranged large cloves.

Cloves are generally arranged in multiple layers (usually 3-6) within a bulb. Their stems stay soft and flexible, making them excellent for creating braids. Examples of Indian garlic varieties are 'G41', 'G1', 'G50', 'G282', *etc.* The two sub-classes of softneck garlic are artichoke and silverskin, each having several cultivars.

Artichoke garlic

Artichoke-type softnecks tend to have early maturing cloves with coarse and thick clove skins that are difficult to peel. They are fairly upright plants with average to slightly pale green leaves, rarely with a flower stalk; scapes occurring occasionally in some climates or on some individual plants. During stressful conditions, asexual propagules called bulbils may be produced within the false stems, resulting in the classification of incomplete bolting types. Varieties in this class have a unique multi-layered (usually 3–6 layers) arrangement of 8-16 milky white cloves in a white bulb that resembles an artichoke when sliced. Some have pink or brown blush or partial striping. Many of its cultivars generally have two layers of cloves with the inner layer cloves being tall, thin and narrow while the outer ones are plumpy and squarish. These have a great storage life of 8-12 months. Though their flavour is not as strong as hardnecks, these have a fine subtle flavour that may vary from mild to hot depending on strain and climate. Its some of the popular cultivars are 'California Early', 'Chef's Italian Red', 'Gomecari', 'Inchelium Red', 'Island Star', 'Italian Softneck', 'Lorz Italian', 'Machashi', 'Mchadidzhauri', 'Red Torch', 'Sicilian Gold', 'Susanville', *etc.*

Silver-skin garlic

Silverskin-type softnecks have nearly vertical, blue-green leaves displaying a bilateral symmetry. Though varieties in this class take longer to grow but have an excellent storage life of more than 12 months. These have an attractive silvery-white peel, and depending on the cultivars the bulbs comprise 8-24 or more cloves that are usually arranged in 3-6 layers. The leaf colour is blue-green. In flavour these vary from very mild to very hot. Its some of the popular cultivars are 'Belarus', 'Californian Select', 'Italian Locati', 'Italian Sicilian', 'Nootka Rose', 'Prim', 'Pristinski', 'Silverwhite', *etc.*

Garlic breeding

Garlic is an asexually propagated crop, and methods of improvement through cross-pollination are not viable, though true seed production in garlic (*Allium sativum* and *A. longicuspis*) has been witnessed by certain authorities, therefore, Etoh (1986) made efforts in it to induce flowering *cum* seed formation. Jenderek and Hannan (2004, 2005) evaluated reproductive characteristics and produced F_1 bulbs in a few fertile clones. Jenderek and Zewdie (2005) reported that due to centuries-old vegetative propagation, this crop has become highly heterozygous. Further, they stated that number and weight of cloves, and over all bulb weight are the main factors contributing to its bulb yield.

Barandiaran *et al.* (1998) were the first to report transfer of uidA gene into different garlic tissues including regenerable calli through biolistic particle delivery. Garlic tissues showed a high endogenous nuclease activity, preventing exogenous DNA expression. Since then, genetic transformation in garlic has been reported through indirect as well as direct transformation system (Sawahel, 2002). Park *et al.* (2002) for the first time generated transgenic plants resistant to chlorsulfuron ~ a sulfonylurea herbicide.

Indian garlic breeding programme

Despite the garlic being propagated vegetatively, there is considerable variation in morphological traits (Shashidhar and Dharmatti, 2005; Khar *et al.*, 2006) such as number of green leaves, plant height, bulbing period, maturity period, bulb weight & diameter *cum* yield and number of cloves per bulb (Lawande *et al.*, 2009). Cultivated garlic is usually non-flowering with limited variability. Therefore, breeding methods for development of garlic are limited to clonal selection and mutagenesis among conventional methods, and somaclonal variation among biotechnological approaches.

Most Indian varieties have been developed through clonal selection and only a very few (1 or 2) through introduction (Lawande *et al.*, 2009). Maximum number of varieties have been developed by NHRDF, followed by agricultural universities, *viz.*, GAU, PAU, MPKV, Rahuri, *etc.* Most of the varieties developed in India are soft-necked type suitable for growing under tropical and subtropical climates though some temperate varieties have also been released at the national level and prominent among them is 'Agrifound Parvati'. Other important temperate varieties are 'VLG-1' (VPKAS, Almora), 'SKUAG 1' (SKUAST, Srinagar), 'DARL 52' and 'Solan Local' (YSPUHF, Solan). Moreover, varieties selected by farmers over the years are also available in the market, *e.g.*, 'Jamnagar Local', 'Ooty Local', 'Jeur Local', *etc.* (Lawande *et al.*, 2009).

Some garlic varieties developed in India:

Variety	*Institution Notification year*	*Traits*	*Region*	*Maturity days*	*Yield t/ha*
Short-day type					
Agrifound White (G-41) (dogr126)	N H R D F - 1989	**Bulbs:** Compact, silvery white with creamy flesh; diameter 3.5-4.5 cm. Cloves: bigger, elongated, 20-25 in numbers. **TSS:** 41% ; Good storer. Susceptible to purple blotch or Stemphylium blight in rabi season. (eagri.org)	Madhya Pradesh and Maharashtra (dogr126)	130-135 (dogr126)	12-14 (dogr126)
Yamuna Safed (G-1) (dogr126)	N H R D F - 1991	**Bulbs:** compact, silvery white skin with creamy flesh; diameter 4-4.5 cm; Cloves: sickle shaped, 25-30 in numbers, diameter 0.8-1 cm. **TSS**: 38%; good storer. Tolerant to insect pests and diseases like purple blotch, Stemphylium blight and onion thrips. (eagri.org)	All India (dogr126)	120-130 (dogr126)	15-17 (dogr126)
Yamuna Safed-2 (G-50) (dogr126)	N H R D F - 1996	**Bulbs:** compact attractive white creamy flesh, diameter 3.5-4.0 cm. Cloves: 35-40 in number, diameter 0.75-14 cm. TSS: 38-40%; Good storer (eagri.org)	Northern India (dogr126)	120-130 (dogr126)	15-20 (dogr126)
Yamuna Safed-3 (G-282) (dogr126)	N H R D F - 1999	**Bulbs:** creamy white and bigger sized (5-6cm dia). Cloves: 15-16 in number, diameter 1.2-1.5 cm. Leaves: wider than other varieties. **TSS:** 38.42% ; medium storer (eagri.org)	Chhattisgarh, Gujarat,Haryana, Madhya Pradesh, Maharashtra, Punjab, Rajasthan and Uttar Pradesh (dogr126)	120-130 (dogr126)	17-20 (dogr126)

Variety	*Institution Notification year*	*Traits*	*Region*	*Maturity days*	*Yield t/ha*
Yamuna Safed-4 (G-323) (dogr126)	NHRDF – 2006 (eagri.org)	**Leaf:** wide. **Bulb**: compact, attractive creamy white colour, creamy flesh, diameter 4-5 cm, **Cloves:** no. of cloves 18-23, diameter 0.75-1.0 cm. **Storage**: Storage quality is better than Yamuna Safed-3. (eagri.org)	North and central India (dogr126)	165-175 (dogr126)	16-17 (dogr126)
Yamuna Safed-5 (G-189) (dogr126)	NHRDF	**Bulb:** white (dogr126) and big in size - about 4.5-5.0 cm in diameter	Andaman and Nicobar Island, Arunachal Pradesh, Bihar, Delhi, Gujarat, Haryana, Jharkhand, Manipur, Meghalaya, Mizoram, Nagaland, Punjab, Rajasthan, Sikkim, Tarai region of UP, Tripura and Uttarakhand (dogr126)	150-160 (dogr126)	17-18 (dogr126)
Godavari (dogr126)	MPKV - 1987	**Bulb:** Purple (dogr126), white (Lawande *et al.,* 2009). Bulbs medium size with 25-30 cloves. Cloves are pink in colour. Tolerant to mites (mpkv.ac.in)	Maharashtra (dogr126	140-145 (dogr126)	10-11 (dogr126)
Shweta	MPKV	**Bulb**: White	Maharashtra	130-135	10-11
T-56-4 (Lawande *et al.,* 2009)	PAU	**Bulb:** White (Lawande *et al.,* 2009)			8-10 (Lawande *et al.,* 2009)
Bhima Omkar (dogr126)	DOGR - 2009	**Bulb**: medium size compact white bulbs (eagri.org)	Delhi, Gujarat, Haryana and Rajasthan (dogr126)	120-135 (dogr126)	8-14 (dogr126)

Variety	*Institution Notification year*	*Traits*	*Region*	*Maturity days*	*Yield t/ha*
Bhima purple (dogr126)	D O G R - 2009	Attractive purple skinned bulbs. Good store (dogr126)	Andhra Pradesh, Bihar, Delhi, Haryana, Karnataka, Maharashtra, Punjab and Uttar Pradesh (dogr126)	120-135 (dogr126)	6-7 (dogr126)
Phule Baswant (dogr126)	MPKV	**Bulb:** White (dogr126). Attractive violet coloured large bulbs (mpkv.ac.in). Moderately resistant to purple blotch, thrips, mites under field conditions (mpkv.ac.in)	Madhya Pradesh and Maharashtra (dogr126)	135-140 (dogr126)	10 -11 (dogr126)
GG 4 (dogr126)	JAU in 2009	**Bulb**: White (dogr126)	Gujarat and Maharashtra (dogr126)	130-140 (dogr126)	8-10 (dogr126)
GG7 (Anand Kesari) (aau.in)	AAU - 2019	**Leaves:** dark green, concave shape in cross section, medium density leaves with erect foliage. Bulb: medium with radial distribution of cloves. Cloves: with purple colour of dry external scales and purple scale colour of clove. **Storer**: good. Low incidence of thrips. (aau.in)	Whole of Gujarat (aau.in)		8 (aau.in)
Phule Nilima (Sel. 10-3) (phytojournal.com; mpkv.ac.in)	MPKV - 2014	**Leaves:** semi-erect, dark green, slightly concave. **Bulb:** large (3.5-5 cm), large, purple/violet in colour, medium compact, external cloves sparsely present, cloves > 20 in number. **Cloves:** 1-2 cm diameter, purple skin, yellowish flesh. Moderately resistant to purple blotch disease; minimum infestation of thrips and mites under field condition during rabi season (phytojournal.com; mpkv.ac.in)	Maharashtra (phytojournal.com)	140-150 (phytojournal.com; mpkv.ac.in)	17-18 (phytojournal.com; mpkv.ac.in)

Variety	*Institution Notification year*	*Traits*	*Region*	*Maturity days*	*Yield t/ha*
Intermediate type					
Ooty 1 (dogr126)	TNAU	Bulb: Dull white (dogr126)	Tamil Nadu (dogr126)	120-130 (dogr126)	15-17 (dogr126)
DARL52 (Lawande *et al.*, 2009)	DARL - 2003	**Bulb:** White (Lawande *et al.*, 2009)	Temperate regions (Lawande *et al.*, 2009)		12-15 (Lawande *et al.*, 2009)
Long day type					
VL Garlic 1 (dogr126)	ICAR-VPKAS	**Bulb:** White (dogr126)	Bihar,Himachal Pradesh, Jammu & Kashmir, Punjab, Uttarakhand and Uttar Pradesh (dogr126)	180-190	Hills:14-15 Plains: 9-10 (dogr126)
VL Lahsun 2 (dogr126)	ICAR-VPKAS (dogr126)	**Bulb:** White (dogr126)	Himachal Pradesh, Jammu & Kashmir and Uttarakhand (dogr126)	190-200 (dogr126)	Mid hills:14-16 Above mid hills: 24-26 (dogr126)
Agrifound Parvati (G313) (dogr126; eagri.org)	NHRDF - 1992	**Bulbs:** purple, bigger size (5-6.5 cm), creamy white colour with pinkish tinge. **Cloves:** 10-16 in number, diameter 1.5-1.8 cm. **Storer:** medium.Tolerant to common disease (eagri.org)	Hills of Himachal Pradesh, Uttarakhand, Jammu & Kashmir and high altitudes of North eastern states like Sikkim etc. (dogr126; eagri.org)	165-175 (dogr126; agri.org)	17-22 (dogr126; eagri.org)
Agrifound Parvati-2 (G-408) (dogr.icar.gov)	NHRDF	**Bulb:** White (dogr.icar.gov)	Hills of Himachal Pradesh, Uttarakhand, Jammu &Kashmir and high altitudes of North eastern states like Sikkim etc. (dogr.icar.gov)	165-175 (dogr.icar.gov)	17-22 (dogr.icar.gov)
SKUAG 1 (Lawande *et al.*, 2009)	SKUAST, Srinagar		Temperate regions (Lawande *et al.*, 2009)		

Variety	*Institution Notification year*	*Traits*	*Region*	*Maturity days*	*Yield t/ha*
Solan Local (Lawande *et al.*, 2009)	YSPUHF, Solan		Temperate regions (Lawande *et al.*, 2009)		

NHRDF - National Horticultural Research & Development Foundation , Nasik
MPKV - Mahatma Phule Krishi Vidyapeeth, Rahuri
CCSHAU - Chaudhary Charan Singh Haryana Agricultural University, Hisar
TNAU - Tamil Nadu Agricultural University, Coimbatore,
ICAR-Indian Council of Agricultural Research Institute, New Delhi,
DOGR –Directorate of Onion & Garlic Research (DOGR), Pune
VPKAS - ICAR-Vivekananda Parvatiya Krishi Anusandhan Sansthan, Almora
JAU - Junagadh Agricultural University, Junagadh
DARL – Defence Agricultural Research Laboratory, Pithoragarh, Uttarakhand
PAU - Punjab Agricultural University, Ludhiana
AAU – Anand Agricultural University, Dahod, Gujarat
BSKKV - Dr . Balasaheb Sawant Konkan Krishi Vidyapeeth, Dapoli, Dist. Ratnagiri.

Varieties 'Yamuna Safed-3' and 'Agrifound Parvati' have bigger cloves suitable for export (dogr.icar.gov)
Sources: dogr126; *eagri.org*; aau.in; Lawande *et al.* (2009); phytojournal.com; mpkv.ac.in

Single clove garlic/solo garlic/pearl garlic

Single clove garlic is in fact not a single clove variety of garlic but a product of specific planting practices. Single clove garlic has been grown at the foothills of the Himalayan mountains for about 7,000 years (en.wikipedia.org/wiki/Solo_garlic). The rounded single cloved bulb, approximately 2.5-5cm in diameter, is white-skinned often with purple stripes and usually easy to peel. It is quite similar in appearance to a pickling onion. It has the flavour of the garlic clove but is somewhat milder and slightly perfumed. Single cloved types of other garlic species such as *Allium nigrum* and *Allium ampeloprasum* are also available (en.wikipedia.org/wiki/Solo_garlic). In India, Madhya Pradesh and some parts of Rajasthan are the major solo garlic producing areas; the biggest supplying districts are Neemuch and Mandsaur of MP.

Growing methods

By Dmitry Makarov (Dmit) - Own work, CC BY-SA 3.0, https://commons.wikimedia.org/w/index.php?curid=7249313

Solo garlics can be found naturally with the crop of normal garlic probably when some premature garlics become solo garlic. Sorting out solo garlics from normal garlic crop is quite labour intensive as to recover some 1-2kg of solo garlics one will have to sort out some one metric ton of normal garlic bulbs. Commercial production comes from areas where garlic is likely to produce a solo bulb due to environmental factors. Climate in such areas, combined with careful cultivation, leads to a large percentage of the garlic crop failing to split into multiple cloves as these are grown in *Rabi* season (November-December) and harvested during spring (February-March) (en.wikipedia.org/wiki/Solo_garlic). In comparison to Chinese solo garlic, Indian single clove garlics are smaller in size and more cylindrical in shape. Their skin is mostly white with light purple lines as appearing on some garlic bulbs.

Using bulbils to grow solo garlics

Small bulbs of solo garlic can also be obtained by planting the bulbils of any variety of garlic. In varieties producing scapes, the flower stalk of some plants are allowed to grow and mature. These scapes usually form bulbils at the top which on maturity are harvested along with underground bulbs. These teardrop-shaped tiny bulbils (cloves) are planted to produce a bit larger, round and single-cloved bulbs after one season. These round bulbs can be peeled and eaten. If these round bulbs are planted for a second year, most of them will grow into a regular garlic bulb with the usual cloves. They are genetically identical to their parent plant.

Nutritional values

Per 100 grammes of garlic contains several nutrients in rich amounts [20% or more of the Daily Value (DV)] including vitamins C and B6 including thiamine and pantothenic acid, the dietary minerals manganese and phosphorus, as well as Ca, Fe and Zn (en.wikipedia.org/wiki/Garlic).

Nutritional value of raw garlic

Nutritional value per 100g of raw garlic is shown in the following table:

Nutritional value per 100g	
Energy	623 kJ (149 kcal)
Carbohydrates	33.06g
Sugars	1g
Dietary fiber	2.1g
Fat	0.5g
Saturated Polysaturated fat Monosaturated	0.09g 0.25g 0.01g
Protein	6.36g
Vitamins	**Quantity (%DV†)**
Thiamine (B1)	17% (0.2mg)
Riboflavin (B2)	9% (0.11mg)
Niacin (B3)	5% (0.7mg)
Pantothenic acid (B5)	12% (0.596mg)
Vitamin B6	95% (1.235mg)
Folate (B9)	1% (3μg)
Choline	5% (23.2mg)
Vitamin C	38% (31.2mg)

Minerals	Quantity (%DV†)
Calcium	18% (181mg)
Iron	13% (1.7mg)
Magnesium	7% (25mg)
Manganese	80% (1.672mg)
Phosphorus	22% (153mg)
Potassium	9% (401mg)
Sodium	1% (17mg)
Zinc	12% (1.16mg)
Other constituents	**Quantity**
Water	59g
Selenium	14.2μg
Units: μg = micrograms; mg = milligrams; IU = International units	
†Percentages are roughly approximated using US recommendations for adults. Source: USDA FoodData Central	
Table source: *https://en.wikipedia.org/wiki/Garlic* *Source: www.myfitnesspal.com*	

Raw garlic provides B vitamins, especially vitamin B6 supplying 6% daily value (DV) in a serving, and niacin, pantothenic acid, riboflavin (B2), thiamine, antioxidant vit. C (5% DV per 3 cloves), vit. K for blood coagulation, and minerals such as Mn with 8% DV per serving for the nervous system and brain as well as smaller amounts of Cu, K and Fe (Gord, 2019).

Garlic composition on dry matter basis

Per 100g garlic on dry matter basis contains 4.55mg moisture, 73.22mg carbohydrate and 15.33mg crude protein, apart from 0.72mg crude fat, 2.1mg crude fibre, 4.08mg ash, 10.19mg K, 26.3mg Ca, 10.19mg P, 5.29mg Fe, 0.001mg Mg, 0.34mg Zn and 0.001mg Mn, however, lead and cobalt were recorded below detection level (Yusuf *et al.*, 2018).

Approximate composition of garlic powder (mg/100g)

Yusuf** et al. **(2018).		***Odebunmi** et al. **(2009)***		***Marina** et al. **(2014)***	
Nutrients	**Composition**	**Nutrients**	**Composition**	**Nutrients**	**Composition**
Moisture	4.55	Moisture	66.57	Moisture	64.58%
Crude protein	15.33	Protein	7.87	Protein	7.87%
Crude fat	0.72	Fat	0.52	Ether extract	0.52%
Crude fibre	2.10	Crude fibre	0.73	Crude fibre	2.3%
Ash	4.08	Ash	1.33	Ash	2.46%
Carbohydrate	73.22	Dry matter	33.43%	N-free extract	22.27%

Approximate mineral composition of garlic on dry matter basis (Yusuf *et al.*, 2018)

Mineral elements	*Concentration (mg/100g)*
Potassium	10.1 9
Calcium	26.30
Phosphorus	10.19
Iron	5.29
Magnesium	0.001
Zinc	0.34
Manganese	0.001
Lead	Nil
Cobalt	Nil
Table source: *Yusuf et al. (2018)*	

Factors affecting nutritional value

Being water soluble, vitamins B and C in garlic are easily destroyed during cooking though light steaming may help preserve vit. C, however, fat soluble vit. K during steaming remains stable though frying reduces it also, therefore garlic should be added in the end of the food process (Gord, 2019). Boiling garlic may not reduce minerals but some of the Mn, Ca and other minerals may leach out to some extent. Fermented black garlic contains far better nutritional value as compared to fresh garlic.

Phytochemicals

Phytochemicals are naturally-occurring, non-nutrient and bioactive plant compounds found in vegetables, fruits, spices and other food items. Garlic has a variety of bioactive compounds, including organosulphur compounds, saponins, phenolic compounds, and polysaccharides (Diretto *et al.*, 2017; Szychowski *et al.*, 2018; Bradley *et al.*, 2016; Wang *et al.*, 2018). It is rich in alkaloids, tannin and flavonoids and has an acidic pH of 3.91 (Yusuf *et al.*, 2018). The carotenoids, saponin, steroids and cardenolides are present in trace amounts. Saponins are produced by plants as a defence mechanism to stop attacks by foreign pathogens which make them natural antibiotics (Okwu and Emenike, 2006). Garlic is not only a rich source of beneficial sulphur compounds, bioactive substances and antioxidants, but is also among the richest sources of phenolic compounds such as β-resorcylic acid, followed by pyrogallol, gallic acid, rutin, protocatechuic acid, as well as quercetin (Nagella *et al.*, 2014). Garlic polysaccharides were reported to contain 85% fructose, 14% glucose and 1% galactose (Hang, 2005).

As per pharmacy180.com, other constituents reported in garlic are fatty oils, mucilage and albumin, and in the essential oil (0.06–0.1%) allyl propyl disulphide, diallyl disulphide and allicin. γ-Glutamyl peptides are isolated from the garlic, whereas amino acids present in the bulb are leucine, methionine, S-propyl-L-cysteine, S-propenyl-L-cysteine, S-methyl cysteine, S-allyl cysteine sulphoxide (alliin), S-ethyl cysteine sulphoxide and S-butyl-cysteine sulphoxide.

Approximate phytochemical composition of garlic powder

Garlic is acidic (pH = 3.91) and contains alkaloids, tannins, carotenoids, saponin, flavonoids, steroids and cardenolides. The approximate phytochemical composition of garlic powder is given in the following table.

Phytochemical	***Composition***
Alkaloids (mg/100g)	4.21
Tannins (mg/100g)	3.54
Carotenoids(μg/100g)	0.64
Saponin (mg/100g)	0.80
Flavonoids (mg/100g)	5.56
Steroids (mg/100g)	0.04
Cardenolides (mg/100g)	0.02
pH	3.91
Titratable acidity (% as lactic acid)	0.97
Salt (% NaCl)	2.39
Table source: *Yusuf et al. (2018)*	

Bioflavonoids

Garlic contains bioflavonoids such as cyanidin, quercetin and allistatin I & II (Etewa and Abaza, 2011). These bioflavonoids and antioxidant vitamins present in garlic protect against oxidative agents and free radicals (El-Kelawy *et al.*, 2017). When compared to ginger, garlic has higher antimicrobial activity which may be due to the presence of allistatin I & II (Edward and Ohaegbu, 2012), which together with garlicin (a well-known antibiotic) and scordinin A and B, also possess broad-spectrum high activity against fungi (Al-sultan and Al-Juwary, 2005; Abdullah and Mahdi, 2006). Allistatin I & II are also present in other *Allium* species such as *A. stracheyi* (Rajan *et el.*, 2014) and *A. cepa* (Arshad *et al.*, 2017; Alamgir, 2017) as well as other plants like *Zingiber officinale* (Helen *et al.*, 2020). Many scholars stated allistatin as a non-sulphur-containing bioflavonoid (Arshad *et al.*, 2017; Farooqui and Farooqui, 2017; Gregrova *et al.*, 2013; Ebadi *et al.*, 2014), while a few researchers relate it to well-studied allicin (a sulphur-containing allyl compound) (Cundell, 2014).

HS/GC/MS analysis of white and green parts

Koca *et al.* (2016) reported that aroma components of white and green parts of garlic were analyzed by 'headspace/gas chromatography/mass spectrometry (HS/GC/MS)'. Acetic acid, methyl ester, methyl-thiirane, allyl mercaptan, ethyl iso-allocholate, 1-propene, 3,3'-thiobis, 2-propen-1-ol, methyl 2-propenyl disulphide, methyl 1-propenyl disulphide, diallyl disulphide, methoxy methyl isothiocyanate, methyl allyl trisulphide and di-2-propenyl trisulphide were determined in both white and green parts of garlic. Methyl hydrogen disulphide, hexanal and 3,4-dimethylthiophene were only found in white part of garlic, however, dimethylsulphide, allyl methylsulphide and dimethyldisulphide were found in green part of garlic.

Major volatile components detected in white part of garlic were methyl hydrogen disulphide (25.42±16.68%) and diallyl disulphide (37.20±21.91%), whereas diallyl disulphide (32.62±3.61%) and 1-propene,3,3'-thiobis (11.85±6.14%) predominated in garlic leaves.

Organosulphur compounds

The flavour and taste associated with garlic are due to the presence of over 200 active constituents which constitute its medicinal properties. Such compounds include sulphur-containing compounds such as diallyl thiosulphonate (allicin), diallyl sulphide (DAS), diallyl disulphide (DADS), diallyl trisulphide (DATS), E/Z-ajoene, S-allyl-cysteine (SAC), and S-allyl-cysteine sulphoxide (alliin) (Yoo *et al.*, 2014a, b; Kodera *et al.*, 2017). One of the most considered medicinal plants garlic has a higher concentration of sulphur than any other *Allium* species (Kushwaha *et al.*, 2018). Garlic cloves are reported to contain several phytochemicals including sulphur-containing compounds such as ajoenes (*E*-ajoene, *Z*-ajoene), allicin, vinyldithiins [2-vinyl-(4H) -1,3-dithiin, 3-vinyl-(4H)-1,2-dithiin], sulphides (DADS, DATS) and others that account 82% of the overall garlic sulphur content (Al-Snafi, 2013).

Alliin (allicin, ajoenes, vinyldithiins and sulphides) is most plentiful sulphur compound found in garlic while allicin is most biologically active compound present in garlic (Kushwaha *et al.*, 2018). Alliin is odourless, crystallized from water acetone and practically insoluble in absolute alcohol, chloroform, acetone, ether and benzene (pharmacy180.com). Fresh garlic contains alliin and an enzyme called alliinase (alliin alkylsulphenate- lyase), each present in separate cells. Alliin is transformed to allicin following subsequent hydrolysis by alliinase enzyme after crushing or cutting off the garlic and breaking down the parenchyma, however, main odoriferous molecules of freshly milled garlic homogenates are S-propyl-cysteine-sulphoxide (PCSO), allicin and S-methyl cysteine-sulphoxide (MCSO)

(El-Saber *et al.*, 2020). Allicin, a yellow liquid responsible for the odour of garlic is very unstable and breaks down quickly producing allyl methyl sulphide that goes into the bloodstream and gets expelled through the skin's pores as garlicky sweat. It also enters the lung and causes garlic smelling breath. Allicin is miscible with alcohol, ether and benzene, and decomposes on distilling (pharmacy180.com). S-alk(en)yl-l-cysteine sulphoxides are the secondary metabolites obtained from cysteine which accumulate in the plants of *Allium* genus. Garlic formulations consist of several organosulphur compounds including N-acetylcysteine (NAC), SAC and S-ally-mercapto cysteine (SAMC) which are derived from alliin.

Plentiful sulphur compounds in garlic are also responsible for turning garlic colour to green or blue during pickling and cooking. Under these conditions, *i.e.*, acidity or heat the sulphur-containing compound alliin reacts with common amino acids to make pyrroles, the clusters of carbon-nitrogen rings (Imai *et al.*, 2006; Cho *et al.*, 2009a). These rings can be linked together into polypyrrole molecules. Ring structures absorb particular wavelengths of light and thus appear coloured. The two-pyrrole molecule looks red, the three-pyrrole molecule looks blue, and the four-pyrrole molecule looks green (like chlorophyll, a tetrapyrrole). Like chlorophyll, the pyrrole pigments are safe to eat. Upon cutting, similar to a colour change in onion caused by reactions of amino acids with sulphur compounds (Lee *et al.*, 2012), garlic can turn green (Cho *et al.*, 2009b; Lukes, 1986).

Pharmacological activity

Garlic contains phytochemicals including allicin, alliin and dithiin which fight diseases. Garlic contains diverse bioactive compounds and studies have shown that garlic and its bioactive constituents exhibit antioxidant, anti-inflammatory, antibacterial, antifungal, immunomodulatory, cardiovascular protective, anticancer, hepatoprotective, digestive system protective, anti-diabetic, anti-obesity, neuroprotective and renal protective properties. SAC has antioxidant, anti-inflammation, regulated redox, pro-energetic, antiapoptotic, and signalling capacities (Liu *et al.*, 2015b) while SAMC shows an anticancer activity through preventing the cancer cell multiplication (Cao *et al.*, 2017). Garlic has shown effective results for anti-microbial, cardiovascular and anticancer activity, and curative results have been found for treatment of hypoglycemia, liver dysfunction, heavy metal poisoning and hyperthyroidism, and its allicin has been found as a strong scavenging and antioxidant compound, and the studies indicate that other compounds such as polar compounds of phenolic and steroidal origin may play stronger roles (Kushwaha *et al.*, 2018).

High alkaloid content in garlic dry matter is probably responsible for its much-acclaimed medicinal values (Yusuf *et al.*, 2018). Purified alkaloids and their synthetic derivatives are medicinal agents as analgesic, anti-malarial, antiseptic and bactericidal (Evans *et al.*, 2002). Saponins are natural antibiotics that have cholesterol lowering effects, *vis-a-vis* also able to kill or inhibit cancer cells (Okwu, 2005; Nwinuka *et al.*, 2005; Okwu and Emenike, 2006; Okwu and Nnamdi, 2008). Therefore, appreciable amount of saponins in garlic could be responsible for its antimicrobial properties. The presence of tannins might have accounted for the sharp taste of garlic and its ability to hasten the healing of wounds and inflamed mucous membranes (Okwu and Emenike, 2006). Cardenolides contribute to the medicinal values of plants in which they occur and have been known for treatment of congestive heart failure (Oloyede, 2005).

Flavonoids protect against allergies, inflammation, platelet aggregation and microbial infection (Okwu and Omodimiro, 2005). Carotenoids are largely responsible for colour of garlic (Tripathi and Mishra, 2009). In general, the presence of these phytochemicals account for the much-touted medicinal properties of garlic in various disease conditions such as atherosclerosis, arthritis, nausea, asthma, bacterial infections and cancer (Das and Saha, 2008; Kaur and Arora, 2009).

Overall, garlic is an excellent natural source of bioactive sulphur-containing compounds and has promising applications in the development of functional foods or nutraceuticals for the prevention and management of certain diseases (Shang *et al.*, 2019).

The pharmacological activity of garlic and its related compounds are shown in the following table:

Activities	***Bioactive Compound***	***Mechanism of Action***	***References***
Antibacterial	Allicin	Chemical interaction with enzymes containing thiol	Gruhlke *et al.*, 2010
Antifungal	DADS	Irreversible ultrastructural changes in the fungal cells, loss of structural integrity and affecting the germination ability	Kuda *et al.*, 2004
	DATS		
Antiviral	Allicin	Chemical interaction with enzymes containing thiol	Zhen *et al.*, 2006
	DATS	Enhancing natural killer-cell (NK-cell) activity that destroys virus-infected cells	
Antiprotozoal	Allicin	Preventing the parasite's RNA, DNA and protein synthesis.	Zhen *et al.*, 2006
	DATS		
	Ajoene	Inhibiting the human glutathione reductase and *T. cruzi* trypanothione reductase	Gallwitz *et al.*, 1999

Activities	*Bioactive Compound*	*Mechanism of Action*	*References*
Antioxidant	Allicin, DADS, and DATS	Modulation of ROS, increasing glutathione and cellular antioxidant enzymes	Gruhlke *et al.*, 2010
	Alliin	Controlling ROS generation and preventing mitogen-activated protein kinase (MAPK)	Chen *et al.*, 2016
	DAS	Suppressing the enzymatic activity of cytochrome P450-2E1, reducing the generation of reactive oxygen and nitrogen species	Abdel-Daim *et al.*, 2018
Anti-inflammatory	Allicin	Enhancing the immune cell activity f, inhibiting the SDF1α chemokine and transendothelial migration of neutrophils	Abdel-Hafeez *et al.*, 2015
	DAS	Diminishing the expression of the inflammatory cytokines (e.g., NF- κB, IL-1β, and TNF-α), and ROS generation by suppressing CYP-2E1 hepatic enzyme	Abdel-Daim *et al.*, 2020
	Thiacremonone	Blocking the NF-κB activity	Jin *et al.*, 2013
Anti-cancer	Allicin, alliin, DADS, DAS	Enhancing p38 expression and cleaved caspase 3.	Zhang *et al.*, 2015a
	Z-Ajoene	Stimulating apoptosis in human leukemic cells, promoting the peroxide production, caspase-3-like, and caspase-8 activities	Bayan *et al.*, 2014
Immunomodulatory	Allicin	Suppressing BuChE and AChE	Liu *et al.*, 2015a
Anti-obesity	Ajoene	Decreasing the fat accumulation in 3T3-L1 adipocytes and dramatically decreases the body weight gain	Han *et al.*, 2011
	1,2-Vinyldithiin	Decreasing the C/EBPα, PPARγ2, and LPL expression and the PPARγ effect in human adipocytes	Keophiphath *et al.*, 2009
Antidiabetic	Allyl propyl disulfide, allicin, cysteine sulfoxide, and S-allyl cysteine sulfoxide, alliin	Decreasing the insulin secretion from pancreatic cells, increasing liver metabolism, and thus enhancing the short-acting insulin production	Zhai *et al.*, 2018
Hypolipidemic, hypocholesterolaemic	Different garlic preparations	Decreasing serum TC, TG, and LDL levels and moderately elevating HDL cholesterol	Qidwai and Ashfaq, 2013

Activities	*Bioactive Compound*	*Mechanism of Action*	*References*
Anti-Atherosclerotic, antithrombotic	Different garlic preparations	Preventing ADP-activated platelets binding to immobilized fibrinogen and platelet aggregation, inhibiting GPIIb/IIIa receptor and increasing cAMP	Drobiova *et al.*, 2011
Antihypertensive	Gamma-glutamylcysteine	Inhibiting the angiotensin-converting enzyme	Bayan *et al.*, 2014
Table source: *El-Saber et al. (2020)*			

The biological activities of garlic products, their active components, study type, and effects and mechanism have been summarized in the following table. The data in the table have been collected from the studies from the Web of Science Core Collection, reviewing the main bioactive compounds and biological functions of garlic, with special attention paid to the relevant mechanisms of actions (Shang *et al.*, 2019).

Product	Compounds	Study type	Subjects/cell lines	Main effects	Possible mechanisms	Ref.
Antioxidant Activity						
Aged garlic extract		In vitro	Human endothelial cells	Protecting cells against oxidative stress	Inducing the expression of several antioxidant enzymes, HO-1 and GCLM subunit, through Nrf2- ARE pathway	Liu *et al.* (2018)
	Saponins	In vitro	Mouse-derived C2C12 myoblasts	Protecting cells against the growth inhibition and DNA damage induced by H_2O_2	Scavenging intracellular reactive oxygen species	Kang *et al.* (2016)
Anti-inflammatory Activity						
	Ethyl linoleate	In vitro	Lipopolysaccharide-stimulated RAW 264.7 macrophages	Reducing the production of NO and prostaglandin E-2	Down regulating the expression of iNOS and COX2	Park *et al.* (2014)
	Garlic 14-kDa protein	In vitro	Lipopolysaccharide-stimulated J774A.1 macrophages	Inhibiting the inflammatory mediators such as NO, TNF-α, and IL-1β	Inhibiting the transcription factor NF-κB signaling pathway	Rabe *et al.*, (2015)
Aged garlic extract		In vivo	Apolipoprotein E-knockout mice	Inhibiting inflammation	Reducing the level of TNF-α and interleukin IL-1 receptor-associated kinase 4 Increasing the activity of AMPK in the liver	Morihara *et al.* (2017)
	Allicin	In vivo	BALB/c mice	Protecting against the inflammatory response induced by schistosome infection		Metwally *et al.* (2018)
Garlic tablets (equal to 2.5 g fresh garlic daily)		Clinical trial	40 post-menopausal obese or overweight patients	Alleviating osteoarthritis	Reducing resistin	Dehghani *et al.* (2018)

Product	Compounds	Study type	Subjects/cell lines	Main effects	Possible mechanisms	Ref.
Antimicrobial Activity						
Garlic "Rosato" and "Caposele"		In vitro	*Aspergillus versicolour, Penicillum citrinum* and *Penicillium expansum*	Inhibiting the growth of bacteria		Fratianni *et al.* (2016)
Aged garlic extract	Allicin	In vitro	*Burkholderia cepacian*	Inhibiting the growth of bacteria		Wallock-Richard *et al.* (2014)
	Garlic oil	In vitro	*Staphylococcus aureus, Escherichia coli* and *Bacillus subtilis*	Inhibiting the growth of bacteria		Guo (2014)
		In vitro	*Penicillium funiculosum*	Inhibiting the growth of bacteria	Penetrating into cells and organelles Destroying the cell structure Leading to the leakage of cytoplasm and macromolecules	Li *et al.* (2014)
		In vitro	*Candida albicans*	Disrupting the normal metabolism of bacteria	Inducing key genes involved in oxidative phosphorylation, the cell cycle, and protein processing in the endoplasmic reticulum	Li *et al.* (2016)
Raw garlic		Clinical trial	15 patients with *H. pylori* infection	Inhibiting *Helicobacter pylori* in the stomach		Zardast *et al.* (2016)

Product	*Compounds*	*Study type*	*Subjects/cell lines*	*Main effects*	*Possible mechanisms*	*Ref.*
Modulating Immune System						
Fresh garlic	Polysaccharides/ Fructan	In vitro	RAW 264.7 macrophages	Exerting immunomodulatory effect	Regulating the expressions of IL-6, IL-10, TNF-α, and interferon-γ	Li *et al.* (2017)
	Garlic oil	In vivo	Wistar rats	Normalizing several immunological parameters of rats, such as the serum total immunoglobulin concentration and T-cell subtype $CD4^+$ Combination of garlic oil and levamisole could balance the T-helper 1/ T-helper 2 response		Hassouna *et al.* (2015); Mohamed et al. (2016)
	Selenizing garlic polysaccharides	In vivo	14-day-old chickens	Promoting lymphocyte proliferation Enhancing interferon-γ and IL-2 Increase the serum antibody titer		Qiu *et al.* (2014)
Aged garlic extract		Clinical trial	56 healthy human participants	Reducing the occurrence and severity of cold and flu Improving the immune system functions		Percival (2016)
Cardiovascular Protection						
Aged garlic extract		In vitro	Isolated rat aortic rings	Leading to endothelial-dependent vasodilation	Stimulating the production of NO	Takashi-ma *et al.* (2017)

Product	*Compounds*	*Study type*	*Subjects/cell lines*	*Main effects*	*Possible mechanisms*	*Ref.*
Aged black garlic extract	Polyphenols	In vivo	Isolated hearts of male Sprague–Dawley rats	Relaxing coronary arteries before and after ischemia-reperfusion in rat Preventing the decrease of myocardial contractility		Garcia-Villalon *et al.* (2016)
Aged garlic extract	S-1-propylenecysteine	In vivo	Spontaneous hypertension rats	Improving peripheral blood circulation Reducing the systolic blood pressure		Ushijima *et al.* (2018)
Fermented garlic extract by *Bacillus subtilis*		In vivo	Spontaneous hypertension rats	Reducing the systolic blood pressure	Modulating the sGC-cGMP-PKG pathway	Park *et al.* (2016)
Fermented garlic extract		In vivo	Monocrotaline-induced pulmonary hypertension rats	Alleviating pulmonary hypertension	Decreasing the expression of vascular endothelial cell adhesion molecule-1 and MMP- 9 increasing the expression of PKG and eNOS	Park *et al.* (2017)
Garlic	Alliin	In vivo	Female Wistar albino rats	Increasing the activity of captopril on inhibiting ACE and hypertension		Asdaq and Inamdar (2010)
1.5% black garlic extract		In vivo	High-fat diet-fed male Sprague–Dawley rats	Modulating the metabolism of lipid and cholesterol Decreasing the levels of blood total lipids, triglyceride, and cholesterol	Reducing the mRNA expression of sterol regulatory element binding protein-1c	Ha *et al.* (2015)
Raw garlic	Allyl methyl sulfide/Allyl methyl sulfoxide	In vivo	Male Sprague–Dawley rats	Reduces cardiac hypertrophy remodeling induced by isoproterenol	Increasing Na^+/K^+-ATPase protein level	Khatua *et al.* (2017)

Product	Compounds	Study type	Subjects/cell lines	Main effects	Possible mechanisms	Ref.
Raw garlic		In vivo	Streptomycin-induced diabetic rats	Protecting the heart function Activating sirtuin 3-manganese superoxide dismutase pathway	Deacetylating manganese superoxide dismutase	Sultana *et al.* (2016)
Garlic extract		In vivo	Insulin-resistant obese rats	Protecting heart rate variability, cardiac dysfunction, and mitochondrial dysfunction		Supakul *et al.* (2014)
		In vivo	Rat model of gentamicin-induced chronic renal failure	Protecting the heart tissue	Reducing oxidative stress. Controlling Na^+/K^+-ATPase activity and Ca^{2+} levels	Gomaa *et al.* (2018)
Aged garlic extract	SAC	In vivo	Rats with myocardial dysfunction induced by isoproterenol	Protecting against cardiotoxicity		Avula *et al.* (2014)
Aged garlic extract		In vivo	Apolipoprotein E-knockout mice	Inhibiting inflammatory response to prevent atherosclerosis	Reducing serum level of C-reactive protein and thromboxane B-2, protein level of TNF-α and IL-1 receptor-associated kinase 4. Increasing AMPK activity in the liver	Morihara *et al.* (2017)
		In vivo	Apolipoprotein E-knockout mice	Inhibiting the vascular inflammation and lipid deposition in the early stage of atherosclerosis development		Morihara *et al.* (2016)

Product	Compounds	Study type	Subjects/cell lines	Main effects	Possible mechanisms	Ref.
High temperature and high pressure-processed garlic		In vivo	High-cholesterol diet-fed Sprague–Dawley rats	Reducing the levels of total cholesterol, low-density lipoprotein cholesterol and triglyceride		Sohn *et al.* (2012)
Garlic		Cohort study	30 patients with diabetic dyslipidemia	Decreasing the level of cholesterol and low-density lipoprotein. Increasing the level of high-density lipoprotein		Siddiqui *et al.* (2016)
Aged garlic		Clinical trial	41 patients with hypercholesterolemia	Reducing the activity of myeloperoxidase and lipid hydroperoxide in serum. Decreasing the concentration of F2-isoprostanes in plasma and urine		Ho *et al.* (2016)
Enzymatic browning processed garlic		Clinical trial	44 patients with hypertension	Reducing systolic blood pressure and diastolic blood pressure		Han *et al.* (2011)
Anticancer Activity						
Garlic extract		In vitro	Bladder cancer EJ cells	Inducing G_2/M-phase cell cycle arrest. Inhibiting cell growth, cell migration and invasion	Activating the ATM pathway and CHK2; Inhibiting the expression of MMP-9 Reducing the binding activity of transcription factors AP-1, specificity protein-1 and NF-κB motifs. Increasing the expression of heat shock protein A6	Shin *et al.* (2017)

Product	Compounds	Study type	Subjects/cell lines	Main effects	Possible mechanisms	Ref.
Aged garlic extract		In vitro	Colourectal cancer cell lines (SW480 and SW620) ECV304 cells and the transformed rat lung endothelial cells	Decreasing invasive activity. Inhibiting cell proliferation. Decreasing invasive activity. Inhibiting the tube formation of endothelial cells	Inhibiting cell motility	Matsuura *et al.* (2006)
		In vitro	DLD-1 human colon cancer cells (ATCC CCL-221)	Inhibiting cell proliferation	Down regulating the expression of cyclin B1 and CDK1. Inhibiting of activation of NF-κB	Jikihara *et al.* (2015)
Crude garlic extract	Lipid bioactive compounds	In vitro	Human liver cancer (Hep-G2), Colon cancer (Caco-2), Prostate cancer (PC-3), Breast cancer (MCF-7), Mouse macrophage cell (TIB-71) lines	Inhibiting the growth rate of Hep-G2, PC-3, MCF-7, and TIB-71 cells by 80%–90% at 72 h ($P < 0.05$).	Inhibiting cell proliferation and inducing cell cycle arrest. Inducing apoptosis	Bagul *et al.* (2015)
	Allicin	In vitro	Human gastric adenocarcinoma cell line	Inhibiting cell proliferation	Inducing cell cycle arrest at S-phase	Mansingh *et al.* (2018)
	DATS	In vitro	Human gastric carcinoma cell line (SGC-7901)	Inhibiting cell proliferation. Blocking cell cycle. Increasing apoptotic cell death	Accumulating Bax, p53, and cytochrome C and decreasing the expression of Bcl-2	Jiang *et al.* (2017)
		In vitro	Human breast cancer cell line (MDA-MB-231)	Inhibiting angiogenesis		Wei *et al.* (2017)

Product	***Compounds***	***Study type***	***Subjects/cell lines***	***Main effects***	***Possible mechanisms***	***Ref.***
	Z-ajoene	In vitro	Glioblastoma multiforme cells	Inhibiting the growth of the cancer stem cells population		Jung *et al.* (2014)
		In vitro	Human breast cancer cells (MDA-MB-231)	Inhibiting cell growth. Inducing cell apoptosis	Targeting the folding of proteins in the endoplasmic reticulum of cancer cells	Kaschula *et al.* (2016)
	SAC	In vitro	Human epithelial ovarian cancer cell line (A2780)	Inhibiting cell proliferation. Inducing G_1/S-phase cell cycle arrest. Increasing apoptosis. Reducing the migration of cells	Decreasing the expression of pro-caspase-3, Parp-1 Bcl-2 and increasing active caspase-3 and Bax Reducing the expression of Wnt5a, phosphorylation protein kinase B and c-Jun proteins	Xu *et al.* (2014)
	SPRC	In vitro	Human pancreatic ductal adenocarcinoma cells (Panc-1)	Inhibiting cell proliferation. Inducing apoptosis	Inducing G_2/M-phase cell cycle arrest. Regulating the level of JNK protein	Wang *et al.* (2015)
	SAMC	In vitro	Human colourectal carcinoma cell line (SW620)	Inhibiting cell proliferation. Inducing cell apoptosis	Regulating JNK and p38 MAPK pathways	Zhang *et al.* (2014)
		In vitro	Hepatoma cell lines (Hep3B and Huh-7)	Reducing the cell viability. Shortening the S phase and increasing the G_0/G_1 phase		Xiao *et al.* (2018)

Product	Compounds	Study type	Subjects/cell lines	Main effects	Possible mechanisms	Ref.
	Alliin	In vitro	Gastric adenocarcinoma cells	Regulating cell apoptosis	Generating reactive oxygen species. Decreasing mitochondrial membrane potential by Bax/Bcl-2 Up-regulating cytochrome C	Mansingh *et al.* (2018)
Aged garlic extract		In vivo	Adult male Wister albino rats treated with cisplatin	Improving the renal histological, ultrastructural and biochemical changes, such as hemorrhage, glomerular atrophy, tubular necrosis and degeneration		Nasr and Saleh (2014)
		In vivo	Fibrosarcoma tumors implanted BALB/c mice	Improving the immune responses of mice to fibrosarcoma. Inhibiting tumor growth	Increasing the ratio of CD_4^+/CD_8^+ Producing interferon-γ in splenocytes	Tabari and Ebrahim-pour (2014)
Garlic and lemon aqueous extract		In vivo	BALB/c mice xenograft model of breast cancer EMT6/P cells	Reducing tumor size. Inhibiting angiogenesis. Inducing apoptosis. Activating the immune system	Inhibiting the expression of vascular endothelial growth factor. Increasing interferon-γ, IL-2, and IL-4 levels	Talib (2017)

Product	Compounds	Study type	Subjects/cell lines	Main effects	Possible mechanisms	Ref.
	Allicin	In vivo	Female Swiss albino mice	Alleviating liver injury induced by tamoxifen	Changing the decrease of superoxide dismutase, glutathione and total protein and the increase of aspartate aminotransferase, alkaline phosphatase and alanine aminotransferase levels	Suddek (2014)
	DADS	In vivo	FVB/N mice	Preventing colourectal tumorigenesis induced by azoxymethane and dextran sulfate	Inhibiting inflammation. Inhibiting glycogen-synthase kinase-3β. Reducing the nuclear localization of NF-κB	Saud *et al.* (2016)
	DATS	In vivo	Female BALB/c-nude mouse xenograft model of human gastric carcinoma SGC-7901 cells	Inhibiting tumor growth. Promoting tumor apoptosis	Regulating the expressions of MMP-9 and E-cadherin protein	Jiang *et al.* (2017)
	SPRC	In vivo	Xenograft model of pancreatic ductal adenocarcinoma Panc-1 cells	Inhibiting tumor growth	Regulating the level of JNK protein	Wang *et al.* (2015)
	SAMC	In vivo	Mouse xenograft model of hepatoma Huh-7 cells	Inhibiting tumor growth	Interacting with the Wnt-pathway co-receptor LRP6 on the cell membrane	Xiao *et al.* (2018)
Raw, crushed garlic		Cohort study	17 volunteers from Beltsville, Maryland	Up-regulating seven genes including *AHR, ARNT, HIF1A, JUN, NFAM1, OSM* and *REL*		Charron *et al.* (2015)

Product	*Compounds*	*Study type*	*Subjects/cell lines*	*Main effects*	*Possible mechanisms*	*Ref.*
Garlic extract		Cohort study	Patients who received chemotherapy for hematological malignancies	Protective effect on febrile neutropenia in lower-risk subgroup		Gatt *et al.* (2015)
Hepatoprotective Activity						
Black garlic extract		In vitro	Rat clone-9 hepatocytes	Inhibiting apoptosis, lipid peroxidation, oxidative stress, and inflammation		Lee *et al.* (2018)
Garlic extract		In vivo	Wistar rats	Attenuating the liver damage induced by alloxan Improving plasma biochemical factors of hepatic function, such as urea, creatinine, aspartate transaminase, and alanine transaminase		Aprioku and Amah-Tariah (2017)
Single clove garlic extract		In vivo	Male rabbits	Protecting against CCl_4-induced acute liver injury		Naji *et al.* (2017)
LAFGE		In vivo	C57/BL6 J mice	Reducing the liver lipid level. Ameliorating the hepatic steatosis		Lee *et al.* (2017)
		In vivo	rats	Inhibiting liver cell apoptosis. Protecting liver from acetaminophen-induced liver injury	Suppressing MAPK phosphorylation Down regulating p53	Lee *et al.* (2016a)

Product	Compounds	Study type	Subjects/cell lines	Main effects	Possible mechanisms	Ref.
	Garlic oil	In vivo	1,3-Dichloro-2-propanol-treated rats	Protecting liver	Enhancing the activities of hepatic antioxidant enzymes. Blocking metabolic activation of 1, 3-dichloro-2-propanol. Reducing the apoptosis in liver	Ko *et al.* (2016)
	DADS	In vivo	Wistar rats	Protecting mice from nonalcoholic fatty liver disease induced by long-term high-fat diet.	Reducing the release of pro-inflammatory cytokines in the liver. Increasing antioxidant activity by inhibiting the expression of cytochrome P450 2E1	Lai *et al.* (2014)
	LAFGE	Clinical trial	36 adults with mildly high level of serum gamamyl glutamyl transpeptiase	Improving the levels of gamamyl glutamyl transpeptias and alanine aminotransferase without adverse effects		Kim *et al.* (2017)
Digestive System Protection						
Black garlic extract		In vitro	Small intestine	Stimulating gastrointestinal peristalsis. Promoting gastrointestinal emptying and facilitates defecation.		Chen *et al.* (2018)

Product	Compounds	Study type	Subjects/cell lines	Main effects	Possible mechanisms	Ref.
	DADS DAS	In vitro	Interferon-γ-stimulated intestinal cells	Reducing interferon-inducible protein-10, IL-6. Inhibiting NO and the expression of STAT-1		Fasolino *et al.* (2015)
		In vivo	Male ICR mice	Improving the colitis induced by dinitrobenzenesulfonic acid		Fasolino *et al.* (2015)
Garlic and cabbage extract		In vivo	Sprague–Dawley rats	Reducing the length of gastric ulcer, the total gastric acid, gastric juice volume, total bacteria count, and histopathological changes caused by aspirin. Improving the pH value of gastric juice		Ben *et al.* (2014)
Aged garlic extract		In vivo	Male albino rats	Healing the gastric mucosal injury induced by indomethacin. Reducing the total microbial amount in stomach		Badr and Al-Mulhim (2014)
		In vivo	Male Wistar rats	Preventing the indomethacin-induced ulcer	Reducing oxidative stress. Elevating the level of prostaglandin E-2, glutathione, and NO in gastric tissue	El-Ashmawy *et al.* (2016)

Product	Compounds	Study type	Subjects/cell lines	Main effects	Possible mechanisms	Ref.
	Allicin	In vivo	Dextran sulfate sodium-induced colitis mice	Alleviating the ulcerative colitis induced by dextran sulfate sodium	Inhibiting the activation of AP-1/NF-κB/signal transducer and activator of transcription-1. Inhibiting the phosphorylation of p38, JNK, and extracellular signal-regulated kinase 1/2 -regulated PPAR-γ	Shi *et al.* (2017)
Raw garlic		Clinical trial	15 patients with *H. pylori* infection	Decreasing the bacterial urease activity. Reducing the residing of *Helicobacter pylori* in the stomach		Zardast *et al.* (2016)
Anti-Diabetic Activity						
Garlic		In vivo	Diabetic rats	Protecting against diabetic retinopathy. Improving weight, blood glucose, and morphological changes of retinal tissue		Al-brakati (2016)
		Clinical trial	768 patients with type 2 diabetes mellitus	Reducing fructosamine and glycosylated hemoglobin		Wang *et al.* (2017)
Anti-Obesity Activity						
	LAFGE	In vivo	High-fat diet-fed male C57BL/6J mice	Reducing the weight. Reducing the epididymal, retroperitoneal, and mesenteric adipose tissue mass	Inhibiting the lipogenesis by down-regulating the mRNA and protein expression of PPAR-γ, C/EBPα, and lipogenic proteins	Lee *et al.* (2016b)

Product	Compounds	Study type	Subjects/cell lines	Main effects	Possible mechanisms	Ref.
Methanolic extract of black garlic		In vivo	High-fat diet-fed male Wistar rats	Reducing the weight. Regulating lipid metabolism	Upregulating the expression of AMPK, forkhead box protein O1, perilipin, and adiponectin in the adipose tissue. Down-regulating cluster of differentiation 36, plasminogen activator inhibitor 1, resistin, and TNF-α	Chen *et al.* (2014)
	Garlic oil	In vivo	High-fat diet-fed male Sprague–Dawley rats	Counteracting the influence of high-fat diet on the body weight and adipose tissue weight		Yang *et al.* (2018)
Neuroprotection						
Aged garlic extract	FruArg	In vitro	Lipopolysaccharide-activated murine BV-2 microglial cells	Alleviating neuroinflammation	Inhibiting the production of NO. Regulating the expression of multiple protein targets related to oxidative stress	Zhou *et al.* (2014)
Garlic extract		In vivo	Female Wistar rats	Reducing the concentration of Pb in the blood and brain. Preventing the Pb-induced apoptosis of neurons		Ebrahimzadeh-Bideskan *et al.* (2016)

Product	*Compounds*	*Study type*	*Subjects/cell lines*	*Main effects*	*Possible mechanisms*	*Ref.*
Aged garlic extract		In vivo	Adult male Wistar rats	Attenuating the damage of working memory	Improving the loss of cholinergic neurons. Increasing the level of vesicular glutamate transporter 1 and glutamate decarboxylase in the hippocampal area	Thorajak *et al.* (2017)
Ethanol extract of garlic		In vivo	Diabetic Wistar rats	Improving memory	Increasing the activity of Na^+/K^+ ATPase, Ca^{2+} ATPase, and glutamine synthetase in the hippocampus	Semuyaba *et al.* (2017)
	Z-ajoene	In vivo	Male gerbils	Preventing I/R-induced delayed neuronal death and gliosis region of the hippocampus	Reducing lipid peroxidation in the CA1	Yoo *et al.* (2014)
	SAC	In vivo	Male albino Wistar rats	Ameliorating the cognitive impairment	Reducing oxidative stress, neuroinflammation, astrogliosis, and acetylcholinesterase activity	Zarezadeh *et al.* (2017)
Renal Protection						
Aqueous extract of garlic		In vivo	Type 1 diabetic rats	Reducing the oxidative stress in the kidneys		Nasiri *et al.* (2017)
		In vivo	Wistar rats	Improving the renal plasma biochemical factors induced by alloxan		Aprioku and Amah-Tariah (2017)

Product	Compounds	Study type	Subjects/cell lines	Main effects	Possible mechanisms	Ref.
	DATS	In vivo	Male albino rats	Protecting the kidney from oxidative stress injury induced by As	Activating the Nrf2-ARE pathway	Miltonpr-abu *et al.* (2017)

Abbreviations: ACE, angiotensin-converting enzyme; *AHR*, aryl hydrocarbon receptor; allicin, diallyl thiosulfonate; alliin, S-allyl-cysteine sulfoxide; AMPK, adenosine monophosphate-activated protein kinase; AP-1, activator protein 1; ARE, antioxidant response element; *ARNT*, aryl hydrocarbon receptor nuclear translocator; ATM, ataxia-telangiectasia mutated; Bax, Bcl-2 associated X; Bcl-2, B cell lymphoma-2; CDK1, cyclin-dependent kinase 1; cGMP, cyclic guanosine monophosphate; CHK2, checkpoint kinase 2; COX2, cyclooxygenase-2; DADS, diallyl disulfide; DAS, diallyl sulfide; DATS, diallyl trisulfide; eNOS, endothelial nitric oxide synthase; FruArg, N-α-(1-deoxy-D-fructos-1-yl)-L-arginine; GCLM, glutamate-cysteine ligase modifier; *HIF1A*, hypoxia-inducible factor 1α; HO-1, heme oxygenase-1; IL, interleukin; iNOS, inducible NO synthase; JNK, Jun *N*-terminal kinase; *JUN*, proto-oncogene c-Jun; LAFGE, fermented garlic extract by *Lactobacillus plantarum* BL2; MMP-9, matrix metalloproteinase-9; *NFAM1*, nuclear factor of activated T cells (NFAT) activating protein with immunoreceptor tyrosine-based activation motif 1; NF-κB, nuclear factor-kappa B; NO, nitric oxide; Nrf2, nuclear factor erythrobia-2 related factor 2; *OSM*, oncostatin M; Parp-1, poly (ADP-ribose) polymerase-1; PKG, protein kinases G; PPAR-γ, peroxisome proliferator-activated receptor-γ; p38 MAPK, p38 mitogen activated protein kinase; *REL*, V-relavian reticuloendotheliosis viral oncogene homolog; SAC, S-allyl-cysteine; sGC, soluble guanylyl cyclase; SAMC, S-allylmercaptocysteine; SPRC, S-propargyl-L-cysteine; STAT-1, signal transducer and activator of transcription-1; TNF-α, tumor necrosis factor-α; Wnt5a, wingless-type MMTV integration site family member 5A.

Table source: *Shang et al. (2019)*

Factors affecting phytochemicals

Fresh or crushed garlic yields the sulphur-containing compounds allicin, ajoene, diallyl polysulfides, vinyldithiins, and S-allylcysteine; as well as nonsulphur compounds such as enzymes, saponins, flavonoids and Maillard-reaction products when cooked (en.wikipedia.org/wiki/Garlic). When through chewing, crushing or chopping the garlic plant cells get damaged, the enzymes present in the cytosol (cell fluids) of cell vacuoles trigger the breakdown of several sulphur-containing compounds (Jones and Jill Hughes, 2004), and in turn the phytochemical compounds responsible for the sharp or hot taste *cum* strong smell of garlic are produced. Some of the compounds are unstable and continue to react over time. Among the alliums, in fact, garlic has by far the highest concentration of initial reaction products which make the garlic more potent than onion, shallot or leeks (en.wikipedia.org/wiki/Garlic).

Fresh garlic preparation has more phytochemical content as compared to garlic extract and garlic powder (iopscience.iop.org). The most significant antibacterial property remains in the fresh garlic preparations with a concentration of 50% (iopscience.iop.org). Moreover, organosulphur compounds in raw garlic generally have higher digestibility than those in cooked garlic (Torres-Palazzolo *et al.,* 2018), though saponins are found more stable in the cooking process (Lanzotti *et al.,* 2013). When garlic is chopped, crushed, sliced or chewed, alliinase enzymes are activated and through a series of conversions form a sulphur phytonutrient called allicin (Gord, 2019) which is a strong scavenging and antioxidant compound which affects the cardiovascular system. Heated garlic has a lower level of allicin concentration. Heating garlic cloves whole or immediately after crushing can destroy the sensitive enzyme (alliinase) that is responsible for producing allicin but the health benefits can be partially conserved by crushing garlic and allowing it to stand for 10 minutes prior to cooking as this extends the enzyme time to be released and for allicin to form which after formation becomes relatively heat stable (Gord, 2019). The saponin in purple garlic is about 40 times higher than in white garlic, and several saponin compounds such as desgalactotigonin-rhamnose, proto-desgalactotigonin, proto-desgalactotigonin- rhamnose, voghieroside D1, sativoside B1-rhamnose, and sativoside R1 were found to exist only in purple garlic (Diretto *et al.,* 2017).

The 'black garlic' ~ a processed garlic product has increased polyphenol and flavonoid contents as well as better antioxidant properties as compared to the fresh garlic (Kimura *et al.,* 2017). Compared to regular garlic, black garlic has a smaller amount of allicin, the compound that gives regular garlic some of its health benefits. It was found that 38 components of garlic changed after thermal treatment during the processing of black garlic (Liang *et al.,* 2015). In addition, the polysaccharide degraded and the content of reducing sugar increased during thermal processing of black garlic (Lu *et al.*, 2018). Also, increasing temperature and decreasing humidity enhance the contents of polyphenols and the total flavonoids in black garlic (Sun and Wang, 2018; Kim *et al.*, 2013).

Smell and spiciness of garlic

Among several sulphur compounds contributing to the smell and taste of garlic, allicin has been found to be most responsible for 'hot' sensation of raw garlic as this chemical opens thermo-transient receptor potential channels that are responsible for bu r n ing sense of heat in foods (en.wikipedia.org/wiki/Garlic) though while cooking the garlic the allicin is almost removed so the spiciness is mellowed (Macpherson *et al.*, 2005). Allicin along with its decomposition products diallyl disulfide and diallyl trisulfide are major contributors to the characteristic odour of garlic along with other allicin-derived compounds such as vinyldithiins and ajoene (en.wikipedia.org/wiki/Garlic).

Uses

Culinary use

Garlic is used widely as a spice and medicinal herb all around the world. It is used in chutneys, pickles, dips, curry powder, sauces, ketchups, *etc.* For thousands of years, garlic has been used as a common food seasoning or condiment around the world for its pungent and spicy flavour that mellows and sweetens considerably with cooking. It enhances the flavour of most vegetables and meats. Though all parts of the plant except the root cluster and bulb's basal plate are edible, the most commonly used part of the plant is its cloves (bulb) which are used for consumption (raw, roasted, fried or cooked) or for medicinal purposes. Garlic cloves can be infused into edible oils that can be used to season all categories of vegetables, breads, pasta and meats. Like most of the vegetables and herbs, garlic cloves and shoots too can be pickled in oil or vinegar. Garlic along with onion and other spices is used in stir-frying (*tadka*) of many Indian dishes. Lightly smoked garlic can be used in stuffings, soups, stews, *etc.* Raw or lightly smoked garlic, along with ginger, tomato, coriander leaves and/or green pepper can be used to prepare chutney. Garlic may be applied to different kinds of bread usually in a medium of butter or oil to create a variety of classic dishes such as garlic bread, garlic toast, bruschetta, crostini, *etc.* The flavour varies in intensity and aroma with different cooking methods. It is often paired with onion, tomato or ginger. Garlics (alone or with other ingredients such as onion and/or ginger) can be soaked in vinegar for a few days to create garlic vinegar pickle.

The leaves, aerial bulbs and flowers if appearing on the head (spathe) are also edible. The tender and edible immature scapes (also known as garlic spears, stems or tops) generally have a milder taste than the cloves. They are often used in stir-frying or braised like asparagus. These are most often consumed when still tender and immature. Immature garlic is sometimes pulled and sold as 'green garlic'. When green garlic is allowed to grow past the 'scallion' stage but is not fully mature, it may produce garlic 'round' ~ a bulb like a boiling onion though not separated into cloves like the mature bulbs. It imparts a garlic flavour and aroma in food minus the spiciness. Green garlic is often chopped and stir-fried or cooked in soup or hot pot. Garlic powder has somewhat different taste from fresh garlic but can be used as a garlic substitute.

Medicinal use

Garlic has several health benefits mainly due to its diverse bioactive compounds such as organic sulphides, saponins, phenolic compounds and polysaccharides (Bose *et al.,* 2014; Diretto *et al.* 2017; Szychowski *et al.,* 2018). Garlic is carminative, aphrodisiac, expectorant, stimulant and used in fevers, coughs, febrifuge in intermittent fevers, respiratory diseases such as chronic bronchitis, bronchial asthma, whooping cough, tuberculosis; in atherosclerosis and hypertension; as well as a complement in the diet of hyperlipidemic patients and for the prophylaxis of the vascular changes induced by ageing (pharmacy180.com). Garlic or its constituents exhibit various biological activities, such as antibacterial, antifungal, antiviral, antitumor, and antidiabetic effects.

In Ayurvedic system of medication the garlic is regarded as a rejuvenative, detoxicant and aphrodisiac. It is a pungent and warming herb that can be used to clear bacterial infection. It increases perspiration thus helping in lowering the fever. It is also used as an antibiotic, to cleanse blood, to reduce blood pressure, cholesterol and sugar levels. It is also used as a protection against common cold, influenza, bronchitis, whooping cough, worms, dysentery and typhoid. Its consumption is associated with a lower risk of stomach and prostate cancer. It is externally used for skin problems especially acne and fungal infections. It is one of the ingredients of the preparations known as 'Lasuna' (Himalaya Drug Company) and 'Lashunadi bati' (Baidyanath) (pharmacy180.com).

Though garlic has numerous health benefits its global consumption is declining with exception of China and India. Some people are unwilling to eat raw garlic because of its pungent taste and smell (Tanamai *et al.,* 2004), while some avoid eating it due to spiritual and religious belief, especially the followers of Jainism and Buddhism, and some Hinduism in India. Moreover, raw garlic can cause gastrointestinal discomfort in some people. Food scientists have developed aged garlic preparations to reduce these discomforts.

Industrial uses

Variety of garlic products are available in the market such as dehydrated garlic flakes or pieces, garlic powder, garlic paste, aged garlic extract, black garlic, garlic pickles and vinegars, garlic salt, garlic oleoresin (dark viscous liquid, having much strong flavour than dehydrated and fresh garlic cloves), garlic oil, *etc.* It is also included as an ingredient to impart flavour in various products such as wafers, mayonnaise, butter, edible oils, *etc.*

Other uses

Apart from having culinary and medicinal uses, garlic plants are said to repel rabbits and moles. Garlic displays antimicrobial effects at temperatures as high as 120°C, hence used as a fish and meat preservative. These plants are usually hardy and not affected significantly by any pest or disease. The sticky juice within the bulb cloves is used as an adhesive in mending glass and porcelain.

Adverse effects

Garlic can cause gastrointestinal distress causing halitosis (bad breath) and body odour (pharmacy180.com) described as a pungent garlicky smell to sweat. This is caused by allyl methyl sulphide (AMS). Also, some people suffer from allergies to garlic and other species of *Allium*. Symptoms can include irritable bowel, diarrhoea, mouth *cum* throat ulcerations, nausea, breathing difficulties and in rare cases anaphylaxis. Garlic-sensitive patients show positive tests to diallyl disulphide, allylpropyldisulphide, allylmercaptan and allicin ~ all of which are present in garlic. People who suffer from garlic allergies are often sensitive to many other plants including onions, chives, leeks, shallots, garden lilies, ginger and bananas.

Several reports of serious burns resulting from garlic being applied topically for various purposes, including naturopathic uses and acne treatment, indicate that care must be taken for these uses usually by testing a small area of skin using a very low concentration of garlic. On the basis of numerous reports of such burns even in children, topical use of raw garlic as well as insertion of raw garlic into body cavities is discouraged. In particular, topical application of raw garlic to young children is not advisable. Possible side effects of long-term garlic supplementation include gastrointestinal discomfort, sweating, dizziness, allergic reactions, bleeding, and menstrual irregularities.

If higher-than-recommended doses of garlic are taken with anticoagulant medications this can lead to a higher risk of bleeding. Garlic may interact with warfarin, antiplatelets, saquinavir, antihypertensives, calcium channel blockers, the quinolone family of antibiotics such as ciprofloxacin and hypoglycemic drugs, as well as other medications. Alliums might be toxic to cats or dogs.

Spiritual and religious belief

In both Hinduism and Jainism garlic and onion are thought to stimulate and warm the body and to increase one's desires. Some Hindus generally avoid using garlic and the related onion in the preparation of foods while some may only observe this for religious festivities and events. Followers of the Jain religion avoid eating garlic and onion on a daily basis. In some Buddhist traditions, garlic along with other five pungent spices is understood to stimulate sexual and aggressive drives to the detriment of meditation practice. In Mahayana Buddhism, monks and nuns are not allowed to consume garlic or other pungent spices such as chillies which are deemed as being 'earthly pleasures' and are viewed as promoting aggression due to their spiciness and pungency.

Black garlic

The process of making black garlic is of East Asian origin and has been used as a traditional and functional food for centuries there. Many ancient cultures used black garlic for medicinal and culinary purposes. Black garlic has a sweet taste and chewy texture without the offensive odour of fresh garlic. Black garlic with deep brownish-black colour is a type of aged garlic obtained from fresh garlic that has been fermented for 15 to 90 days at a controlled 60-90^{0}C temperature and 80-90% humidity (typically 85% humidity at 70^{0}C for 40 days) (en.wikipedia.org/wiki/Black_garlic). The processing period for aged black garlic could be changed according to the temperature, with shorter periods at higher temperatures (Toledano-Medina *et al.*, 2016). Different temperature and incubation periods make a difference in the concentration of active components in garlic.

Heat produces key intermediate compounds through the Maillard reaction which give the black garlic its colour and distinctive taste (en.wikipedia.org/wiki/Black_garlic). The reaction produces hundreds of flavour-making compounds giving black garlic its unique taste. No additives or preservatives are used and there is no bur n ing of any kind. The enzymes that give fresh garlic its sharpness break down. These conditions are thought to facilitate the Maillard reaction, the chemical process that produces new flavour compounds responsible for the deep taste of seared meat and fried onions. During the aging process, the odorous, harsh and irritating compounds in fresh raw garlic are converted naturally into stable and safe compounds. The cloves tu r n black and develop a sticky date-like or jelly-like texture (Stanek, 2016) with sweet and sour taste. Bacterial endophytes capable of fermentation and with strong heat resistance have been identified in common garlic and black garlic, and these may have relevance in black garlic production (Qiu *et al.*, 2018).

Source: Stank, 2016

Nutritional and phytochemical content

It has been found through various sources over internet that fermented black garlic contains better nutritional value than fresh garlic. Fermentation of fresh garlic dramatically raises its health promoting properties and its bio-availability. Despite the fact that black garlic has less of the active compound allicin than fresh one, it does show off higher concentrations of many nutrients, antioxidants and other beneficial compounds. Black garlic possesses an abundant amount of antioxidant compounds such as polyphenols, flavonoids, tetrahydro-β-carboline derivatives and organosulphur compounds including S-allyl-cysteine and S-allyl-mercaptocysteine, as compared with fresh garlic. These higher concentrations may be at least partly responsible for the many health benefits that black garlic provides. These potential benefits include contributing to anti-inflammatory effects, repair of age-damaged skin, cardioprotective effects, lowered cholesterol levels, prevention of diabetic complications, boosting the immune system, lowering the risk of cancer, therapeutic management of liver injury and increased levels of antioxidants.

When garlic undergoes fermentation not only physiochemical characteristics are altered but bioactive compound concentration is also found improved (Kimura *et al.*, 2017). Choi *et al.* (2014) recorded decreased moisture and pH levels of garlic along with the fermentation process, whereas reducing sugar and total acidity were found accumulated. Choi *et al.* (2014) in black garlic recorded altered colour spectra *cum* amino acid composition when compared with fresh garlic, *vis-a-vis* black garlic has elastic and chewy texture with sweet taste without any offensive flavour of garlic.

Comparison of components of Black Garlic and Raw Garlic

In the process of producing black garlic, the physiochemical properties of fresh garlic change during the aging process as the process increases the concentrations of flavonoids, pyruvate, total phenol and SAC. Black garlic which is dark brown in colour has a soury-sweet taste with jelly-like texture, *vis-a-vis* concentrations of phytochemical components, free sugars and minerals are also found increased than those in fresh garlic (Ryu and Kang, 2017).

Tables below show the comparisons of nutritional and phytochemical components, antioxidant and anti-inflammatory activities of black garlic as well as fresh garlic.

Nutrition facts, minerals and amino acids

Nutrition Facts (100g)		
	Black Garlic	**Raw Garlic**
Water	33.6g	58.6g
Protein	14.2g	6.4g
Fat	0.3g	0.5g
Carbohydrate	39.8g	1.0g
Food Fibre	9.8g	2.1g
Ash	2.28g	1.5g
Sodium	0.0932g	0.017g
Vitamin B1	0.00002g	0.0007g
Calories	219 kcal	149 kcal
Cholesterol	0.0	0.0
Mineral Content (100g)		
Ref: Sunatec Japan	**Black Garlic**	**Raw Garlic**
Sodium (Na)	36mg	17.0mg
Potassium (K)	930mg	401.0mg
Magnesium (Mg)	52mg	25.0mg
Calcium (Ca)	13mg	181.0mg
Iron (Fe)	2.1mg	1.7mg
Zinc (Zn)	1.4mg	1.2mg
Information for Raw Garlic is taken from the "USDA National Nutrient Database for Standard Reference". (NDB No: 11215, Scientific Name: Allium sativum)		
Table source: *www.blackgarlichasarrived.com/nutrition-facts-and-mineral-content*		

Amino Acid Content (100g)/Dry From		
	Black Garlic	**Raw Garlic**
	***Essentials**	
Isoleucine	371 mg	217 mg
Leucine	779 mg	308 mg
Lysine	385 mg	273 mg
Methionine	210 mg	76 mg
Histidine	287 mg	113 mg
Phenylalanine	461 mg	183 mg
Tryptophan	102 mg	66 mg
Valine	610 mg	291 mg
	***Non-essentials**	
Arginine	558mg	634mg
Alanine	714mg	132mg
Aspartic Acid	1020mg	489mg
Glutamic Acid	2327mg	805mg
Glysine	697mg	200mg
Proline	544mg	100mg
Serine	458mg	190mg
Tyrosine	292mg	81mg
Cystine	223mg	65mg
*Essentials are amino acids that the adult human body cannot produce and must be ingested. Non-essentials are those that the body can synthesize.		

Free sugars and minerals

Components		*Content in black garlic*	*Content in fresh garlic*	*Change in black garlic compared to fresh garlic*	*References*
Free sugars (mg/100 g)	Arabinose	1.6 ± 0.3	-	Increase	Choi *et al.* (2008)
		114.5 ± 15.6	51.1 ± 6	Increase	Kang (2016)
	Galactose	13.1 ± 1.7	-	Increase	Choi *et al.* (2008)
	Glucose	181.7 ± 8.8	91.6 ± 2.7	Increase	Choi *et al.* (2008)
		210 ± 5	-	Increase	Jeong *et al.* (2016)
		221.9 ± 11.5	16.7 ± 0.7	Increase	Kang (2016)
	Fructose	2043.7 ± 5	63.9 ± 3.4	Increase	Choi *et al.* (2008)
		4002 ± 71	707 ± 8	Increase	Jeong *et al.* (2016)
		3383.2 ± 44.0	31.4 ± 1.0	Increase	Kang (2016)
	Sucrose	119.1 ± 3.5	76.3 ± 0.1	Increase	Choi *et al.* (2008)
		242.9 ± 18.1	181.7 ± 1.6	Increase	Kang (2016)
	Maltose	7.8 ± 0.4	1.7	Increase	Choi *et al.* (2008)
		48.2 ± 4.4	11.7 ± 0.8	Increase	Kang (2016)

Components		*Content in black garlic*	*Content in fresh garlic*	*Change in black garlic compared to fresh garlic*	*References*
Minerals (mg/100 g)	Al	25.4 ± 0.8	0.5	Increase	Choi *et al.* (2008)
	Ca	13.1 ± 0.4	7.5	Increase	Choi *et al.* (2008)
		15.2 ± 0.4	12.2 ± 0.3	Increase	Kang (2016)
	Cu	2 ± 0.1	1.7	Increase	Choi *et al.* (2008)
		0.27	0.32	Increase	Kang (2016)
	Fe	4.4 ± 0.2	3.3 ± 0.2	Increase	Choi *et al.* (2008)
		2.3 ± 0.5	1.6	Increase	Kang (2016)
	K	738.2 ± 11.2	434.9 ± 4.2	Increase	Choi *et al.* (2008)
		1018.4 ± 23.2	907.4 ± 20.6	Increase	Kang (2016)
	Mg	27.2 ± 0.8	15.7 ± 0.1	Increase	Choi *et al.* (2008)
		48.7 ± 1.1	43.2 ± 1.0	Increase	Kang (2016)
	Mn	0.7	0.5	Increase	Choi *et al.* (2008)
		0.5	0.4	Increase	Kang (2016)
	Na	13.0 ± 0.1	9.1	Increase	Choi *et al.* (2008)
		35.5 ± 0.8	22.0 ± 0.5	Increase	Kang (2016)
	Zn	1.6 ± 0.1	1.5	-	Choi *et al.* (2008)
		1.3	1.2	Increase	Kang (2016)
	P	143.7 ± 5.5	93.3 ± 0.3	Increase	Choi *et al.* (2008)
		2.6 ± 0.1	2.2 ± 0.1	Increase	Kang (2016)
	Se	0.13	0.05	Increase	Kang (2016)
	S	189.7 ± 4.3	183.1 ± 4.2	-	Kang (2016)
The dash (-) represents no change in free sugars and minerals during the aging processes. Comparison between black garlic and fresh garlic was performed within a single study.					
Table source: *Ryu and Kang (2017). www.ncbi.nlm.nih.gov/pmc/articles/PMC6152780/*					

Phytochemicals

Comparison of some phytochemical components between Black garlic and Fresh garlic:

Components	*Black garlic content*	*Fresh garlic content*	*Change in black garlic compared to fresh garlic*	*Analytical Methods*	*Basis for Comparison*	*References*
Allicin (mg/100 g)	-	362 ± 1	Decrease	colorimetric method	allicin	Jeong *et al.* (2016)
	20	345	Decrease	HPLC		Zhang *et al.* (2015b)

Components	*Black garlic content*	*Fresh garlic content*	*Change in black garlic compared to fresh garlic*	*Analytical Methods*	*Basis for Comparison*	*References*
Flavonoid (mg/100 g)	0.8	0.1	Increase	colorimetric method	quercetin	Choi *et al.* (2008)
	1570 ± 211	322 ± 7	Increase		rutin	Choi *et al.* (2014)
	195 ± 8	125 ± 13	Increase		rutin	Jung and Sohn (2014)
Pyruvate (mmol /100 g)	27.8 ± 0.3	18.8 ± 0.3	Increase	Colorimetric method		Choi *et al.* (2008)
	245.7 ± 2.4	48.7 ± 1.2	Increase			Jeong *et al.* (2016)
Thiosulfate (mmol /100 g)	9.12 ± 0.05	0.65 ± 0.03	Increase	Colorimetric method		Jeong *et al.* (2016)
	0.3	10.5 ± 0.4	Decrease			Kang (2016)
	0.8	0.1	Increase		OD value	Choi *et al.* (2008)
Total phenol (mg/100 g)	1.6 ± 0.1	0.6 ± 0.1	Increase	Colorimetric method	caffeic acid	Choi *et al.* (2008)
	4835 ± 114	1391 ± 162	Increase		garlic acid	Choi *et al.* (2014)
	1000 ± 100	367 ± 22	Increase		garlic acid	Jang *et al.* (2008)
	22.3 ± 0.8	3.7 ± 0.2	Increase		garlic acid	Kim *et al.* (2012)
	1023 ± 19	255 ± 12	Increase		tannic acid	Jung *et al.* (2014)
SAC (mg/100 g)	8.5 ± 0.1	2	Increase	HPLC-FLD		Bae *et al.* (2014)
	19.4	2.4	Increase	HPLC		Sasaki *et al.* (2007)
	9.8 ± 0.2	2.2	Increase	HPLC-FLD		Bae *et al.* (2012)
	11.4 ± 0.9	2.3	Increase	HPLC		Bae *et al.* (2012)

Table source: *Ryu and Kang (2017). www.ncbi.nlm.nih.gov/pmc/articles/PMC6152780/*

	Components	*Original concentration*
Water-soluble sugar	Increased 1.88-7.91-fold	450mg/g
Polyphenol	Increased 4.19-fold	13.91mg GAE/g
Flavonoid	Increased 4.77-fold	3.22mg RE/g
Amadori & Heyns	Increased 40-100-fold	10μg/g

	Components	*Original concentration*
Fructan	Decreased 0.15-0.01-fold	580mg/g
Leucine	Increased 1.06-fold	58.62mg/100 g
Isoleucine	Increased 1.67-fold	50.04mg/100 g
Cysteine	Decreased 0.58-fold	81.06mg/100 g
Phenylalanine	Increased 2.43-fold	55.64mg/100 g
Tyrosine	Decreased 0.18-fold	449.95mg/100 g
GAE = gallic acid equivalents; RE = rutin equivalents.		
Table source: *www.blackgarlichasarrived.com/nutrition-facts-and-mineral-content*		

Antioxidant and anti-inflammatory

As compared to fresh garlic, the black garlic has strong antioxidant effects but low anti-inflammatory (Jeong *et al.*, 2016), anti-coagulation (Jung and Sohn, 2014), anti-allergic (Kim *et al.*, 2012) and immunomodulatory activities (Li *et al.*, 2017). These differences could result from the concentration of components changing during black garlic processing (Ryu and Kang, 2017).

Measurement	*Content in ABG*	*Content in FRG*	*Change in ABG Compared to FRG*	*Basis for Comparison*	*References*
Antioxidant activity					
DPPH radical scavenging activity	0.1 mg/ml	0.4 mg/ml	Increase	IC50	Jeong *et al.* (2016)
	4.1 mg/ml	114.9 mg/ml	Increase	IC50 (Japanese garlic)	Sasaki *et al.* (2007)
	7.3 mg/ml	88.5 mg/ml	Increase	IC50 (Chinese garlic)	Sasaki *et al.* (2007)
	97%	10.4%	Increase	900 μg/mL	Jang *et al.* (2008)
	67.4 ± 0.2%	35.7 ± 0.6%	Increase	1 mg/mL	Shin *et al.* (2008)
	1.3 mg/ml	0.7 mg/ml	Decrease	IC50	Jung and Sohn (2014)
	82.5 ± 0.5%	35.1 ± 0.7%	Increase	2 mg/mL	Kim *et al.* (2012)
ABTS radical scavenging activity	0.2 mg/ml	0.3 mg/ml	Increase	IC50	Jeong *et al.* (2016)
	1 mg/ml	1.1 mg/ml	Increase	IC50	Jung and Sohn (2014)
Hydroxy radical scavenging activity	75 ± 0.7%	60.7 ± 0.2%	Increase	2 mg/mL	Kim *et al.* (2012)

Measurement	*Content in ABG*	*Content in FRG*	*Change in ABG Compared to FRG*	*Basis for Comparison*	*References*
Antioxidant activity					
Nitrite radical scavenging activity	32.9%	55.2%	Decrease	5 mg/mL	Jang *et al.* (2008)
	0.1 mg/ml	0.2 mg/ml	Increase	IC50	Jung and Sohn (2014)
Fe2+-chelating activity	18.2 ± 0.7%	30.6 ± 1.4%	Decrease	2 mg/ml	Kim *et al.* (2012)
Reducing power	2.6	0.4	Increase	OD	Kim *et al.* (2012)
SOD activity	64.4%	16.8%	Increase	100 mg/ml	Jang *et al.* (2008)
Anti-inflammatory activity					
Inhibition of COX-2 activity	39.1 ± 3.8%	80.5 ± 7.8%	Decrease	250 μg/ml	Jeong *et al.* (2016)
Inhibition of 5-LO activity	29.5 ± 2.1%	97.4 ± 9.5%	Decrease	250 μg/ml	Jeong *et al.* (2016)

DPPH: 2,2-diphenyl-1-picrylhydrazyl; ABTS: 2,2-azinobis-(3-ethylbenzothiazoline-6-sulfonate); IC50: the concentration of a sample that gives half inhibition. Comparison between black garlic and fresh garlic was performed within a single study.

Table Source: *Ryu and Kang (2017). www.ncbi.nlm.nih.gov/pmc/articles/PMC6152780/*

In addition to the antioxidant and anti-inflammatory effects, black garlic has anti-cancer, anti-diabetic, anti-obesity, anti-allergic, hepatoprotective, cardioprotective, neuroprotective and anti-thrombotic effects.

Flavour

Black garlic has a flavour different from fresh garlic. It does not release a strong offensive flavour owing to the reduced content of allicin. Black garlic is soft and slightly sticky with intensely sweet and a very rich savory flavour which is quite different from normal fresh garlic. Although the flavour of black garlic depends on the variety of fresh garlic used as a raw material, it usually develops notes of licorice, tamarind and caramel (en.wikipedia.org/wiki/Black_garlic). Garlic with higher sugar content produces a milder and more caramel-like flavour, whereas garlic with low sugar content produces a sharper, slightly more acidic flavour (en.wikipedia.org/wiki/Black_garlic). During heating, the garlic turns black in colour well before the full extent of its sweetness is able to develop and its softness increases with water content.

Uses

Black garlic can be used alone on bread, pizzas, pastas, risottos, aioli, eggs, with cheese, red wine or dark chocolate, in soups or sauces, in seafood, with meat or fish, crushed into mayonnaise, added to vinaigrette, with a vegetable dish, and even in dessert dishes. The cloves may also be crushed, and added with water to make the paste or liquid.

Being rich in nutrients, black garlic can improve human body function, eliminate fatigue, provide immunity to influenza, improve physical strength, help in relieving constipation, protect the liver, improve the prostate activity and help promote the sleep. It is anti-aging, helps in the prevention of cancer, promotes recovery of diabetic patients, lowers blood pressure, prevents blood clots and helps lower cholesterol levels.

Adverse effects

Although the side effects of black garlic are much less as compared to fresh garlic, its internal use may show the side effects such as bad breath, burning sensation in the mouth or stomach, gases, nausea, vomiting, body odour, diarrhoea, asthma, *etc.* When applied topically, it may cause skin burn and severe skin irritation. Moreover, administration of black garlic for a long time could induce problems in the control of bleeding during a surgery because of its anti-coagulation effects.

Pregnant women and nursing mothers should consult the doctor before taking black garlic either internally or topically. Oral consumption of large doses or small amount for long-term are not safe and may even be fatal for children. Topical use on children skin may cause damage similar to burns.

Taking black garlic may cause a drop in blood sugar. In diabetic patients, garlic decreases blood sugar levels significantly and may even lead to diabetic coma. Garlic is not recommended for patients with chronic bleeding disorders or after surgery because it may increase the risk of bleeding. Black garlic may cause gastrointestinal distress, particularly in people having stomach or digestive issues. Garlic lowers blood pressure and it may benefit people with high blood pressure but in case of low blood pressure it may cause an alarming situation.

Propagation

Garlic by nature propagates vegetatively. Individual cloves or top bulbils of well-developed and mature plants are planted during autumn. A 200-500kg cloves having 8-10mm diameter are required to plant one hectare area.

The cloves should be carefully separated from the basal plate and must be planted within 24 hours of cracking so that the root nodules do not dry away (Farmingindia, 2017). Cracking is the process of separating the cloves from bulbs.

Usually the mature, uninjured cum even at basal plate, healthy and larger cloves having 8-10mm diameter are used as planting materials.

Using bulbils

In the scape producing varieties the growers usually remove the scapes so that plant can divert its energy in developing larger bulbs. However, if the scapes of these plants are allowed to grow and mature, they usually form bulbils at the top which can be harvested at the time of harvesting when bulbs mature.

The tiny teardrop-shaped cloves of these bulbils can be planted to produce a larger, round and single cloved bulbs after one season and if these round bulbs are again planted for a second year most of these will grow into a regular garlic bulb with the usual cloves.

Bulbils have main advantage of being free from infection of soil-pathogens as is case with underground cloves (bulbs). Several garlic growers have found bulbil-grown garlic to be larger and more vigorous than conventional clove-grown garlic. Another reason to grow from bulbils is that some garlic varieties produce much more cloves in bulbils as compared to fewer numbers of cloves in an underground bulb. Even though it takes two full seasons to produce mature garlic from bulbil though some varieties can multiply faster similar to underground cloves.

Bulbils are usually planted just like normal cloves, but since they are very small, their green sprouts will also be very small so these may be lost with the weeds.

Therefore, it would be better to plant them in at least 15cm deep containers (25-30cm is better) for the first year to get single-clove bulbs (rounds) which as the planting material may be used for the next season to have optimum size of bulbs. Deep containers are best to allow good root growth. They should be planted just 2-2.5cm deep and 2.5-5cm apart from each other in a container. A plastic pot 15cm in diameter should give plenty of room for 10 evenly spaced bulbil cloves.

It would be better to place the container into the soil in a manner that rim of the pot is just above the ground. Mulching with straw or leaves will prevent the bulbils from freezing and thawing excessively. At the time of sprouting, mulch must be removed to allow the tiny sprouts to grow better.

The grower should make sure that these containers are consistently watered with timely weeding and hoeing. At about the same time when regular clove-planted garlic crop is ready to harvest, these bulbil crops which have grown into 'rounds' of about 1.25-2.5cm in size are also harvested. Now these 'rounds' are also planted as to those of normal cloves in the beds.

Land Preparation

As detailed through Farmingindia (2017), the planting area must be ploughed and disked properly making it free of clods and other debris. It is advisable to use a moldboard plough since it enhances the drainage property of the soil and also pushes the crop residue deeper down the earth to enhance better decomposition. The land is tilled 3-4 times to get rid of the soil clod and organic manure being incorporated at the last ploughing. For *rabi* crops, flat beds of 4-6m length and 1.5-2m width are formed. However, flat beds are avoided during *kharif* or rainy seasons so as to prevent waterlogging. In case of *kharif* crop, broad bed furrows (BBF) with a height of 15cm are made. The top width is kept about 1.2m and each furrow 45cm deep. BBFs are suitable for drip and sprinkler irrigation. The rows must be made at a distance of 15cm from each other.

Factors influencing growth and development

Variety selection

Several high yielding and disease resistant varieties with good storage life have been developed by various research institutes for different seasons (*rabi*, *kharif* or for both) and climatic conditions. The latitude where the garlic is grown affects the choice of the type as garlic can be day-length sensitive. There are various subspecies or varieties of garlic, most notably long-day/hardneck and short-day/softneck types as detailed under the heading '**Subspecies and varieties**' above.

Planting

Dipping the planting cloves in 0.1% carbendazim solution just before sowing reduces the incidence of fungal diseases. Garlic plants can be grown closely leaving enough space for the bulbs to grow and mature, and for instant kitchen use in containers of sufficient depth. Cloves are planted perpendicular to the ground with planting distance of 10cm or slightly more. About 500 kg cloves of 8-10 mm diameter are required to plant one hectare. Garlic is planted through the use of dibbler (in case of flat beds) or in the furrows (indiaagronet.com).

- **Dibbling:** The field is divided into small beds of convenient size to facilitate irrigation and further tillage activities. Cloves may be dibbled 5-7.5cm deep, keeping their growing ends upwards. They are laced 7.5-10cm apart from each other in rows of 10-15cm apart and then they are covered with loose soil. May-July on the hilly regions and October-November in the subtropical and tropical regions are the normal planting seasons for garlic.

- **Furrow planting:** The furrows are made 15cm apart with hand hoe or a cotton drill. In these furrows, cloves are dropped by hand 7.5-10cm apart. They are covered lightly with loose soil, followed with light irrigation.

Climate

While garlic can be grown year-round in mild climates but in colder climates, cloves are best planted about six weeks before the soil freezes. The goal is to have the bulbs producing only roots but no shoots above the ground. Hardneck garlic is generally grown in cooler climates which produces relatively large cloves whereas softneck garlic is generally grown closer to the equator and produces small tightly-packed cloves. Garlics should be planted only at sunny situations.

Garlic cannot tolerate extreme cold or a very hot conditions. Bulb development and vegetative growth of garlic requires a cool and moist climate, whereas for maturity it prefers warm and dry climate (Farmingindia, 2017; eagri.org).

Exposure of young plants to temperatures lower than 20^{0}C for 1 or 2 months would accelerate the bulb formation though a prolonged exposure to lower temperature may reduce the yield of the bulbs. A cooler growing period gives higher yield as compared to warmer growth conditions. Long-day varieties require 13-14h day length for optimum growth whereas short-day types only 10-12h.

Soil

In mild climates, garlic can be grown year-round in a wide range of soil conditions and pH levels though these prefer to grow in a well-drained, loose, loamy soil having a pH of 6-8 and not the clayey waterlogging type. Soils with rich organic content having high amount of nutrient contents aid in proper bulb formation. A heavy and waterlogging type soil would result in deformed bulbs and those having poor drainage facility make the bulbs discoloured (Farmingindia, 2017).

Season

The time of planting differs from region-to-region. In India, garlic is planted as both *kharif* (May-July) and *rabi* (October-November) crop. It is planted as a *rabi* crop in Andhra Pradesh, Bihar, Haryana, Uttar Pradesh, Orissa, Punjab, Uttarakhand, Rajasthan, Bengal and hilly regions; though as *rabi* and *kharif* both in Tamil Nadu, Karnataka, Maharashtra, Gujarat, Madhya Pradesh and Chhattisgarh looking into the climatic conditions (Farmingindia, 2017).

Planting and harvesting time of garlic in different regions of India:

State	Season	Time of planting	Time of harvesting
Chhattisgarh, Gujarat, Madhya Pradesh and Maharashtra	*Kharif*	June-July	October-November
	Rabi	September-November	March
Tamil Nadu	*Kharif*	June-July	October-November
	Rabi	October-November	March-April
Karnataka	*Kharif*	June-July	October-November
	Rabi	September-October	March
Andhra Pradesh	*Rabi*	September-October	March
Bihar, Haryana, Punjab, Rajasthan, Uttar Pradesh and Uttarkhand	*Rabi*	October-November	March-April
Odisha and West Bengal	*Rabi*	October-November	March
Hilly areas (Long day types)	*Rabi*	September	May
Source: dogr125			
Proper season for planting in higher hills in northern part of India.	March-April It can also be, however, planted during September-October.		

Manures and fertilizers recommended by DOGR

Based on the field experiments carried out through 'All India Network Research Project on Onion and Garlic' at different locations, DOGR recommends application of 75:40:40:40kg NPKS/ha along with a combination of two or three organic manures (FYM, poultry manure and Vermicompost) equivalent to 75kg N/ha (7.5t poultry manure or 7.5t vermicompost or 15t FYM/ha) for Bihar, Chhattisgarh, Gujarat, Karnataka, Madhya Pradesh, Maharashtra, Odisha, Rajasthan and Tamil Nadu (Ooty) (dogr129).

For Haryana, Uttarkhand and Uttar Pradesh, there is recommendation of 100:50:50:50kg NPKS + 20t FYM/ha (dogr129). Recommended organic manures should be applied before last ploughing and mixed well in the soil. One third of nitrogen and full recommended dose of phosphorus, potassium and sulphur should be applied as basal at the time of planting though remaining two-third of nitrogen should be applied in two equal splits at 30 and 45 days after planting (dogr129).

If the soil test shows a deficiency of any micronutrient besides NPKS, the deficient micronutrient should also be applied to correct the deficiency. If the plant is diagnosed with micronutrient deficiency during growth stages, the deficiency should be corrected by either foliar or soil application of concerned nutrients immediately (dogr129).

- ZnSO @ 10kg/ha as basal is recommended in areas having Zn deficiency (dogr129).

- Borax @ 10kg/ha is recommended for areas having boron deficiency (dogr129).
- FYM @ 15t/ha is recommended in areas having multi-micronutrient deficiency along with foliar application of micronutrient mixture (Fe: 2.5%, Zn: 0.3%, Mn: 1%, Cu: 1.0%, B: 0.2%) at 45 and 60 days after transplanting for increasing onion productivity (dogr129).

Biofertilizers are efficient strains of microorganisms that improve plant nutrient uptake.When applied to seed or soil, the microbes colonize the rhizosphere or the interior of the plant and promote growth by increasing availability of primary nutrients to the host plant by biological nitrogen fixation, phosphorus solubilization and stimulating plant growth through the synthesis of growth-promoting substances (dogr129). Based on the experiments carried out at DOGR, biofertilizers *Azospirillum* and 'phosphorus solubilizing bacteria' (PSB) @ 5 kg/ha each is recommended for garlic. Recommended biofertilizers should be mixed with 100-200kg FYM having 50% moisture and stored overnight before planting (dogr129).

Irrigation

Since garlic is a bulb crop producing shallow roots, it requires a good amount of moisture. However, too much water would result in splitting of the bulbs and too little water or moisture level will cause yield loss by producing under-developed bulbs (Farmingindia, 2017). Best way is to irrigate the crop frequently but shallowly. First irrigation is applied just after planting and then at 7-10 days intervals depending on the moisture content in the soil. Irrigation is usually stopped 15 days before bulb harvest.

Alternating the irrigation period with a dry spell causes the outer scales of garlic to split. Waterlogging results in development of diseases such as purple blotch and basal rot. A continuous irrigation until maturity causes secondary roots to develop, *vis-à-vis* production of new sprouts and further growth (Farmingindia, 2017).

The best way to irrigate garlic is by use of sprinkler and drip irrigation. It helps in improving the yield considerably. In case of drip irrigation, the discharge flow rate of the emitters must be 4 l/h which will help improving the yield by 15-25% better than flood irrigation system, however, the discharge rate through sprinklers should be 135 l/h (Farmingindia, 2017).

Fertigation

In fetigation both irrigation water as well as crop nutrients are applied through drip irrigation only in the root zone of the plants. This saves the water to a great extent together with reducing nitrogen losses through leaching into ground water (dogr.icar.gov.in; vikaspedia.in; Farmingindia, 2017). Fertigation is an efficient method of using drip irrigation to apply fertilizers in garlic farming. However, at the time of planting a basal dose of 75kg nitrogen per hectare is applied (Farmingindia, 2017).

Weed management

Garlic requires routine care during the growing period to keep the rows free of competing weeds, especially when the plants are young (dogr.icar.gov.in). First hoeing and weeding is done with hand hoe usually one month after sowing (eagri.org) and second one month after (indiaagronet.com; eagri.org). This loosens the soil and helps in the setting of bigger and well filled bulbs. Weeding or hoeing should not be done at a later stage because this may damage the stem and impair the keeping quality. Pendimethalin @ 3.5 litres or Goal @ 0.25 kg/ha + 1 hand-weeding gives good control of broad leaved weeds (eagri.org).

Whenever emerging garlic scapes are noticed, these are removed immediately to divert whole energy towards bulb growth and development. These scapes can be eaten raw or cooked.

Crop rotation and mixed cropping

Being a shallow-rooted crop, garlic is not able to utilize all the supplied nutrients that leach down with water and settle in the sub-soil. These nutrients can be utilized better through growing deep-rooted leguminous crops grown together. Alternating garlic with leguminous crops not only boosts the yield of garlic but also improves the soil fertility. Alternating crops like groundnut with garlic can ensure better profit for the farmers as well (Farmingindia, 2017).

Harvest

Garlic is one of the few plants with two harvests, i.e. first for immature scapes, and second for the bulbs. Harvesting the scapes is an integral part of garlic farming which are sheared off so that the plant can spend its entire energy only in bulb development. These scapes can be eaten and sold in the market as greens. Depending on the variety, garlic bulbs generally become ready for harvest within 120-150 days of sowing when the leaves start yellowing and become dry (Farmingindia, 2017).

Plants are pulled out or uprooted from the soil, followed by small bundling with the help of rope and then kept in a field under the shade for 2-3 days. Then the dry stalks are sheared off near the bulb and the roots are trimmed, bulbs are properly cleaned and then sun-dried for a week (en.wikipedia.org/wiki/Garlic). This process is important for the hardening of the bulbs. Before storing they are graded according to the size and weight. Also there is common practice to hang the bulbs in an open and well-ventilated room without shearing the dried plant part from the neck of the bulb and whenever the need be or for selling purpose the bulbs are taken out devoid of stalk, cleaned and sold.

The yield of bulbs vary from 100 to 200q/ha depending upon variety and regions.

Postharvest

Many operations are performed for getting mature and quality bulbs from the field to maintain the reliability of the consumer. Losses occur if proper post-harvest management practices are not followed. These practices differ from place-to-place. Proper curing, sorting *cum* grading, storage and transportation are essential to minimize the looses, *vis-à-vis* for satisfaction of consumers (eagri.org).

Drying and curing

The garlic bulbs are generally harvested at 100% neck-fall stage. To remove excess moisture from the outer skin *cum* neck region, the garlic bulbs are field cured for 3-5 days under the sun covering the bulbs with the tops to avoid damage by sunlight. This way fungal infection in storage is minimized. Curing is an additional process of drying so that bulbs attain proper coloration, excess moisture is removed, bulbs become compact and are paved for dormancy. The method and period of curing may vary depending on the varieties as well as weather conditions at the time of harvesting. Sun-curing is followed by shade curing for 7-15 days either with tops intact or shearing the tops leaving only 2.5 cm above the bulbs and removing the roots gently. Windrow method curing is most common as here with intact tops the bulbs are hanged on wire racks before the windows or without tops on the floor with proper bottom ventilation. However, commercial growers cure these in a room or chamber by passing 27-35^{0}C hot air where it takes about 48 hours if the humidity is between 60-75% (eagri.org).

Sorting, grading and packing

After curing but before marketing or storage the garlic bulbs are either **graded** manually or run over a grader. Thick-necked, splitted, injured, diseased or hollow-cloved bulbs are sorted out and then graded manually or through graders. Government of India has prescribed certain grade designations for different qualities of garlic for export (eagri.org).

Since garlic needs less ventilation compared to onion, there is a need to develop suitable packaging to reduce drying loss (eagri.org). In India, garlic bulbs are packed in open-mesh jute bags, nylon-netted bags or leno bags of 40-60 or even 90kg capacity for domestic use or local sale especially in Andhra Pradesh, Karnataka or elsewhere where garlic is grown commercially but 18 and 25kg perforated 10-ply corrugated cardboard boxes for export. Nylon-netted bags used for packing and further storage cause minimum losses in storage. In foreign countries, plastic-wooven bags are quite commonly used as such bags are attractive having good strength. However, 5kg small packs are also used for sale in the Malls and Super Markets.

Three weeks before harvesting if garlic crop is sprayed with 3000 ppm MH, storage bulb weight loss is reduced and life prolonged, spraying of 0.1% carbendazim before harvesting *cum* disinfection of premises for handling and storage of garlic also reduces post-harvest losses especially in terms of bulb decay, and application of phosphorus and potassium in the field also reduces weight loss of garlic bulbs in store whereas nitrogen application increases it (eagri.org).

Storage

Though garlic is commercially stored in a dry and low-humidity environment at 0–0.5°C, it is domestically stored warm and dry (above 18°C) to keep the bulbs dormant. Usually the garlic is stored at room temperature in ordinary well-ventilated chambers for up to 8 months and these store well when kept with intact tops, however, traditionally these are hung with soft-neck varieties often braided in strands (plaits or grappes) (en.wikipedia.org/wiki/Garlic). Peeled cloves can be stored in wine or vinegar in the refrigerator. Garlic are also kept in oil to produce flavoured oil but it requires measures to be taken to prevent the garlic from spoiling which may include rancidity and growth of *Clostridium botulinum* so to minimize or check the bacterial growth through acidification with a mild solution of vinegar. Garlic granules are drying of cloves at low temperatures to preserve the enzymatic activity which can be rehydrated to reactivate it (en.wikipedia.org/wiki/Garlic).

'Yamuna Safed' (G 1), 'Yamuna Safed-2' (G 50) and 'G 323' are better storer (eagri.org). Since garlic bulbs store well under a wide range of temperatures therefore cold storage facility is not required, however, if these are cold-stored at 4.4°C for prolonged period bulbs start sprouting, hence cold storage is of no use in this case, however, as such or irradiated bulbs store well at 0.5-2.2°C with 60-70% relative humidity (eagri.org). Storage above 70% RH at any temperature will generate root initiation and its further growth *cum* mould development (Mahajan, 2019) as storage loss of 12.5% has been recorded in garlic stored at 1-5 °C with 75% relative humidity compared to 42.4% loss at ambient temperature though UV light treatment for 30 minutes reduces loss to 8% when stored for 150 days (eagri.org). In small bundles with attached dried foliage these store well in heaps in the well-ventilated storage structures when bulbs are properly exposed all around in the heaps. Bulb exposure to 2-6 krad Co-60 gamma rays within 8 weeks of harvesting reduces weight loss, controls sprouting and prolongs bulb life up to one year easily though doses higher than 10 krad reduces diallyl disulphide responsible for typical garlic flavour (eagri.org). Smoke treatment also reduces storage loss and improves bulb quality.

Physiological disorders

Its nutritional deficiencies symptoms are almost the same as elaborated under the previous chapter 'Onion'.

Pre-mature sprouting, rubberification and splitting

Sometimes, especially when there are winter rains or excessive soil moisture and nitrogen supply, premature sprouting of bulbs in the field is noticed towards the start of maturity stage of bulbs. However, this disorder is not of permanent nature and varies from variety-to-variety. Pre-mature sprouting is also caused by early-planting or delayed harvesting during rainy season. Irrigation after long spell of drought or delayed harvesting during rainy season may sometimes cause splitting in some varieties (eagri.org; nhrdf.com).

For the past few years, rubberification problem is noticed in Rajkot area as well as in Nilgiri hills (T.N.) which is increasing day-by-day. Actual cause for the disorder was since not identified, a trial was planned in collaboration with TNAU at HRS, Ooty during 1995-96 to study the effects of insecticides, fungicides, micronutrients and growth regulators. It was observed that rubberification was totally controlled by application of micronutrients, i.e. zinc sulphate and ammonium molybdate. It was also controlled by application of neem-cake insecticides and GA (eagri.org; nhrdf.com).

Rubberification and premature sprouting of bulbs are noticed usually in those low-lying areas of watershed where there is heavy deposition of nutrients along with silt during heavy rains. These problems are severe in garlic fields which are more frequently irrigated than the usual requirement. With application of higher levels of nitrogen, there is an increased level of pre-mature sprouting of bulbs which results in bulb splitting and rubberification. Rubberification incidence also increases when higher level of nitrogen is applied in the form of urea. Mettupalayam-type short-duration garlics are more susceptible than long duration Singaporean types (eagri.org; nhrdf.com).

Aerial bulbils

Aerial bulbil formation in garlic is quite common in areas where lower temperature prevails for longer duration or where during growing season there is more temperature variation. It, however, does not affect much on yield or quality of bulbs. 'Yamuna Safed-3' (G-282) is quite prone for this type of effect.

Insects–pests and diseases

There are only a few insect-pests or pathogens affecting garlic plants. Major insect-pests and diseases of garlic along with their symptoms and management are furnished here under.

Mites and insects

Bulb mites (*Rhizoglyphus* spp.; *Tyrophagus* spp. *and Glycohaqus* spp.)

Rhizoglyphus callae, R. robini, Glycohaqus domesticus and *Tyrophagus* spp. may attack the bulbs in clusters in storage (eagri.org; nhrdf.com) as these inhabit damaged areas under the garlic root plate. Bulb mite is creamy-white, <1mm in length, with prominent legs and is pearl-like. Its infestation to plants causes secondary invasion by certain pathogens which cause bulb rot (Plantvillage). Its attack is symptomised as reduced plant stand, stunted growth and bulb rotting in field as well as in storage. Its infestation is checked by avoiding planting successive crops of onion or garlic at the same location, keeping the field fallow to ensure that any residual organic matter decomposes completely as crop residues can harbour mite populations so garlic seed-cloves should be treated with hot water prior to planting (Plantvillage). Methyl bromide fumigation @ 1kg/1000 cubic feet for 2h at 26^0C to the surface dry bulbs after harvesting and before storage is quite effective against mites (eagri.org; nhrdf.com).

Eriophyid mite (*Aceria tulipae*)

Identification with naked eye is quite difficult as this banana-shaped mite is microscopic. In its infestation the leaves do not open fully, plants show stunting, mottling, twisting and curling, especially along the edges of leaves so whenever such symptoms are first noticed the crop should be sprayed wirth Dicofol (0.2%) or sulphur (0.05%) and if necessary this should be repeated again after a fortnight (Farmingindia, 2017; dogr133).

Red spider mite (*Tetranychus cinnabarinus*, syn. *T. urticae*)

Leaves due to its attack show white, yellow and bronze spots or discoloration due to small patches of eggs, nymphs, adult mites, their faeces *cum* silky spider webbing on the upper surface of the leaves apart from the damage symptoms so as soon as the symptoms appear Dicofol (0.2%) or sulphur (0.05%) should be sprayed and if necessary it should be repeated again after one fortnight (dogr133). Red spider mite infestation usually occurs at maturity stage of garlic, especially there where insecticides have been used excessively.

Leafminers (*Lyriomyza* spp.)

Its infestation causes mature larvae dropping from the leaves into soil to pupate, entire lifecycle can take as little as 2 weeks in warm weather and the insect may go through 7 to 10 generations per year (Plantvillage). Its infestation shows white, thin *cum* winding trails on leaves, heavy mining can result in white blotches on leaves and leaves dropping from the plant prematurely and the early infestation causes yield to be reduced drastically. Adult leafminer is a small black *cum* yellow fly which lays its eggs in the leaves which when hatch the larvae start feeding on the leaves interior through mining (Plantvillage). If any plant shows signs of leafminer damage such plants are immediately removed from the field and plants burnt. Insecticides are used only when leafminer damage has been identified so that the populations of their natural enemies are unnecessarily not controlled (Plantvillage).

Onion maggot (*Delia antiqua*)

Mature larvae are about 1cm long with feeding hooks. Females can lay several hundred eggs during their 2-4 week lifespan. Insect overwinters as pupae in the soil (Plantvillage). Its infestation causes stunting or wilting of seedlings, plants usually breaking at the soil level if an attempt is made to pull it up and if infestation occurs when plants have started bulbing the bulbs will be deformed and become susceptable to storage rots. Adult insect is a greyish fly which lays white and elongate eggs around the base of the plant. White larvae that emerge from the eggs are tiny and bore into the plant. Management of onion maggots is heavily reliant on good sanitation and all the onion/garlic bulbs should be removed at the end of the season as maggots will die without a food source. Commercial onion growers often rely on the application of appropriate granular insecticides but in some cases even insecticidal sprays are also required. Home gardeners should try to remove any volunteer wild *Allium* plants as these can act as a host and floating row covers may also help to protect plants and prevent females from laying eggs on plants (Plantvillage).

Thrips {Onion thrips (*Thrips tabaci*) and Western flower thrips (*Frankliniella occidentalis*)}

Onion thrips and western flower thrips are slender, yellow to dark brown and small (1.5mm) creatures with 4 wings having an extensive host range and can be introduced to garlic from other plants whereas the nymphs are even smaller and lighter in colour (Plantvillage). Their infestation causes twisting and curling of leaves up to 45 days of transplanting, white to silvery patches on leaves and in extreme cases whole plant becomes blemished and turns white (Farmingindia, 2017; dogr133). Its infestation causes discolored *cum* distorted tissues and scarring of leaves though severly infested plants may have a silvery appearance (Plantvillage). Planting one or two outer rows of maize or wheat can act as a good barrier for thrips (Farmingindia, 2017; dogr133). Natural enemies include some species of predatory mites, pirate bugs and lacewings but planting garlic in close proximity to grain fields should also be avoided as thrips populations build up on these plant in the spring (Plantvillage). Overhead irrigation of plants may help reduce thrips population. When thrips population has attained 30 thrips per plant the crop should be sprayed with carbosulfan (0.2%), fipronil (0.1%) or profenofos (0.1%) (Farmingindia, 2017; dogr133). Malathion spraying @ 1 ml/l or methyldemeton @ 0.75 ml/l of water gives good control if sprayed at frequent intervals (eagri.org). Four sprays of fluvalinate (0.04%) at fortnightly interval at Karnal and malathion (0.1%) at Nasik provided best control of thrips. Phorate @ 1kg ai/ha at planting and 30 days after planting controls thrips effectively (eagri.org).

Lesion nematode (*Pratylenchus penetrans*)

Lesion nematode has one of the widest host ranges of any nematode. The nematode enters the plant through the root epidermis and feeds on cell contents (Plantvillage). Its infestation causes plant stunting as root system has round and irregular lesions and lacking fine roots. Due to its extensive host range, crop rotation is not very effective though HWT is quite effective (Plantvillage).

Stem and bulb nematode (*Ditylenchus dipsaci*)

This is a tiny worm almost invisible to the naked eye which invades the tissues of garlic so reducing its yield drastically (eagri.org). It enters through stoma or plant wounds, creating galls or malformations in plant growth, thus allowing the entrance of secondary pathogens for further infection (dogr133). However, this nematode is uncommon on garlic in India but its severely infested plants rot basally and when pulled gently the plant comes out leaving the rotten roots in the ground (eagri.org). Its infestation causes twisting and stunting in plant growth, bulb discoloration, swollen stems, swelling *cum* sponginess at the base in mature plants, rotting of tissues of stems & leaves and frequently broken by longitudinal splits (dogr133). In no case the diseased bulbs (cloves) should be planted. Since this nematode survives and reproduces inside infested plants so all such fields as well as the concerned tools should be properly sanitized (dogr133). Pre-planting soil fumigation with dichloro-propene-dichloropropane mixture @ 500 kg/ha is recommended (eagri.org).

Fungal Diseases

Botrytis neck and bulb rot (*Botrytis porri*)

It is a disease of onion, garlic, leek and shallot. *Botrytis allii* and *Botrytis aclada* cause this disease in onion and *Botrytis porri* causes bulb and neck rot in garlic. Fungus spreads through air-borne conidia. It is also seed-borne. Sclerotia are formed in the soil and persist until the next season. In both onion and garlic, initial infections may not be noticeable and symptoms may develop only when leaves become necrotic and senesce. The greatest incidence of infection occurs in persistently cool (10° to 24.3°C) and moist weather. Infections are favoured by excessive use of nitrogen fertilization, N application after bulb initiation, and poor drying of the tops & curing of the necks. Healthy onions that are stored properly and harvested at the right time are seldom affected (en.wikipedia.org/wiki/Garlic). *Botrytis allii* grows on *Allium* species, including onion (*A. cepa*), aggregating onion (*A. cepa* var. *aggregatum*), shallot (*A. cepa* var. *ascalonicum*), garlic (*A. sativum*) and leek (*A. porrum*). It may also infect wild *Allium* species and can grow saprophytically on decaying crop residues such as cereal, pea and bean straw. It is capable of colonizing and producing spores on sterilized poppy straw (*Papaver somniferum*) (Chilvers and du Toit, 2006).

Symptoms appear either in the field towards the end of the season or during bulb storage. Plants infected in the field may be stunted with dying and dead outer leaves. Initial symptoms usually begin at the neck where affected tissue softens, becomes water-soaked and turns brown. In a humid atmosphere, a gray and felt-like growth (where spores are produced) appears on rotting scales and mycelia may develop between scales. In its infection, the dark-brown-to-black sclerotia (the resting bodies of the pathogen) may eventually develop in the neck or between scales (en.wikipedia.org/wiki/Garlic).

Excessive and late application of nitrogenous fertilizer is always dangerous to this crop. Late-season irrigation is not desirable in this crop as garlic this way invites the pathogen. Before lifting it should be ensured that foliage and necks are properly dry and before storage the neck tissue is well cured. While lifting damaged bulbs (bruising and wounding) as well as diseased and insect infested bulbs should be discarded. Bulbs at this time may be stored at 5^0C temperature with minimum relative humidity but good air circulation. Stable temperatures should be maintained when moving bulbs into and out of storage as well as while shipping to markets. Wide temperature fluctuations can cause condensation on the bulbs which also favours development of this disease. Seed and bulb treatment with fungicides prior to planting can reduce incidence of *Botrytis* neck rot. However, late use of fungicides during last month of harvest is most effective in preventing neck infection since late-season infection poses the greatest risk for bulb rot in storage. Chemical control may not be effective once the symptoms develop, and cultural control is the most reliable way to avoid infection (en. wikipedia.org/wiki/Garlic).

Cercospora leaf spot/blight (*Cercospora duddiae* Welles)

This disease appears on leaves as small, ash-coloured and irregular-shaped spots scattered on leaf lamina where at later stage the spots coalesce gradually and result in blightening of foliage (eagri.org). Spraying of ziram or captan @ 2g/l of water or copper oxychloride @ 3g/l of water at fortnightly intervals gives good control (eagri.org).

Downy mildew (*Peronospora destructor*)

Emergence of this fungul disease is favoured by cool temperatures *cum* leaf wetting, and the symptoms of its infection are pale leaf spots, paling *cum* elongated patches on leaves with gray-purple fuzzy growth on leaf surface, *vis-a-vis* collapsing of leaf tips (Plantvillage). Planting of infected sets should strictly be avoided. Crop rotation of non-*Allium* crops should be followed for 3-4 years. This crop should be planted only in well-draining areas and that too at proper distance. All the plant debris should be destroyed and proper sanitation should be maintained. Application of sulphur fungicides in a proper way should be done as the garlic has waxy leaves (Plantvillage).

Garlic bulb canker or skin blotch (*Embellisia alli*)

This fungus in garlic crop causes bulb canker in many countries. Until interception of *E. allii* in garlic bulbs imported from China by the Indian Plant Quarantine Authorities, it was not known to occur in India (Latha *et al*, 2007). In the same year it was isolated for the first time in India from garlic bulbs collected from Himachal Pradesh (Singh and Khar, 2007). The fungus is reported to be overwintering in the plant debris, soil and infected bulbs and its infection in garlic is symptomized as irregular lesions of dark layer underneath the epidermis (David, 1991). The dark areas are developed due to powdery mycelia and conidia of the organism scattered on the outer scales of the developing garlic or usually near the base of the bulb, on the back and sides of the cloves within a head or on the edges of cloves, and the plants in the field show yellowing due to its infection (Ram *et al.*, 2007). This disease is reported to be a problem in wet years. For controlling this malady only the healthy bulbs should be kept in storage by maintaining a quite low humidity and for marketing the bulbs should be wrapped in breathable materials and in case of affected ones the discoloured outer scales of the bulbs should be peeled (Embellisia/OSU Plant Clinic 2007). Under field conditions, there should be good drainage and the crop stubbles should be burnt in managing this disease.

Onion powdery mildew [*Leveillula taurica* [Lev.) G. Arnaud, (1921)]

This disease develops in warm and dry environments (en.wikipedia.org/wiki/Leveillula_taurica) by producing distinct pale-yellow circular or oblong lesions of variable sizes on abaxial surface of leaf associated with powdery mass (eagri.org). On the lesions white mycelium can be found with conidiophores bearing either lanceolate or rounded condia. All the crop residues from the previous onion crop should be removed before subsequent planting, deep tillage, rotating to a non-host crop the year following an onion crop, crops with adequate moisture will be more resistant to the pathogen, and crops with overhead irrigation rarely see powdery mildew development in the field (en.wikipedia.org/wiki/Leveillula_taurica). Though disease rarely progresses enough to justify the use of a fungicide, sulphur fungicides @ 2g/ litre of water if sprayed at regular intervals fortnightly after disease appearance gives good control (eagri.org).

Purple bloch (*Alternaria porri*)

Mainly it is a soil-borne disease which also spreads through infected bulbs, plant debris, etc. (Farmingindia, 2017). Disease emergence is favoured by wet foliage, with sporulation occurring in nights during periods of high humidity (Plantvillage). Initially small but elliptical white cum sunken lesions or spots on leaves appear, lesions often turn purplish-brown and get surrounded by chlorotic margin, if the spots enlarge chlorotic margins extend above and below the actual lesion, lesions usually girdle leaves causing them to fall over and the lesions may also start at the tips of older leaves (dogr133).

Long rotations with non-hosts and the reduction of leaf wetness by planting in well-draining soil coupled with timed irrigations to allow plants to dry adequately during the day will keep this disease at the minimum. Several fungicides are effective in controlling this disease but the crop rotation should be followed for optimum control (Plantvillage). This disease can be kept at bay by spraying Mancozeb @ 0.25%, Tricyclazole @ 0.1%, Hexaconazole @ 0.1% or Propiconazole @ 0.1% at 10-15 days intervals from 30 days after planting or as soon as disease appears (dogr133).

Rust or leek rust (*Puccinia porri*)

It is favoured by high humidity but low rainfall and its spores can be transported over long distances by wind. The disease is symptomized by small white flecks on leaves and stems which later on develop into circular or elongated orange pustules though under severe infestations it can cause leaves to yellow and die (Plantvillage).

To keep this disease at the minimum, only disease-free planting material should be used in well-draining soil, the field should be kept weed-free, air circulation in the field should be improved and either of the fungicides such as penconazole, bupirimate with triforine, mancozeb, myclobutanil, *etc.* should be applied (Plantvillage).

Stemphylium blight (*Stemphylium vesicarium*)

Plant debris and soil are the major transmitting agents (Farmingindia, 2017). Small yellow to orange streaks or flecks develop in the middle of the leaves when the plants are at 3-4 leaf stage. These streaks soon develop into elongated, spindle-shaped to ovate-elongate and diffused spots surrounded by characteristic pinkish margin.

These spots progress from the tip to the base of the leaves, coalescing into extended patches and blight the entire leaves on the plants gradually (dogr133). To control the disease Mancozeb @ 0.25%, Tricyclazole @ 0.1%, Hexaconazole @ 0.1% or Propiconazole @ 0.1% is sprayed at 10-15 days intervals from 30 days after planting or as soon as disease appears (dogr133).

White rot (*Sclerotium cepivorum*)

It gets transmitted through soil, garlic debris and diseased garlic sets (Farmingindia, 2017). Once disease is established in the field, it is unusable for garlic production as fungus can survive in the soil for 20 years. This is one of the most devastating diseases of Allium crops worldwide causing extensive crop losses (Plantvillage) as well as bulb rot in the storage. It is symptomised as yellowing and dying back of leaf tips, roots normally get destroyed, superficial white fluffy mycelium may be present on decaying scales and brown or black sclerotia develop on surface or within tissues (dogr133).

Seed treatment with hot water before sowing should be done, crop rotation should be followed, infected plants should be destroyed (dogr133) and soil and plant material transfer between sites should be avoided (Plantvillage). Soil solarization during summer at high temperature reduces the disease incidence, but during cropping period carbendazim (0.1%) should be applied (dogr133).

Viral Diseases

Irish yellow spot virus (IYSV)

It is transmitted through onion thrips and transplantation of infected plant (Farmingindia, 2017). Its infection causes straw-colouration, spindle-shaped spots with poorly defined ends on the leaves, and coalescing of spots to form larger patches on the leaves as are clearly visible on older leaves (dogr133). To keep this disease at bay, volunteers and weeds in and around garlic field are eliminated, virus-free planting material is used, crop stress should be avoided and thrips as virus vector should be controlled effectively (dogr133).

Leek yellow stripe virus (LYSV)

Aphids are the main vector of this virus (Farmingindia, 2017). Its infection causes light yellow striping on the distal part the leaves which leads to dwarfing of the entire plant, developed bulbs become smaller and malformed which results into significant yield loss, *vis-a-vis* post harvest storage loss (dogr133). This disease can be managed by use of virus-free planting material and foliar spraying of insecticides such as Profenofos @ 0.1%, Carbosulfan (0.2%) or Fipronil (0.1%) to control the vector-aphid (dogr133).

Garlic mosaic virus (GarMV)

This virus is usually asymptomatic and is transmitted through aphids. This virus on garlic is often found alongside other viruses such as onion yellow dwarf virus producing mild mosaic pattern with plant stunting *cum* reduced bulb size (Plantvillage). Chlorotic mottling and stripes on first emerging leaf may be seen followed by pale-yellow broken stripes, resulting in typical mosaic pattern on matured leaves along with yellowish dots, whitish leaf margin or twisting of some leaves in a few cultivars, cloves being fewer in a bulb and the symptoms being mild on younger leaves than on matured ones (eagri.org). For planting only virus-free cloves produced from meristem tip culture under virus-free conditions should be managed (Plantvillage). Since virus is transmitted through aphids, spraying of monocrotophos @ 0.5 ml/litre of water or methyl demeton @ 0.75 ml/litre of water is useful (eagri.org).

Onion yellow dwarf virus (OYDV)

This virus in garlic is transnmitted through aphids, infected cloves and seed bulbs (Farmingindia, 2017). Its symptoms on garlic plants are mild chlorotic to bright yellow striping, mosaic, curling of leaves and stunted growth and this disease can be controlled by use of virus-free planting material and for controlling the vector-aphid the crop should be sprayed with Profenofos @ 0.1%, Carbosulfan (0.2%) or Fipronil (0.1%) (dogr133).

Storage diseases

Storage diseases can be controlled by avoiding mechanical injuries to the bulbs during harvesting, through proper bulb curing after lifting and by storing the bulbs under ideal conditions. Gargi and Roy (1988) recommend 0.1% Bavistin spraying fortnightly to standing garlic crop. Garlic stored for table purposes may be fumigated with Formalin (Rath and Mohanty, 1986).

Blue or green mould (*Penicillium hirsutum* and *Penicillium allii*)

Stored garlic can be affected by blue or green mould (*Penicillium* spp.) decay if storehouse has high humidity (above 70%). The pathogens usually invade bulbs of onion and garlic through wounds, bruises or uncured neck tissues. Many species of *Penicillium* are responsible for causing blue mould. These fungi are common on plant debris and senescent plant tissues. For onions, secondary infections by other pathogens are also common and then it becomes a serious problem. Infection may first appear as soft or water-soaked spots, followed by white patches (of mycelium) which turn blue or green with sporulation. Its sporulation and germination are delayed at low temperature and at -4^0C these activities of fungi are inhibited entirely. In refrigerated cloves, one may only see the white mycelium during early stages.

To reduce the risk of blue mould, the bulbs after lifting should immediately be cured so that there is no extra moisture on the surface and necks become dry and thin. Bruising or wounding of bulbs during harvest and handling should always be avoided. Any injured bulb at sorting should be removed and only well cured and intact bulbs should be sent to storage where humidity should not be more than 70% in any case. Moreover, storage is always better with intact plant which can be hanged in a well ventilated chamber having 5^0C temperature if it is only for short duration.

Penicillium rot (*Penicillium corymbiferum*)

Affected bulbs are light in weight and individual cloves are soft, spongy and powdery-dry. In extreme stage of decay, the cloves break down in a green or gray powdery mass (dogr133). The bulbs are stored under dry conditions in storage structures to reduce rot development.

Black mould (*Aspergillus* spp.)

Bulbs show black discoloration at the neck and streaks of black mycelium and conidia beneath the outer dry scales and at advanced stages all the scales are infected and the bulbs shrivel (dogr133). To get rid of this problem, the bruising or injury on the bulbs should be avoided at any level, *viz.* lifting, curing, handling, storing and transportation. The bulbs are properly dried before storing and these are stored in a well ventilated chamber under dry conditions.

Fusarium basal rot (*Fusarium oxysporum cepae*)

Its infection causes shattering of the cloves, bulbs appear spongy or sunken and/or brown *cum* watery when cut open, and white to light pink growth appears over the cloves. Injured and abnormal bulbs should be discarded from storing. Bulb must be stored under dry conditions with good ventilation. Crop rotation for 3-4 years with non-host crops such as wheat and cowpea should be followed.

Bacterial rot (*Erwinia* spp. and *Pseudomonas* spp.)

Watery and foul-smelling fluid squeezes from the neck of diseased bulbs when pressed. This problem can be solved by proper bulb handling *cum* curing and storing. Bulbs must always be stored in a chamber having dry conditions with good ventilation.

Other diseases

Garlic may also suffer from leek rust (*Puccinia porri*) and pink root (*Phoma terrestris*). Pink root is a typically non-fatal disease though stunts the roots and turns them pink to red. The larvae of the leek moth (*Acrolepiopsis assectella*) attack garlic by mining into the leaves or bulbs (Landry, 2007).For details about certain insect-pests and diseases attacking garlic are the same as given under the previous chapter onion (*Allium cepa*) along with Integrated Pest Management.

References

Abdel-Daim, M.M., A.I. Abushouk, S.G. Bungau, M. Bin-Jumah, A.F. El-Kott, A.A. Shati, L. Aleya and S. Alkahtani, 2020. Protective effects of thymoquinone and diallyl sulphide against malathion-induced toxicity in rats. *Environ. Sci. Pollut. Res.*, (doi: 10.1007/s11356-019-07580-y. www.ncbi.nlm.nih.gov/pmc/articles/PMC7146530/).

Abdel-Daim, M.M., H.M. Shaheen, A.I. Abushouk, E.A. Toraih, M.S. Fawzy, W.S. Alansari, L. Aleya, S. Bungau, 2018. Thymoquinone and diallyl sulfide protect against fipronil-induced oxidative injury in rats. *Environ. Sci. Pollut. Res Int.*, **25**:23909–16 (doi: 10.1007/s11356-018-2386-3. www.ncbi.nlm.nih.gov/pmc/articles/PMC7146530/).

Abdel-Hafeez, E.H., A.K. Ahmad, A.M. Kamal, M.Z. Abdellatif and N.H. Abdelgelil, 2015. *In vivo* antiprotozoan effects of garlic (*Allium sativum*) and ginger (*Zingiber officinale*) extracts on experimentally infected mice with *Blastocystis* spp. *Parasitol. Res.*, **114**:3439–44 (doi: 10.1007/s00436-015-4569-x. www.ncbi.nlm.nih.gov/pmc/articles/PMC7146530/).

Abdullah, I.N. and L.M. Mahdi, 2006. The effect of aqueous extract of *Allium sativum* on some hematological characteristics of Broiler chicken. Israel *J. Al-Qadisiyah Pure Sci.*, **6**: 44-52. (https://www.sciencedirect.com/science/article/pii/S2667142522000811).

Al-Brakati, A.Y. 2016. Protective effect of garlic against diabetic retinopathy in adult albino rats. *Res. J. Pharm. Biol. Chem. Sci.*, **7**:2748–59.

Alamgir, A.N.M. 2017. Origin, definition, scope and area, subject matter, importance, and history of development of pharmacognosy (doi:10.1007/978-3-319-63862-1_2). (https://www.sciencedirect.com/science/article/ pii/S2667142522000811)..

Al-Snafi, A. 2013. Pharmacological effects of *Allium* species grown in Iraq.: An overview. *Int. J. Pharm. Health Care Res.*, **1**:132–47. (www.ncbi.nlm.nih.gov/pmc/articles/PMC7146530/).

Al-Sultan, S.A.S.H. and M.R.A. Al-Juwary, 2005. Comparative *in vitro* activity of *Allium sativum* (garlic) aqueous extract with other selected antibiotics against *Brucella melitensis*. Raf. *Jour. Sci.*, **16**:9-16. (https://www.sciencedirect.com/science/article/pii/S2667142522000811).

Aprioku, J.S. and F.S. Amah-Tariah, 2017. Garlic (*Allium sativum* L.) protects hepatic and renal toxicity of alloxan in rats. *Br. J. Pharm. Res.*, **17**:349-59. (doi: 10.9734/JPRI/2017/34909).

Arshad, M.S., M. Sohaib, M. Nadeem, F. Saeed, A. Imran, A. Javed, Z. Amjad and S.M. Batool. 2017. Status and trends of nutraceuticals from onion and onion by-products: A critical review. *Cogent Food Agric.*, **3**. Article 1280254, 10.1080/23311932.2017.1280254. https://www.sciencedirect.com/science/article/pii/S2667142522000811.

Asdaq, S.M. and M.N. Inamdar, 2010. Potential of garlic and its active constituent, S-allyl cysteine, as antihypertensive and cardioprotective in presence of captopril. *Phytomedicine*, **17**:1016–26. (doi: 10.1016/j.phymed.2010.07.012).

Avula, P.R., S.M. Asdaq and M. Asad, 2014. Effect of aged garlic extract and S-allyl cysteine and their interaction with atenolol during isoproterenol induced myocardial toxicity in rats. *Indian J. Pharmac.*, **46**:94–9.

Badr, G.M. and J.A. Al-Mulhim, 2014. The protective effect of aged garlic extract on nonsteroidal anti-inflammatory drug-induced gastric inflammations in male albino rats (Evid. Based Complement). *Altern. Med.* (doi: 10.1155/2014/759642).

Bae, S.E., S.Y. Cho, Y.D. Won, S.H. Lee, H.J. Park, 2012. A comparative study of the different analytical methods for analysis of s-allyl cysteine in black garlic by hplc. *LWT-Food Sci. Technol.*, **46**:532–5. (doi: 10.1016/j.lwt.2011.11.013. www.ncbi.nlm.nih.gov/pmc/articles/PMC6152780/).

Bae, S.E., S.Y. Cho, Y.D. Won, S.H. Lee and H.J. Park, 2014. Changes in s-allyl cysteine contents and physicochemical properties of black garlic during heat treatment.

LWT-Food Sci. Technol., **55**:397–402. (doi: 10.1016/j.lwt.2013.05.006. www.ncbi.nlm.nih.gov/pmc/articles/PMC6152780/).

Bagul, M., S. Kakumanu and T.A. Wilson, 2015. Crude garlic extract inhibits cell proliferation and induces cell cycle arrest and apoptosis of cancer cells *in vitro*. *J. Med. Food.*,**18**:731–7. (doi: 10.1089/jmf.2014.0064).

Barandiaran, X., A. Di Petro. And J. Matin, 1998. Biolistic transfer and expression of a uid A reporter gene in different tissues of *Allium sativum* L. *Plant Cell Rep.*, 17 :737-41.

Bayan, L., P.H. Koulivand and A. Gorji, 2014. Garlic: A review of potential therapeutic effects. *Avicenna J. Phytomed.*, **4**:1–14. (www.ncbi.nlm.nih.gov/pmc/articles/PMC7146530/).

Ben, H.T., N.A. ElSawy, E.A.M. Header, Y.N. Mabkhot and M.S. Mubarak, 2014. Effect of garlic and cabbage on healing of gastric ulcer in experimental rats. *Med. Chem. Res.*, **23**:5110–19. (doi: 10.1007/s00044-014-1092-z).

Block, E. 2010. Garlic and Other Alliums: The Lore and the Science. *Royal Society of Chemistry*, pp. 5–6. (ISBN 9780854041909).

Bose, S., B. Laha and S. Banerjee, 2014. Quantification of allicin by high performance liquid chromatography-ultraviolet analysis with effect of post-ultrasonic sound and microwave radiation on fresh garlic cloves. *Pharmacogn. Mag.*, **10**:288–93. (doi: 10.4103/0973-1296.133279).

Bradley, J.M., C.L. Organ and D.J. Lefer, 2016. Garlic-derived organic polysulfides and myocardial protection. *J. Nutr.*, **146**:403S–409S. (doi: 10.3945/jn.114.208066).

Cao, X., L. Cao, L. Ding and J.S. Bian, 2017. A new hope for a devastating disease: Hydrogen sulfide in Parkinson's disease. *Mol. Neurobiol.*, **55**:3789–99. (doi: 10.1007/s12035-017-0617-0. www.ncbi.nlm.nih.gov/pmc/articles/PMC7146530/).

Charron, C.S., H.D. Dawson, G.P. Albaugh, P.M. Solverson, B.T. Vinyard, G.I. Solano-Aguilar, A. Molokin and J.A. Novotny, 2015. A single meal containing raw, crushed garlic influences expression of immunity- and cancer-related genes in whole blood of humans. *J. Nutr.*, **145**:2448–55. (doi: 10.3945/jn.115.215392).

Chen, Y.C., T.H. Kao, C.Y. Tseng, W.T. Chang, C.L. Hsu, 2014. Methanolic extract of black garlic ameliorates diet-induced obesity via regulating adipogenesis, adipokine biosynthesis, and lipolysis. *J. Funct. Foods*, **9**:98–108. (doi: 10.1016/j.jff.2014.02.019).

Chen, Y., J. Sun, C. Dou, N. Li, F. Kang, Y. Wang, Z. Cao, X. Yang and S. Dong, 2016. Alliin attenuated RANKL-induced osteoclastogenesis by scavenging reactive oxygen species through inhibiting Nox1. *Int. J. Mol. Sci.*, **17**:1516. (doi: 10.3390/ijms17091516. www.ncbi.nlm.nih.gov/pmc/articles/PMC7146530/).

Chen Y.A., J.C. Tsai, K.C. Cheng, K.F. Liu, C.K. Chang and C.W. Hsieh, 2018. Extracts of black garlic exhibits gastrointestinal motility effect. *Food Res. Int.*, **107**:102–9. (doi: 10.1016/j.foodres.2018.02.003).

Chilvers, M. I. and L.J. du Toit, 2006. Detection and Identification of *Botrytis* species Associated with Neck Rot, Scape Blight, and Umbel Blight of Onion (PDF). *Plant Health Progress,* **7**:38. (doi:10.1094/PHP-2006-1127-01-DG. Archived from the original (PDF) on 2016-03-04. Retrieved 2013-03-31).

Cho, J, S.K. Lee, B.S. Patil, E.J. Lee and K.S. Yoo, 2009a. Separation of Blue Pigments in Crushed Garlic Cloves: The Color-Forming Potential of Individual Amino Acids". *Acta Horticulturae,* No. 841: 491–4. (doi:10.17660/ActaHort..2009.841.66).

Cho, J., E.J. Lee, K.S. Yoo, S.K. Lee and B.S. Patil, 2009b. Identification of Candidate Amino Acids Involved in the Formation of Blue Pigments in Crushed Garlic Cloves (*Allium sativum* L.). *J. Food Sci.,* **74** (1): C11–C16. (doi:10.1111/j.1750-3841.2008.00986.x. ISSN 1750-3841. PMID 19200080).

Choi, D..-J., S,-J. Lee, M.-J. Kang, H.-S. Cho, N.-J. Sung and J.-H. Shin, 2008. Physicochemical characteristics of black garlic (*Allium sativum* L.). *J. Korean Soc. Food Sci., Nutr.,* **37**:465–71. (doi: 10.3746/jkfn.2008.37.4.465. www.ncbi.nlm.nih.gov/pmc/articles/PMC6152780/).

Choi, I.S., H.S. Cha and Y.S. Lee, 2014. Physicochemical and antioxidant properties of black garlic. *Molecules,* **19**:16811–23. (doi: 10.3390/molecules191016811. www.ncbi.nlm.nih.gov/pmc/articles/PMC6152780/).

Cundell, D. 2014. Herbal phytochemicals as immunomodulators. *Curr. Immunol. Rev.,* **10**:64-81. (10.2174/1573395510666140915213156.https://www.sciencedirect.com/science/article/pii/S2667142522000811).

Das, I. and T. Saha, 2008. Effect of garlic on lipid peroxidation and antioxidation enzymes in DMBA induced skin carcinoma. *Nutrition,* **25**(4): 459-71.

David, J.C. 1991. *Embellisia allii.* IMI descriptions of fungi and bacteria. CAB Abstract (No.108), Sheet 1078..

Dehghani, S., E. Alipoor, A. Salimzadeh, M. Yaseri, M. Hosseini, C. Feinle-Bisset and M.J. Hosseinzadeh-Attar, 2018. The effect of a garlic supplement on the pro-inflammatory adipocytokines, resistin and tumor necrosis factor-α, and on pain severity, in overweight or obese women with knee osteoarthritis. *Phytomedicine,* **48**:70–75. (doi: 10.1016/j.phymed.2018.04.060).

Diretto, G., A. Rubio-Moraga, J. Argandona, P. Castillo, L. Gomez-Gomez and O. Ahrazem, 2017. Tissue-specific accumulation of sulfur compounds and saponins in different parts of garlic cloves from purple and white ecotypes. *Molecules,* **22**:1359. (doi: 10.3390/molecules22081359).

Dogr125:- dogr.icar.gov.in/index.php?option=com_content&view=article&id=125&Itemid=176&lang=en

Dogr126:- dogr.icar.gov.in/index.php?option=com_content&view=article&id=126&Itemid=178&lang=en

Dogr129:- dogr.icar.gov.in/index.php?option=com_ content&view=article&id=129&Itemid=186&lang=en

Dogr133:- dogr.icar.gov.in/index.php?option=com_ content&view=article&id =133&Itemid=204&lang=en

Drobiova, H., M. Thomson, K. Al-Qattan, R. Peltonen-Shalaby, Z. Al-Amin and M. Ali, 2011. Garlic increases antioxidant levels in diabetic and hypertensive rats determined by a modified peroxidase method. *Evid. Based Complement. Altern. Med.*, (703049. doi: 10.1093/ecam/nep011.www.ncbi.nlm.nih.gov/ pmc/articles/PMC7146530/).

eagri.org/eagri50/HORT281/lec16.html.

Ebadi P., M.H. Karimi and Z. Amirghofran, 2014. .Plant components for immune modulation targeting dendritic cells: implication for therapy. *Immunotherapy,* **6**: 1037-53. (10.2217/imt.14.77. https://www.sciencedirect.com/science/ article/ pii/S2667142522000811).

Ebrahimzadeh-Bideskan, A.R., J. Hami, F. Alipour, H. Haghir, A.R. Fazel and A. Sadeghi, 2016. Protective effects of ascorbic acid and garlic extract against lead-induced apoptosis in developing rat hippocampus. *Metab. Brain Dis.*, **31**:1123–32. (doi: 10.1007/s11011-016-9837-7).

Edward, K.C. and C.G. Ohaegbu, 2012. The effect of ginger and garlic on the microbial load and shelf life of Kunun-zaki. J. *Appl. Pharm. Sci.*, **2**:150-3. (10.7324/ JAPS.2012.2538. https://www.sciencedirect.com/science/ article/pii/ S2667142522000811).

El-Ashmawy, N.E., E.G. Khedr, H.A. El-Bahrawy and H.M. Selim, 2016. Gastroprotective effect of garlic in indomethacin induced gastric ulcer in rats. *Nutrition*, **32**:849–54. (doi: 10.1016/j.nut.2016.01.010).

El-Kelawy, H.M., M.A. Mansour, R.E. El-Naggar, E. Nabila and E.M. Elkassas, 2017. Effect of garlic (*Allium sativum*) treatment on hematological, biochemical, hormonal and fertility parameters of male BOUSCAT rabbits. *Egypt. J. Rabbit Sci.*, **27**:341-58. (https://gain.fas.usda.gov/RecentGAINPublications/ Agricultural Biotechnology Annual_Ottawa_Canada_11-20-2018. pdf%0A10.1016/j.resourpol.2020.101869%0A; http://dx.doi.org/10.1016/j .jenvman.2017.06.039%0A; http://www.oecd.org/gov/regulatory-poli; https://www.sciencedirect.com/science /article/pii/S2667142522000811).

El-Saber, B.G., A.G. Magdy Beshbishy, L. Wasef, Y.H.A. Elewa, A. Al-Sagan, M.E. Abd El-Hack, A.E. Taha, Y. M. Abd-Elhakim and H. Prasad Devkota, 2020. Chemical Constituents and Pharmacological Activities of Garlic (*Allium sativum* L.): A Review. *Nutrients,* **24**{12(3)}:872. (doi: 10.3390/nu12030872. PMID: 32213941; PMCID: PMC7146530. www.ncbi.nlm.nih.gov/pmc/articles/ PMC7146530/).

en.wikipedia.org/wiki/Allium_tuncelianum

en.wikipedia.org/wiki/Black_garlic

en.wikipedia.org/wiki/Garlic.

en.wikipedia.org/wiki/Leveillula_taurica

en.wikipedia.org/wiki/Solo_garlic

Etewa, S.E. and S.M. Abaza, 2011. Herbal medicine and parasitic diseases. *Herb. Med. Parasites*, **4**: 3-14. (https://www.sciencedirect.com/science/article/pii/S2667142522000811).

Etoh, T. 1986. Fertility of the garlic clones collected in Soviet Central Asia. *J. Japan. Soc. hort. Sci.*, **55**:312-9.

Evans, M., A. Roberts and A. Rees, 2002. The future direction of cholesterol-lowering therapy. *Curr. Opin. Lipidol.*, **13**: 663-9.

Farmingindia: December 8, 2017. www.farmingindia.in/garlic-cultivation/

Farooqui, T. and A.A. Farooqui, 2017. Neuroprotective Effects of Garlic in Model Systems of Neurodegenerative Diseases. *Elsevier Inc.* (10.1016/B978-0-12-811959-4.00016-X). (https://www.sciencedirect.com/science/article/ pii/S2667142522000811).

Fasolino, I., A.A. Izzo, T. Clavel, B. Romano, D. Haller and F. Borrelli, 2015. Orally administered allyl sulfides from garlic ameliorate murine colitis. *Mol. Nutr. Food Res.*, **59**:434–42. (doi: 10.1002/mnfr.201400347).

Fratianni, F., R. Riccardi, P. Spigno, M.N. Ombra, A. Cozzolino, P. Tremonte, R. Coppola and F. Nazzaro, 2016. Biochemical characterization and antimicrobial and antifungal activity of two endemic varieties of garlic (*Allium sativum* L.) of the Campania region, southern Italy. *J. Med. Food.*, **19**:686–91. (doi: 10.1089/jmf.2016.0027).

Gallwitz, H., S. Bonse, A. Martinez-Cruz, I. Schlichting, K. Schumacher and R.L. Krauth-Siegel, 1999. Ajoene is an inhibitor and subversive substrate of human glutathione reductase and Trypanosoma cruzi trypanothione reductase: Crystallographic, kinetic, and spectroscopic studies. *J. Med. Chem.*, **42**:364–72. (doi: 10.1021/jm980471k.www.ncbi.nlm.nih.gov/pmc/articles/PMC7146530/).

Garcia-Villalon, A.L., S. Amor, L. Monge, N. Fernandez, M. Prodanov, M. Munoz, A.M. Inarejos-Garcia and M. Granado, 2016. *In vitro* studies of an aged black garlic extract enriched in S-allylcysteine and polyphenols with cardioprotective effects. *J. Funct. Foods.*, **27**:189–200. (doi: 10.1016/j.jff.2016.08.062).

Gargi, N. and A.N. Roy, 1988. Prevention and control of some post harvest fungal diseases of garlic bulbs. *Pesticides,* **22**:11-5.

Gatt, M.E., J. Strahilevitz, N. Sharon, D. Lavie, N. Goldschmidt, Y. Kalish, A. Gural and O.B. Paltiel, 2015. A randomized controlled study to determine the efficacy of garlic compounds in patients with hematological malignancies at risk for chemotherapy-related febrile neutropenia. Integr. *Cancer Ther.*, **14**:428–35. (doi: 10.1177/1534735415588928).

Gomaa, A.M.S., A.T. Abdelhafez and H.A Aamer, 2018. Garlic (*Allium sativum*) exhibits a cardioprotective effect in experimental chronic renal failure rat model by reducing oxidative stress and controlling cardiac Na+/K+-ATPase activity and Ca2+ levels. *Cell Stress Chaperones*, **23**:913–20. (doi: 10.1007/s12192-018-0898-x).

Gord, K.. 2019. Which Is Healthier, Raw or Cooked Garlic? www.livestrong.com/article/340975-what-are-the-health-benefits-of-raw-garlic-vs-cooked-garlic/

Gregrova, A., H. Cizkova, I. Bulantova, A. Rajchl and M. Voldrich, 2013. Characteristics of garlic of the Czech origin, *Czech J. Food Sci.*, **31**:581-8. (10.17221/539/2012-cjfs. https://www.sciencedirect.com/science/article/pii/S2667142522000811).

Gruhlke, M.C., I. Nwachwukwu, M. Arbach, A. Anwar, U. Noll and A.J. Slusarenko, 2010. Allicin from garlic, effective in controlling several plant diseases, is a reactive sulfur species (RSS) that pushes cells into apoptosis. *Proc. of the Modern Fungicides and Antifungal Compounds VI.* 16th International Reinhardsbrunn Symposium; Friedrichroda, Germany, held on. 25–29 April 2010. (www.ncbi.nlm.nih.gov/pmc/articles/PMC7146530/).

Guo, Y.J. 2014. Experimental study on the optimization of extraction process of garlic oil and its antibacterial effects. *Afr. J. Tradit. Complement. Altern. Med.*, **11**:411–4. (doi: 10.4314/ajtcam.v11i2.27).

Ha, A.W., T. Ying and W.K. Kim, 2015. The effects of black garlic (*Allium satvium*) extracts on lipid metabolism in rats fed a high fat diet. *Nutr. Res. Pract.*, **9**:30–6. (doi: 10.4162/nrp.2015.9.1.30).

Han, C.H., J.C. Liu, K.H. Chen, Y.S. Lin, C.T. Chen, C.T. Fan, H.L. Lee, D.Z. Liu and W.C. Hou, 2011. Antihypertensive activities of processed garlic on spontaneously hypertensive rats and hypertensive humans. *Bot. Stud.*, **52**:277–83.

Han, C.Y., S.H. Ki, Y.W. Kim, K. Noh, da Y. Lee, B. Kang, J.H. Ryu, R. Jeon, E.H. Kim and S.J. Hwang, 2011. Ajoene, a stable garlic by-product, inhibits high fat diet-induced hepatic steatosis and oxidative injury through LKB1-dependent AMPK activation. *Antioxid. Redox Signal.*, **14**:187–202. (doi: 10.1089/ars.2010.3190. www.ncbi.nlm.nih.gov/pmc/articles/PMC7146530/).

Hang, X. 2005. Isolation and identification of garlic polysaccharide. *Food Sci.*, **26**:48–51.

Hassouna, I., H. Ibrahim, F.A. Gaffar, I. El-Elaimy and H.A. Latif, 2015. Simultaneous administration of hesperidin or garlic oil modulates diazinon-induced hemato- and immunotoxicity in rats. *Immunopharmacol. Immunotoxicol.*, **37**:442–9. (doi: 10.3109/08923973.2015.1081932).

Helen O.N., A.O. Job and O. Oghenebrorhie, 2020. Antibiotic and antihelminthic effects of garlic bulbs (*Allium sativum*) and ginger rhizomes (*Zingiber officinale*) as growth promoters on broiler chickens in the tropics. *Adv. Anim. Vet. Sci.*,

8: 1019-27. (10.23959/sfmrl-1000016. https://www.sciencedirect.com/science/article/pii/S2667142522000811).

Hirst, K.K. 2021. Garlic Domestication - Where Did it Come from and When? *ThoughtCo* (Feb. 16). (thoughtco.com/garlic-domestication-where-and-when-169374).

Ho, X.L., S.Y. Tsen, M.Y. Ng, W.N. Lee, A. Low and W.M. Loke, 2016. Aged garlic supplement protects against lipid peroxidation in hypercholesterolemic individuals. *J. Med. Food,* **19**:931–7. (doi: 10.1089/jmf.2016.3693).

hort.extension.wisc.edu/articles/garlic-allium-sativum/

Imai, S., K. Akita, M. Tomotake and H. Sawada, 2006. Model Studies on Precursor System Generating Blue Pigment in Onion and Garlic. *J. agr. Food Chem.*, **54** (3): 848–52. (doi:10.1021/jf051980f. PMID 16448193).

indiaagronet.com/Horticulture/CONTENTS/Garlic.htm

iopscience.iop.org/article/10.1088/1757-899X/1053/1/012041/pdf#:~:text=Garlic%20(Allium%20sativum) %20is%20a,%2C%20allicin%2C%20ajoene%20and%20flavonoids.

Ipek, M., A. Ipek and W. Simon Philipp, 2008. Genetic characterization of *Allium tuncelianum*: An endemic edible *Allium* species with garlic odour. Archived 2011-07-21 at the Wayback Machine. *Sci Hort.*, **115**:409-15. (web.archive.org/web/20110721070234/http://ddr.nal.usda.gov/bitstream/10113/18219/1/IND44029827.pdf).

Jang, E.-K., J.-H. Seo and S.-P. Lee, 2008. Physiological activity and antioxidative effects of aged black garlic (*Allium sativum* L.) extract.. *Korean J. Food Sci. Technol.*, **40**:443–8. (www.ncbi.nlm.nih.gov/pmc/articles/PMC6152780/).

Jenderek, M.M. and R.M. Hannan, 2004. Variation in reproductive characteristics and seed production in the USDA garlic germplasm collection. HortSci., **39**:485-8.

Jenderek, M.M. and Y. Zewdie, 2005. Within and between family variability for important bulb and plant traits among sexually derived progenies of garlic. *HortSci.*, **40**:1234-6.

Jeong, Y.Y., J.H. Ryu, J.-H. Shin, M.J. Kang, J.R. Kang, J. Han and D. Kang, 2016. Comparison of anti-oxidant and anti-inflammatory effects between fresh and aged black garlic extracts. *Molecules,* **21**:430. (doi: 10.3390/molecules21040430. www.ncbi.nlm.nih.gov/pmc/articles/PMC6152780/).

Jiang, X.Y., X.S. Zhu, W.Z. Huang, H.Y. Xu, Z.X. Zhao, S.Y. Li, S.Z. Li, J.H. Cai and J.M. Cao, 2017. Garlic-derived organosulfur compound exerts antitumor efficacy via activation of MAPK pathway and modulation of cytokines in SGC-7901 tumor-bearing mice. *Int. Immunopharmacol.*, **48**:135–45. (doi: 10.1016/j.intimp.2017.05.004).

Jones, M.G. and Jill Hughes, 2004. Biosynthesis of the flavour precursors of onion and garlic. *J. exper. Bot.*, **55** (404): 1903–18. (doi:10.1093/jxb/erh138. PMID 15234988).

Jung, I. and H. Sohn, 2014. Antioxidation, antimicrobial and antithrombosis activities of aged black garlic *(Allium sativum* L.). *Korean J. Microbiol. Biotechnol.* (doi: 10.4014/kjmb.1407.07002. www.ncbi.nlm.nih.gov/pmc/articles/PMC6152780/).

Jung, Y., H. Park, H.Y. Zhao, R. Jeon, J.H. Ryu and W.Y. Kim, 2014. Systemic approaches identify a garlic-derived chemical, Z-ajoene, as a glioblastoma multiforme cancer stem cell-specific targeting agent. *Mol. Cells*, **37**:547–53. (doi: 10.14348/molcells.2014.0158).

Kamenetsky, R., I.L. Shafir, F. Khassanov, C. Kik, A.W. van Heusden, M. Vrielink-van Ginkel, K. Burger-Meijer, J. Auger, I. Arnault and H.D. Rabinowitch, 2005. Diversity in fertility potential and organo-sulphur compounds among garlics from Central Asia. *Biodivers Conserv.*, **14** (2): 281–95.

Kang, J.S., S.O. Kim, G.Y. Kim, H.J. Hwang, B.W. Kim, Y.C. Chang, W.J. Kim, C.M. Kim, Y.H. Yoo and Y.H. Choi, 2016. An exploration of the antioxidant effects of garlic saponins in mouse-derived C2C12 myoblasts. Int *J. Mol. Med.*, **37**:149–56. (doi: 10.3892/ijmm.2015.2398).

Kang, O.-J. 2016. Physicochemical characteristics of black garlic after different thermal processing steps. *Prev. Nutr. Food Sci.*, **21**:348. (doi: 10.3746/pnf.2016.21.4.348. www.ncbi.nlm.nih.gov/pmc/articles/PMC6152780/).

Kaschula, C.H., R. Hunter, J. Cotton, R. Tuveri, E. Ngarande, K. Dzobo, G. Schafer, V. Siyo, D. Lang and D.A. Kusza, 2016. The garlic compound ajoene targets protein folding in the endoplasmic reticulum of cancer cells. *Mol. Carcinog.*,**55**:1213–28. (doi: 10.1002/mc.22364).

Kaur, G.J. and D.S. Arora, 2009. Antibacterial and phytochemical screening of *Anethum graveolens, Foeniculum vulgare and Trachyspermum ammi. BMC Compl. Alt. Med.*, **9**: 30.

Keophiphath, M., F. Priem, I. Jacquemond-Collet, K. Clement and D. Lacasa, 2009. 1,2-Vinyldithiin from garlic inhibits differentiation and inflammation of human preadipocytes. *J. Nutr.,* **139**: 2055–60. (doi: 10.3945/jn.109.105452. www.ncbi.nlm.nih.gov/pmc/articles/PMC7146530/).

Khar, A., Asha Devi, V.A. Mahajan and K.E. Lawande, 2006. Genetic divergence analysis in elite lines of garlic (*Allium sativum* L.). *J. Mah. Agri. Univ.*, **31**:52-5.

Khatua, T.N., R.M. Borkar, S.A. Mohammed, A.K. Dinda, R. Srinivas and S.K. Banerjee, 2017. Novel sulfur metabolites of garlic attenuate cardiac hypertrophy and remodeling through induction of Na+/K+-ATPase expression. *Front. Pharmacol.*, **8**:18. (doi: 10.3389/fphar.2017.00018).

Kim, H.N., S.G. Kang, Y.K. Roh, M.K. Choi and S.W. Song, 2017. Efficacy and safety of fermented garlic extract on hepatic function in adults with elevated serum gamma-glutamyl transpeptidase levels: A double-blind, randomized, placebo-controlled trial. *Eur.. J. Nutr.*, **56**:1993–2002. (doi: 10.1007/s00394-016-13 18-6).

Kim, J.H., S.H. Nam, C.W. Rico and M.Y. Kang, 2012. A comparative study on the antioxidative and anti-allergic activities of fresh and aged black garlic extracts. *Int. J. Food Sci. Technol.*, **47**:1176–1182. (doi: 10.1111/j.1365-2621.2012.02957.x. www.ncbi.nlm.nih.gov/pmc/articles/PMC6152780/).

Kim, J.S., O.J. Kang and O.C. Gweon, 2013. Comparison of phenolic acids and flavonoids in black garlic at different thermal processing steps. *J. Funct. Foods.*, **5**:80–86. (doi: 10.1016/j.jff.2012.08.006).

Kimura, S., Y.C. Tung, M.H. Pan, N.W. Su, Y.J. Lai and K.C. Cheng, 2017. Black garlic: A critical review of its production, bioactivity, and application. *J. Food Drug Anal.*, **25**:62–70. (doi: 10.1016/j.jfda.2016.11.003).

Ko, J.W., S.H. Park, I.C. Lee, S.M. Lee, I.S. Shin, S.S. Kang, C. Moon, S.H. Kim, J.D. Heo and J.C. Kim, 2016. Protective effects of garlic oil against 1,3-dichloro-2-propanol-induced hepatotoxicity: Role of CYP2E1 and MAPKs. *Mol. Cell. Toxicol.*, **12**:185–95. (doi: 10.1007/s13273-016-0023-0).

Koca, I., B. Karadeniz and B. Tekguler, 2016. Aroma components of green kormen (*Allium scorodoprasum L. spp. rotundum*) and garlic (*Allium sativum*) plants. *Acta Hortic.*, No. 1143, pp. 291-6. (DOI: 10.17660/ActaHortic.2016 .1143.40. doi.org/10.17660/ActaHortic.2016.1143.40; www.actahort.org/ books/1143/1143_40.htm).

Kodera, Y., M. Ushijima, H. Amano, J. Suzuki and T. Matsutomo, 2017. Chemical and biological properties of S-1-propenyl-l-cysteine in aged garlic extract. *Molecules,* **22**:570. (doi: 10.3390/molecules22040570).

Kuda, T., A. Iwai and T. Yano, 2004. Effect of red pepper *Capsicum annuum* var. *conoides* and garlic *Allium sativum* on plasma lipid levels and cecal microflora in mice fed beef tallow. *Food Chem. Toxicol.*, **42**:1695–1700. (doi: 10.1016/j. fct.2004.06.007. www.ncbi.nlm.nih.gov/pmc/articles/PMC7146530/).

Kushwaha, Sadhna, Vandna Pathak and I.P. Tripathi, 2018. Phytochemical screening of some herbs ginger, garlic and onion. *World J. pharmaceut. Res.*, **7**(7):2243-7. (wjpr.s3.ap-south-1.amazonaws.com/article_issue/1573029146.pdf).

Lai, Y.S., W.C. Chen, C.T. Ho, K.H. Lu, S.H. Lin, H.C. Tseng, S.Y. Lin and L.Y. Sheen, 2014. Garlic essential oil protects against obesity-triggered nonalcoholic fatty liver disease through modulation of lipid metabolism and oxidative stress. *J. Agric. Food Chem.*, **62**:5897–5906. (doi: 10.1021/jf500803c).

Lanzotti, V., G. Bonanomi and F. Scala, 2013. What makes *Allium* species effective against pathogenic microbes? *Phytochem. Rev.*, **12**:751–72. (doi: 10.1007/ s11101-013-9295-3).

Latha, S., N. Sathyanaryana and O.R. Reddy, 2007. Interception of *Embellisia allii* in garlic bulbs imported from China. *Ind. Phytopath.*, **60**:125-7.

Lawande, K.E., Anil Khar, V. Mahajan, P.S. Srinivas, V. Sankar and R.P. Singh, 2009. Onion and garlic research in India. *J. Hort. Sci.*, **4**(2): 91-119. Directorate of Onion and Garlic Research, Rajgurunagar, Pune, Maharashtra. (www.researchgate.net/publication/332143333_Onion_and_garlic_research_in_India).

Lee, E.J., Y.H. Rezenom, D.H. Russell, B.S. Patil and K.S. Yoo, 2012. Elucidation of chemical structures of pink-red pigments responsible for 'pinking' in macerated onion (*Allium cepa* L.) using HPLC–DAD and tandem mass spectrometry. *Food Chem.*, **131** (3): 852–61. (doi:10.1016/j.foodchem.2011.09.059).

Lee, H.S., W.C. Lim, S.J. Lee, S.H. Lee, H.J. Yu, J.H. Lee and H.Y. Cho, 2016a. Hepatoprotective effects of lactic acid-fermented garlic extract against acetaminophen-induced acute liver injury in rats. *Food Sci. Biotechnol.*, **25**:867–73. (doi: 10.1007/s10068-016-0143-2).

Lee, H.S., W.C. Lim, S.J. Lee, S.H. Lee, J.H. Lee and H.Y. Cho, 2016b. Antiobesity effect of garlic extract fermented by *Lactobacillus plantarum* bl2 in diet-induced obese mice. *J. Med. Food.*, **19**:823–9. (doi: 10.1089/jmf.2016.3674).

Lee, K.C., C.C. Teng, C.H. Shen, W.S. Huang, C.C. Lu, H.C. Kuo and S.Y. Tung, 2018. Protective effect of black garlic extracts on tert-butyl hydroperoxide-induced injury in hepatocytes via a c-Jun N-terminal kinase-dependent mechanism. *Exp. Ther. Med.*, **15**:2468–74. (doi: 10.3892/etm.2018.5719).

Li, W.R., Q.S. Shi, Q. Liang, X.M. Huang and Y.B. Chen, 2014. Antifungal effect and mechanism of garlic oil on *Penicillium funiculosum*. Appl. *Microbiol. Biot.*, **98**:8337–46. (doi: 10.1007/s00253-014-5919-9).

Li, W.R., Q.S. Shi, H.Q. Dai, Q. Liang, X.B. Xie, X.M. Huang, G.Z. Zhao and L.X. Zhang, 2016. Antifungal activity, kinetics and molecular mechanism of action of garlic oil against *Candida albicans. Sci. Rep.*, **6**:22805. (doi: 10.1038/srep22805).

Li, M., Y.X. Yan, Q.T. Yu, Y. Deng, D.T. Wu, Y. Wang, Y.Z. Ge, S.P. Li and J. Zhao, 2017. Comparison of immunomodulatory effects of fresh garlic and black garlic polysaccharides on RAW 264.7 macrophages. *J. Food Sci.*, **82**:765–71. (doi: 10.1111/1750-3841.13589).

Liang, T.F., F.F. Wei, Y. Lu, Y. Kodani, M. Nakada, T. Miyakawa and M. Tanokura, 2015. Comprehensive nmr analysis of compositional changes of black garlic during thermal processing. *J. agric. Food Chem.*, **63**:683–91. (doi: 10.1021/jf504836d).

Liu, S.G., P.Y. Ren, G.Y. Wang, S.X. Yao and X.J. He, 2015a. Allicin protects spinal cord neurons from glutamate-induced oxidative stress through regulating the heat shock protein 70/inducible nitric oxide synthase pathway. *Food Funct.*, **6**:321–30. (doi: 10.1039/C4FO00761A. www.ncbi.nlm.nih.gov/pmc/articles/PMC7146530)/

Liu, Y., J. Yan, X. Han and W. Hu, 2015b. Garlic-derived compound S-allylmercaptocysteine (SAMC) is active against anaplastic thyroid cancer cell line 8305C (HPACC). *Technol. Health Care*, **23**:S89–S93. (doi: 10.3233/thc-150936. www.ncbi.nlm.nih.gov/pmc/articles/PMC7146530/).

Liu, J., W. Guo, M.L. Yang, L.X. Liu, S.X. Huang, L. Tao, F. Zhang and Y.S. Liu, 2018. Investigation of the dynamic changes in the chemical constituents of Chinese 'laba' garlic during traditional processing. *Rsc. Adv.*, **8**:41872–83. (doi: 10.1039/C8RA09657K).

Lu, X.M., N.Y. Li, X.G. Qiao, Z.C. Qiu and P.L. Liu, 2018. Effects of thermal treatment on polysaccharide degradation during black garlic processing. *Lwt-Food Sci. Technol.,* **95**:223–9.(doi: 10.1016/j.lwt.2018.04.059).

Lukes, T.M. 1986. Factors Governing the Greening of Garlic Puree. *J. Food Sci.*, **51** (6): 1577. (doi:10.1111/j.1365-2621.1986.tb13869.x. ISSN 1750-3841).

Maaß, H.I. and M. Klaas, 1995. Infraspecific differentiation of garlic (*Allium sativum* L.) by isozyme and RAPD markers. *Theoret. and Appl. Gen.*, **91**(1): 89-97. (https://doi.org/10.1007/BF00220863. https://thescipub.com/pdf/ojbsci.2023.33.43.pdf).

Macpherson, L.J., B.H. Geierstanger, Veena Viswanath, Michael Bandell, R. Eid Samer, Hwang SunWook and Patapoutian Ardem, 2005. The Pungency of Garlic: Activation of TRPA1 and TRPV1 in Response to Allicin (PDF). *Curr. Biol.*, **15** (10): 929–34. (doi:10.1016/j.cub.2005.04.018. PMID 15916949. S2CID 163993). Mahajan, V. 2019. Garlic. In: Handbook of Horticulture, Vol. 1, pp. 544-9, ICAR, New Delhi.

Mansingh, D.P., N. Dalpati, V.K. Sali and A.H.R. Vasanthi, 2018. Alliin the precursor of allicin in garlic extract mitigates proliferation of gastric adenocarcinoma cells by modulating apoptosis. *Pharmacogn. Mag.*, **14**:S84–S91.

Marina, S., M.S. Butt, B. Masood, A. Shehzad and S. Tanweer, 2014. Chemical and mineral analysis of garlic: a golden herb. *Pakistan J. Food Sci.*, **24**(1): 108-10.

Mathew, D., F. Yitzhak, D.R. Haim and Rina Kamenetsky. 2011. Effect of long photoperiod on the reproductive and bulbing processes in garlic (*Allium sativum* L.) genotypes. *Environ. and experim. Bot.*, **71**(2): 166–73.

Matsuura, N., Y. Miyamae, K. Yamane, Y. Nagao, Y. Hamada, N. Kawaguchi, T. Katsuki, K. Hirata, S.I. Sumi and H. Ishikawa, 2006. Aged garlic extract inhibits angiogenesis and proliferation of colorectal carcinoma cells(1–3). *J. Nutr.*, **136**:842S–6S. (doi: 10.1093/jn/136.3.842S).

Metwally, D.M., E.M. Al-Olayan, M. Alanazi, S.B. Alzahrany and A. Semlali, 2018. Antischistosomal and anti-inflammatory activity of garlic and allicin compared with that of praziquantel *in vivo. BMC Complement. Altern. Med.*, **18**:135. (doi: 10.1186/s12906-018-2191-z).

Miltonprabu, S., N.C. Sumedha and P. Senthilraja, 2017. Diallyl trisulfide, a garlic polysulfide protects against as-induced renal oxidative nephrotoxicity, apoptosis and inflammation in rats by activating the Nrf2/ARE signaling pathway. *Int. Immunopharmacol.*, **50**:107–20. (doi: 10.1016/j.intimp.2017.06.011).

Mohamed, E.H., A.A.A. Baiomy, Z.S. Ibrahim and M.M. Soliman, 2016. Modulatory effects of levamisole and garlic oil on the immune response of wistar rats:

Biochemical, immunohistochemical, molecular and immunological study. *Mol. Med. Rep.*, **14**:2755–63. (doi: 10.3892/mmr.2016.5551).

Morihara, N., A. Hino, T. Yamaguchi and T. Suzuki, 2016. Aged garlic extract suppresses the devegopment of atherosclerosis in apollipoprotein E-knockout mice. *J. Nutr.*,**146**:460S–63S. (doi: 10.3945/jn.114.206953).

Morihara, N., A. Hino, S. Miki, M. Takashima and J. Suzuki, 2017. Aged garlic extract suppresses inflammation in apolipoprotein E-knockout mice. *Mol. Nutr. Food Res.*, **61**:1700308. (doi: 10.1002/mnfr.201700308).

mpkv.ac.in/Uploads/Research/12-Vegetable_20200108060120.pdf

Nagella, P., M. Thiruvengadam, A. Ahmad, J.Y. Yoon and I.M. Chung, 2014. Composition of polyphenols and antioxidant activity of garlic bulbs collected from different locations of Korea. *Asian J. Chem.*, **26**:897–902. (doi: 10.14233/ajchem.2014.16143A).

Naji, K.M., E.S. Al-Shaibani, F.A. Alhadi, S.A. Al-Soudi and M.R. D'Souza, 2017. Hepatoprotective and antioxidant effects of single clove garlic against ccl4-induced hepatic damage in rabbits. *BMC Complement. Altern. Med.*, **17**:411. (doi: 10.1186/s12906-017-1916-8).

Nasiri, A., N. Ziamajidi, R. Abbasalipourkabir, M.T. Goodarzi, M. Saidijam, H. Behrouj and S.S. Asl, 2017. Beneficial effect of aqueous garlic extract on inflammation and oxidative stress status in the kidneys of type 1 diabetic rats. *Indian J. Clin. Biochem.*, **32**:329–36. (doi: 10.1007/s12291-016-0621-6).

Nasr, A.Y. and H.A.M. Saleh, 2014. Aged garlic extract protects against oxidative stress and renal changes in cisplatin-treated adult male rats. *Cancer Cell Int.*, **14**:92. (doi: 10.1186/s12935-014-0092-x).

nhrdf.com/pPlantProtection_G.html

Nwinuka, N.M., G.O. Ibeh and G.L. Ekeke, 2005.. Proximate composition and levels of some toxicants in four commonly consumed spices. *J. Appl. Sci. envir. Manag.*, **9**(1): 150- 5.

Odebunmi, E.O., O.O. Oluwaniyi and M.O. Bashiru, 2009. Comparative proximate analysis of some food condiments. *J. App. Sci. Res.*, **2**(1): 1-3.

Okwu, D.E. 2005. Phytochemical, vitamins and mineral contents of two Nigerian medicinal plants. *Int. J. Mol. Med. Adv. Sci.*, **1**: 375-81.

Okwu, D.E. and I.V. Emenike, 2006. Evaluation of the phytonutrients and vitamin contents of citrus fruits. *Int. J. Mol. Med. Adv. Sci.*, **2**(1): 1-6.

Okwu, D.E. and F.U. Nnamdi, 2008. Evaluation of the chemical composition of *Dacryodes edulis* and *Raphia hookeri* Mann and Wendl exudates used in herbal medicine in South Eastern Nigeria. *Afr. J. Trad. Comp. Alt. Med.*, **5**(2): 194-200.

Okwu. D.E. and D.O. Omodamiro, 2005. Effect of hexane extract and phytochemical content of *Xylopia aethiopica* and *Ocimum gratissimum* on uterus of guinea pig. *Bio-Res.*, **3**(2): 40- 44.

Oloyede, O.I. 2005. Chemical profile of unripe pulp of *Carica papaya. Pak. J. Nutr.*, **4**(6): 379-81.

Pandey, Anjula; Pavan Kumar Malav; K.M. Rai and S.P. Ahlawat, 2022. Genus *Allium* L. of the Indian Region: A Field Guide for Germplasm Collection and Identification. ICAR-National Bureau of Plant Genetic Resources, New Delhi, India, 109p + i-vi. (http://www.nbpgr.ernet.in/Downloadfile.aspx?EntryId=9197).

Park, M.Y., Yi-NaRi, Lee-HanYong, Kim-SungTae, Kim- MinKyun, Park-JuHyun, Kim-JuKon, Lee-JonSeob, Cheong-JongJoo and Choi-YangDo, 2002. Generation of chlorsulfuron-resistant transgenic garlic plants (*Allium sativum* L.) by particle bombardment. *Mol.-Breed.*, **9**:171-81.

Park, S.Y., R. Seetharaman, M.J. Ko, D.Y. Kim, T.H. Kim, M.K. Yoon, J.H. Kwak, S.J. Lee, Y.S. Bae and Y.W. Choi, 2014. Ethyl linoleate from garlic attenuates lipopolysaccharide-induced pro-inflammatory cytokine production by inducing heme oxygenase-1 in RAW 264.7 cells. *Int. Immunopharmacol.*, **19**: 253–61. (doi: 10.1016/j.intimp.2014.01.017).

Park, B.M., S.A. Cha, H.Y. Kim, D.K. Kang D.K., K. Yuan, H. Chun, S.W. Chae and S.H. Kim, 2016. Fermented garlic extract decreases blood pressure through nitrite and sGC-cGMP-PKG pathway in spontaneously hypertensive rats. *J. Funct. Foods,* **22**:156–65. (doi: 10.1016/j.jff.2016.01.034).

Park, B.M., H. Chun, S.W. Chae and S.H. Kim, 2017. Fermented garlic extract ameliorates monocrotaline-induced pulmonary hypertension in rats. *J. Funct. Foods,* **30**:247–53. (doi: 10.1016/j.jff.2017.01.024).

Percival, S.S. 2016. Aged garlic extract modifies human immunity. *J. Nutr.,* **146**:433S–436S. (doi: 10.3945/jn.115.210427).

pharmacy180.com

phytojournal.com

Plantvillage:- plantvillage.psu.edu/topics/garlic/infos#!

Pooler, M.R. and P.W. Simon, 1993. Characterization and classification of isozyme and morphological variation in a diverse collection of garlic clones. *Euphytica,* **68**:121–30.

POWO, 2021. *Allium longicuspis Regel*. Plants of the World Online. Royal Botanic Gardens, Kew. (https://powo.science.kew.org/taxon/urn:lsid:ipni.org:names:528333-1. Retrieved 20 February, 2021).

POWO, 2022a. *Allium macrochaetum* Boiss. & Hausskn. Plants of the World Online. Royal Botanic Gardens, Kew. (https://powo.science.kew.org/taxon/urn:lsid:ipni.org:names:528363-1. Retrieved 20 February, 2022).

POWO, 2022b. *Allium truncatum* (Feinbrun) Kollmann & D.Zohary. Plants of the World Online. Royal Botanic Gardens, Kew. https://powo.science.kew.org/taxon/urn:lsid:ipni.org:names:942813-1. Retrieved 20 February, 2022.

Qidwai, W. and T. Ashfaq, 2013. Role of garlic usage in cardiovascular disease prevention: An evidence-based approach. *Evid. Based Complement. Altern. Med.* (doi: 10.1155/2013/125649. www.ncbi.nlm.nih.gov/pmc/articles/PMC7146530/).

Qiu, S.L., J. Chen, T. Qin, Y.L. Hu, D.Y. Wang, Q. Fan, C.S. Zhang, X.Y. Chen, X.L. Chen and C. Liu, 2014. Effects of selenylation modification on immune-enhancing activity of garlic polysaccharide. *PLoS One*, **9**:e86377. (doi: 10.1371/journal.pone.0086377).

Qiu, Z., N. Li, X. Lu, Z. Zheng, M. Zhang and X. Qiao, 2018. Characterization of microbial community structure and metabolic potential using Illumina MiSeq platform during the black garlic processing. *Food Res. Intern.* (Ottawa, Ont.), **106**: 428–38. (doi:10.1016/j.foodres.2017.12.081. PMID 29579944).

Rabe, S.Z.T., T. Ghazanfari, Z. Siadat, M. Rastin, S.Z.T. Rabe and M. Mahmoudi, 2015. Anti-inflammatory effect of garlic 14-kDa protein on LPS-stimulated-J774A.1 macrophages. *Immunopharmacol. Immunotoxicol.*, **37**:158–64. (doi: 10.3109/08923973.2015.1005229).

Ram, D., Timila Gyanu Manadhar and Sharada Joshi, 2007. Bulb canker of garlic caused by *Embellisia allii,* newly found in Nepal. Nepal Agric. Res. J., **8**: 127-9.

Ranjan, S., V.S. Jadon and S. Gupta, 2014. Studies on the anti-inflammatory and analgesic activities of *Allium stracheyi*, a medicinal plant of alpine Himalayas of Uttarakhand. BDDD, **1**: 27-34. (https://www.sciencedirect.com/science/article/pii/S2667142522000811).

Rath, G.C. and G.N. Mohanty, 1986. *Fusarium rot* of stored garlic. *Ind. Phytopath.*, **39**:614-5.

Ryu, J.H. and D. Kang, 2017. Physicochemical Properties, Biological Activity, Health Benefits, and General Limitations of Aged Black Garlic: A Review. *Molecules,* **22**(6):919. (doi: 10.3390/molecules22060919. PMID: 28587168; PMCID: PMC6152780. www.ncbi.nlm.nih.gov/pmc/articles/PMC6152780/).

Salunkhe, D.K., S.S. Kadam (eds.), 1998. Handbook of Vegetable Science and Technology: Production, Composition, Storage and Processing, 397p. Marcel Dekker (ISBN 978-0-8247-0105-5).

Sasaki, J.-I., C. Lu, E. Machiya, M. Tanahashi and K. Hamada, 2007. Processed black garlic (*Allium sativum*) extracts enhance anti-tumor potency against mouse tumors. *Med. Aromat. Plant Sci. Biotechnol.*,**1**:278–81. (www.ncbi.nlm.nih.gov/pmc/articles/PMC6152780/).

Saud, S.M., W.D. Li, Z. Gray, M.S. Matter, N.H. Colburn, M.R. Young and Y.S. Kim, 2016. Diallyl disulfide (DADS), a constituent of garlic, inactivates NF-κB and prevents colitis-induced colourectal cancer by inhibiting GSK-3β. *Cancer Prev. Res.*, **9**:607–15. (doi: 10.1158/1940-6207.CAPR-16-0044).

Sawahel, W.A. 2002. Stable genetic transformation of garlic plants using particle bombardment. *Cellular and Mol.r Biol. Letters*, **7**:49-59.

Semuyaba, I., A.A. Safiriyu, E.A. Tiyo and R.F. Niurka, 2017. Memory improvement effect of ethanol garlic (*Allium sativum*) extract in streptozotocin-nicotinamide induced diabetic wistar rats is mediated through increasing of hippocampal sodium-potassium ATPase, glutamine synthetase, and calcium ATPase activities. *Evid. Based Complement. Altern. Med.* (doi: 10.1155/2017/3720380).

Shang, A., S.Y. Cao, X.Y. Xu, R.Y. Gan, G.Y. Tang, H. Corke, V. Mavumengwana and H.B. Li, 2019. Bioactive compounds and biological functions of garlic (*Allium sativum* L.). *Foods,* **8**(7):246. (doi: 10.3390/foods8070246. PMID: 31284512; PMCID: PMC6678835. www.ncbi.nlm.nih.gov/pmc/articles/PMC6678835/#:~:text=Garlic%20contains%20diverse%20bioactive%20compounds,and%20S%2Dallyl%2Dcysteine).

Shashidhar, T.R. and P.R. Dharmatti, 2005. Genetic divergence studies in garlic. *Karnataka J. Hort.,* **1**:12-5.

Shemesh-Mayer, E. and R. Kamenetsky Goldstein, 2018.. Recent advances in sexual propagation and breeding of garlic. *Hort Rev.,* **46**: 1-38. (https://doi.org/10.1002/9781119521082.ch1. https://thescipub.com/pdf/ojbsci.2023.33.43.pdf.).

Shi, L.M., Q.L. Lin, X.H. Li, Y. Nie, S.G. Sun, X.Y. Deng, L. Wang, J. Lu, Y.P. Tang and F.J. Luo, 2017. Alliin, a garlic organosulfur compound, ameliorates gut inflammation through MAPK-NF-κB/AP-1/STAT-1 inactivation and PPAR-7 activation. *Mol. Nutr. Food Res.,* **61**:1601013. (doi: 10.1002/mnfr.201601013).

Shin, J.-H., D.-J. Choi, S.-J. Lee, J.-Y. Cha, J.-G. Kim and N.-J. Sung, 2008. Changes of physicochemical components and antioxidant activity of garlic during its processing. *J. Life Sci.,* **18**:1123–31. (doi: 10.5352/JLS.2008.18.8.1123).

Shin, S.S., J.H. Song, B. Hwang, D.H. Noh, S.L. Park, W.T. Kim, S.S. Park, W.J. Kim and S.K. Moon, 2017. HSPA6 augments garlic extract-induced inhibition of proliferation, migration and invasion of bladder cancer EJ cells; implication for cell cycle dysregulation, signaling pathway alteration, and transcription factor-associated mmp-9 regulation. *PLoS ONE,* **12**:e0171860. (doi: 10.1371/journal.pone.0171860).

Siddiqui, N.A., S. Haider, M. Misbah-ur-Rehman and T. Perveen, 2016. Role of herbal formulation of garlic on lipid profile in patients with type 2 diabetes related dyslipidemia. Pak. Heart J., **49**:146–50.

Singh, R.P. and A, Khar, 2007. First report of *Embellisia alli* from garlic. *National Research Centre for Onion and Garlic News,* **11**:4.

Sohn, C.W., H. Kim, B.R. You, M.J. Kim, H.J. Kim, J.Y. Lee, D.E. Sok, J.H. Kim, K.J. Lee and M.R. Kim, 2012. High temperature- and high pressure-processed garlic improves lipid profiles in rats fed high cholesterol diets. *J. Med. Food,* **15**:435–40. (doi: 10.1089/jmf.2011.1922).

Stanek Amiel, 2016. Chefs Are Going Crazy for Black Garlic (and You Will, Too). *Bon Appetit.* Retrieved 2019-09-30.

Suddek, G.M. 2014. Allicin enhances chemotherapeutic response and ameliorates tamoxifen- induced liver injury in experimental animals. *Pharm. Biol.*, **52**:1009–14. (doi: 10.3109/13880209.2013.876053).

Sultana, M.R., P.K. Bagul, P.B. Katare, S.A. Mohammed, R. Padiya and S.K. Banerjee, 2016. Garlic activates SIRT-3 to prevent cardiac oxidative stress and mitochondrial dysfunction in diabetes. *Life Sci.*, **164**:42–51. (doi: 10.1016/j.lfs.2016.08.030).

Sun, Y.E. and W.D. Wang, 2018. Changes in nutritional and bio-functional compounds and antioxidant capacity during black garlic processing. *J. Food Sci. Technol.*, **55**:479–88. (doi: 10.1007/s13197-017-2956-2).

Supakul, L., H. Pintana, N. Apaijai, S. Chattipakorn, K. Shinlapawittayatorn and N. Chattipakorn, 2014. Protective effects of garlic extract on cardiac function, heart rate variability, and cardiac mitochondria in obese insulin-resistant rats. *Eur. J. Nutr.*, **53**:919–28. (doi: 10.1007/s00394-013-0595-6).

Szychowski, K.A., K. Rybczynska-Tkaczyk, K. Gawel-Beben, M. Swieca, M. Karas, A. Jakubczyk, M. Matysiak, U.E. Binduga and J. Gminski, 2018. Characterization of active compounds of different garlic (*Allium sativum* L.) cultivars. *Pol. J. Food Nutr. Sci.*, **68**:73–81. (doi: 10.1515/pjfns-2017-0005).

Tabari, M.A. and S. Ebrahimpour, 2014. Effect of aged garlic extract on immune responses to experimental fibrosarcoma tumor in BALB/c mice. *Indian J. Cancer.*, **51**:609–13. (doi: 10.4103/0019-509X.175359).

Takashima, M., Y. Kanamori, Y. Kodera, N. Morihara and K. Tamura, 2017. Aged garlic extract exerts endothelium-dependent vasorelaxant effect on rat aorta by increasing nitric oxide production. *Phytomedicine*, **24**:56–61. (doi: 10.1016/j.phymed.2016.11.016).

Talib, W.H. 2017. Consumption of garlic and lemon aqueous extracts combination reduces tumor burden by angiogenesis inhibition, apoptosis induction, and immune system modulation. *Nutrition*, **43**:89–97. (doi: 10.1016/j.nut.2017.06.015).

Tanamai, J., S. Veeramanomai and N. Indrakosas, 2004. The efficacy of cholesterol-lowering action and side effects of garlic enteric coated tablets in man. *J. Med. Assoc. Thail.*, **87**:1156–61.

Thorajak, P., W. Pannangrong, J.U. Welbat, W. Chaijaroonkhanarak, K. Sripanidkulchai and B. Sripanidkulchai, 2017. Effects of aged garlic extract on cholinergic, glutamatergic and gabaergic systems with regard to cognitive impairment in a β-induced rats. *Nutrients*, **9**:686. (doi: 10.3390/nu9070686).

Toledano-Medina, M.A., J. Perez-Aparicio, R. Moreno-Rojas and T. Merinas-Amo, 2016. Evolution of some physicochemical and antioxidant properties of black garlic whole bulbs and peeled cloves. *Food Chem.*, **199**:135–9. (doi: 10.1016/j.foodchem.2015.11.128. www.ncbi.nlm.nih.gov/pmc/articles/PMC6152780/).

Torres-Palazzolo, C., D. Ramirez, D. Locatelli, W. Manucha, C. Castro and A. Camargo, 2018. Bioaccessibility and permeability of bioactive compounds in raw and cooked garlic. *J. Food Compos. Anal.*, **70**:49–53. (doi: 10.1016/j.jfca.2018.03.008).

Tripathi, S. and H.N. Mishra, 2009. Nutritional changes in powdered red pepper upon *in vitro* infection of *Aspergillus flavus. Braz. J. Microbiol.*, **40**(1): 139-44.

Ushijima, M., M. Takashima, K. Kunimura, Y. Kodera, N. Morihara and K. Tamura, 2018. Effects of S-1-propenylcysteine, a sulfur compound in aged garlic extract, on blood pressure and peripheral circulation in spontaneously hypertensive rats. *J. Pharm. Pharmacol.*, **70**:559–65. (doi: 10.1111/jphp.12865).

vikaspedia.in (a): vikaspedia.in/agriculture/crop-production/integrated-pest-managment/ipm-for-spice-crops/ipm-strategies-for-garlic/crop-stage-wise-ipm

Wallock-Richards, D., C.J. Doherty, L. Doherty, D.J. Clarke, M. Place, J.R.W. Govan and D.J. Campopiano, 2014. Garlic revisited: Antimicrobial activity of allicin-containing garlic extracts against Burkholderia cepacia complex. *PLoS ONE,* **9**:e112726. (doi: 10.1371/journal.pone.0112726).

Wang, W., J.W. Cheng and Y.Z. Zhu, 2015. The JNK signaling pathway is a novel molecular target for S-propargyl-l-cysteine, a naturally-occurring garlic derivatives: Link to its anticancer activity in pancreatic cancer in vitro and *in vivo. Curr. Cancer Drug Targets,* **15**:613–23. (doi: 10.2174/1568009615666150602143943).

Wang, J., X.M. Zhang, H.L. Lan and W.J. Wang, 2017. Effect of garlic supplement in the management of type 2 diabetes mellitus (T2DM): A meta-analysis of randomized controlled trials. *Food Nutr. Res.*, **61**:1377571. (doi: 10.1080/16546628.2017.1377571).

Wang, Y.C., M. Guan, X. Zhao and X.L. Li, 2018. Effects of garlic polysaccharide on alcoholic liver fibrosis and intestinal microflora in mice. *Pharm. Biol.*, **56**:325–332. (doi: 10.1080/13880209.2018.1479868).

Wei, Z.H., Y.L. Shan, L. Tao, Y.P. Liu, Z.J. Zhu, Z.G. Liu, Y.Y. Wu, W.X. Chen, A.Y. Wang and Y. Lu, 2017. Diallyl trisulfides, a natural histone deacetylase inhibitor, attenuate HIF-1 synthesis, and decreases breast cancer metastasis. *Mol. Carcinog.*, **56:**2317–31. (doi: 10.1002/mc.22686).

Xiao, J., F.Y. Xing, Y.X. Liu, Y. Lv, X.G. Wang, M.T. Ling, H. Gao, S.Y. Ouyang, M. Yang and J. Zhu, 2018. Garlic-derived compound S-allylmercaptocysteine inhibits hepatocarcinogenesis through targeting LRP6/Wnt pathway. *Acta Pharm. Sin. B.*, **8**:575–86. (doi: 10.1016/j.apsb.2017.10.003).

Xu, Y.S., J.G. Feng, D. Zhang, B. Zhang, M. Luo, D. Su, N.M. Lin, 2014. S-allylcysteine, a garlic derivative, suppresses proliferation and induces apoptosis in human ovarian cancer cells *in vitro. Acta. Pharmacol. Sin.*, **35**:267–74. (doi: 10.1038/aps.2013.176).

Yang, C., L.H. Li, L.G. Yang, H. Lu, S.K. Wang and G.J. Sun, 2018. Anti-obesity and hypolipidemic effects of garlic oil and onion oil in rats fed a high-fat diet. *Nutr. Metab.*, **15**:43. (doi: 10.1186/s12986-018-0275-x).

Yoo, D.Y., W. Kim, S.M. Nam, M. Yoo, S. Lee, Y.S. Yoon, M.H. Won, I.K. Hwang and J.H. Choi, 2014. Neuroprotective effects of Z-ajoene, an organosulfur compound derived from oil-macerated garlic, in the gerbil hippocampal CA1 region after transient forebrain ischemia. *Food Chem. Toxicol.*, **72**:1–7. (doi: 10.1016/j.fct.2014.06.023).

Yoo, D.Y., W. Kim, S.M. Nam, M. Yoo, S. Lee, Y.S. Yoon, M.H. Won, I.K. Hwang and J.H. Choi, 2014a. Neuroprotective effects of Z-ajoene, an organosulfur compound derived from oil-macerated garlic, in the gerbil hippocampal CA1 region after transient forebrain ischemia. *Food Chem. Toxicol.*, **72**:1–7. (doi: 10.1016/j.fct.2014.06.023).

Yoo, M., S. Lee, S. Kim, J.B. Hwang, J. Choe and D. Shin, 2014b. Composition of organosulfur compounds from cool- and warm-type garlic (*Allium sativum* L.) in Korea. *Food Sci. Biotechnol.*, **23**:337–44. (doi: 10.1007/s10068-014-0047-y).

Yusuf, A., S.S. Fagbuaro and Oladipo Kolawole Fajemilehin, 2018. Chemical composition, phytochemical and mineral profile of garlic (*Allium sativum*). *J. Biosci. Biotechn. Discov.*, **3**:105-9. (10.31248/JBBD2018.073). (www.researchgate.net/publication/334561917_Chemical_composition_phytochemical_and_mineral_profile_of_garlic_Allium_sativum).

Zardast, M., K. Namakin, J.E. Kaho, S.S. Hashemi, 2016. Assessment of antibacterial effect of garlic in patients infected with Helicobacter pylori using urease breath test. *Avicenna J. Phytomed.*, **6**:495–501.

Zarezadeh, M., T. Baluchnejadmojarad, Z. Kiasalari, S. Afshin-Majd and M. Roghani, 2017. Garlic active constituent S-allyl cysteine protects against lipopolysaccharide-induced cognitive deficits in the rat: Possible involved mechanisms. *Eur. J. Pharmacol.*, **795**:13–21. (doi: 10.1016/j.ejphar.2016.11.051).

Zhai, B., C. Zhang, Y. Sheng, C. Zhao, X. He, W. Xu, K. Huang and Y. Luo, 2018. Hypoglycemic and hypolipidemic effect of S-allyl-cysteine sulfoxide (alliin) in DIO mice. *Sci. Rep.*, **8**:3527. (doi: 10.1038/s41598-018-21421-x. www.ncbi.nlm.nih.gov/pmc/articles/PMC7146530/).

Zhang, Y., H.Y. Li, Z.H. Zhang, H.L. Bian and G. Lin, 2014. Garlic-derived compound S-allylmercaptocysteine inhibits cell growth and induces apoptosis via the JNK and p38 pathways in human colourectal carcinoma cells. *Oncol. Lett.*, **8**:2591–6. (doi: 10.3892/ol.2014.2579).

Zhang, X., Y. Zhu, W. Duan, C. Feng and X. He, 2015a. Allicin induces apoptosis of the MGC-803 human gastric carcinoma cell line through the p38 mitogen-activated protein kinase/caspase-3 signaling pathway. *Mol. Med. Rep.*, **11**:2755–60. (doi: 10.3892/mmr.2014.3109. www.ncbi.nlm.nih.gov/pmc/articles/PMC7146530/).

Zhang, X., N. Li, X. Lu, P. Liu and X. Qiao, 2015b. Effects of temperature on the quality of black garlic. *J. Sci. Food Agric.* (doi: 10.1002/jsfa.7351. www.ncbi.nlm.nih.gov/pmc/articles/PMC6152780/).

Zhen, H., F. Fang, D.Y. Ye, S.N. Shu, Y.F. Zhou, Y.S. Dong, X.C. Nie and G. Li, 2006. Experimental study on the action of allitridin against human cytomegalovirus in vitro: Inhibitory effects on immediate-early genes. *Antiviral Res.*, **72**:68–74. (doi: 10.1016/j.antiviral.2006.03.017. www.ncbi.nlm.nih.gov/pmc/articles/PMC7146530/).

Zhou, H., Z. Qu, V.V. Mossine, D.L. Nknolise, J.L. Li, Z.Z. Chen, J.L. Cheng, C.M. Greenlief, T.P. Mawhinney and P.N. Brown, 2014. Proteomic analysis of the effects of aged garlic extract and its FruArg component on lipopolysaccharide-induced neuroinflammatory response in microglial cells. *PLoS ONE*, **9**:e113531. (doi: 10.1371/journal.pone.0113531).

Zohary, Daniel and M. Hopf, 2000. Domestication of Plants in the Old World (3rd ed.). *Oxford University Press* (published January 11, 2001), 197p (ISBN 978-0-19-850357-6).

3

Broadleaf Wild Leek (*Allium ampeloprasum* L.)

Subgenus: *Allium*; **section**: *Allium* (Pandey *et al.*, 2022).

Other common names: Leek, Sand leek or Wild leek; in Hindi: Hadgand, Vilatipyaz, Ekputiya lahsun.

Synonyms: *Allium adscendens* Kunth, *A. albescens* Guss., *A. ampeloprasum* var. *babingtonii* (Borrer) Syme, *A.a.* var. *bertolonii* (De Not.) Nyman, *A.a.* var. *bulbiferum* Syme, *A.a.* var. *bulgaricum* Podp., *A.a.* var. *caudatum* Pamp., *A.a.* subsp. *euampeloprasum* Hayek, *A.a.* var. *gasparrinii* (Guss.) Nyman, *A.a.* var. *gracile* Cavara, *A.a.* subsp. *halleri* Nyman, *A.a.* var. *holmense* Asch. & Graebn., *A.a.* f. *holmense* (Asch. & Graebn.) Holmboe, *A.a.* subsp. *porrum* (L.) Hayek, *A.a.* var. *porrum* (L.) J.Gay, *A.a.* var. *pylium* (De Not.) Asch. & Graebn., *A. a.* subsp. *thessalum* (Boiss.) Nyman, *A.a.* var. *wiedemannii* Regel, *A. ascendens* Ten., *A. babingtonii* Borrer, *A. bertolonii* De Not., *A. byzantinum* K.Koch, *A. duriaeanum* Regel, *A. durieuanum* Walp., *A. gasparrinii* Guss., *A. halleri* G.Don, *A. holmense* Mill. ex Kunth, *A. kurrat* Schweinf. ex K.Krause, *A. laetum* Salisb., *A. lineare* Mill., *A. porraceum* Gray, *A. porrum* L., *A.p.* var. *ampeloprasum* (L.) Mirb., *A.p.* subsp. *euampeloprasum* Breistr., *A.p. var. kurrat* (Schweinf. ex K.Krause) Seregin, *A. pylium* De Not., *A. scopulicola* Font Quer, *A. scorodoprasum* subsp. *babingtonii* (Borrer) Nyman, *A. spectabile* De Not., *A. syriacum* Boiss., *A. thessalum* Boiss., *Porrum amethystinum* Rchb., *P. ampeloprasum* (L.) Mill., *P. commune* Rchb., *P. sativum* Mill.

Introduction and origin

The specific epithet *ampeloprasum* comes from the Greek word *ampelos* meaning 'vine' and *parson* meaning 'leek' for a leek found growing in vineyards. *Allium ampeloprasum,* related to *Allium sativum* (common garlic) is an herbaceous perennial native to southern Europe (en.wikipedia.org 'a'; Missouri Botanical Garden), western Asia (en.wikipedia.org a; Sanyat Misra and Misra, 2013), northern Africa and Egypt but naturalized and cultivated in many countries. The species is regarded as native to all the countries bordering on the Black, Adriatic and Mediterranean Seas from Portugal to Romania *via* Egypt (en.wikipedia.org a), *vis-a-vis* Crimea, Ethiopia, Uzbekistan, Iran and Iraq (en.wikipedia.org 'a') which is said to be highly invasive in Russia and Ukraine. It grows in diverse type of soils and has diverse distributional range including Africa, Asia, Europe, Northern America, South America, Australia and India (mostly on higher altitudes, foot-hills of NE states, Uttarakhand, Himachal Pradesh and Jammu & Kashmir) (Pandey *et al.,* 2022). In Britain and Ireland this wildflower is rare and mainly restricted to the coastal areas of the SW England and the West of Ireland (First-Nature).

Wild leek Plant Scape and leaves

Allium ampeloprasum species complex contains wild and cultivated forms with ploidy levels having 4x, 5x and 6x ~ the most frequent being tetraploids (Selvakumar *et al.,* 2021). In cultivation, these require planting distance of 30cm (Sanyat Misra and Misra, 2013) and can easily be grown under full sun in well-drained but moisture-retentive deep sandy-loam soil rich in organic matter. Plants perform best with consistent moisture during the growing season (Missouri Botanical Garden). The climatic and soil requirements are nearly the same as those for onion.

Botany

In its wild form, *Allium ampeloprasum* produces 1-3 or sometimes even more ovoid bulbs (Flora of North America), each up to 3cm across (First-Nature). Some plants may have poorly developed bulbs while others 1-2 large bulbs with several 0.4–1(–3) × 0.4–1(–1.5)cm yellowish to light brown bulbils at base. The long green leaves of wild leeks are quite similar to those of cultivated leeks, and when cut or bruised these plants smell strongly like onions (First-Nature). The plant bears simple, alternate and basal distichous leaves bearing long sheaths attached at the base of bulb with height roughly up to midplant (w3.biosci.utexas.edu).

Wild leek inflorescence at different stages			
Half opened 	**Mature** 	**Aged** 	**Seedhead** 

Scapes being round in cross-section, each typically up to 50cm long but exceptionally over a metre (First-Nature) or up to 1.8m tall (Flora of North America), bearing dense umbel of about 100-500 urn-shaped purple to whitish flowers (First-Nature; Sanyat Misra and Misra, 2013) in summer, i.e. from April to July; depending on location these are planted (Flora of North America). Bisexual flowers (w3.biosci.utexas.edu) are up to 6mm across with white, pink or red tepals (First-Nature), yellow or purple anthers and yellow pollen (Flora of North America). At first the inflorescence appears in an almost perfect globe, but as the flowers age they droop forming an attractive waterfall effect (First-Nature). Though the umbel usually does not produce bulbils but some variants are there producing a few flowers with bulbils therein (Flora of North America). In the

Mediterranean region, by the time the 'wild leek' is in full bloom, its leaves are very likely to have shrivelled and become unrecognizable (First-Nature).

Fruit capsule is loculicidal, typically sterile with 6 aborted ovules, sterile fruit ± dehiscent ± pyramidal and 3–sided, 3-3.5mm long; hidden by persistent, papery tepals but no seeds (w3.biosci.utexas.edu). It has tetraploid chromosome number 2n= 4x=32.

Varieties

Since its genetic improvement in India has not been done, hence it does not have any indigeous variety, however, popular cultivar 'Renova' was derived through selection from cv. 'Regions' whose leaves are longer and darker though there is no bulbing. In mainland Britain two varieties are recorded, var. *ampelopsasum* and var. *babingtonii,* while var. *bulbiferum* is recorded in the Channel Islands (First-Nature). *Allium ampeloprasum* var. *babingtonii* (Babington's leek) is fairly common in southern Portugal (First-Nature). Bulbs of wild and endemic Babington's leek to SW Britain and Ireland is widely available in the nursery trade having mild taste similar to that of a cross between leek and garlic (Roper, 2018). From the crosses and family selection in the var. 'St. George', early winter maturing variety 'Miner' was developed. Likewise, 'Regius 930/57' was evolved by crossing 'St. George' × 'Special'. Other exotic varieties are 'American Flag Elephant', 'Ficus', 'Goliath', 'King Richard', 'London Flag', 'Mussel Burg', 'Splendid', 'Titan', 'Varna', 'Winterens', etc. While crossing, inbreeding depression onward to F2 generatons has been observed. In leek, many hybrids with significant heterosis have been recorded. *A. porrum* and the wild *A. ampeloprasum* having same ploidy level cross easily, producing fertile F1. Crosses with *A. cepa* was also successful though with *A. kurrat* the F1 was found infertile (Currah, 1986).

Major vegetable groups

Cultivated leeks are believed to have derived from the wild types. *Allium ampeloprasum* consists of biennial, onion-like plants now divided into five different horticultural/vegetable groups: i) Leek or garden leek (*A. ampeloprasum* var. *porrum*, syn. *A. porrum*), ii) Elephant garlic or Levant garlic (*A. ampeloprasum* var. *ampeloprasum*), iii) Pearl onion or baby onions (*A. ampeloprasum* var. *sectivum* 'Pearl-Onion Group'), iv) Kurrat or Egyptian leek (*Allium ampeloprasum* var. *kurrat*, syn. *A. kurrat*), and v) Persian leek (*Allium ampeloprasum* ssp. *persicum*). *A. ampeloprasum* is found in a broad geographic range that includes western Iran through the Mediterranean to Portugal and all these vegetables can be freely crossed and interbred (Onions and other allium plants).

Leek/Garden leek/Gourmet's onion (*A. ampeloprasum* var. *porrum, syn. A. porrum*)

Introduction and origin

Leek is the national emblem of Wales and the name leek is from Middle English leke, leek, lek, originating from the old English lēac which in Old English means 'onion' (en.wikipedia.org b). Leek is variously classified in various literatures as *Allium porrum*, *A. ampeloprasum*, *A.a.* var. *porrum*, *A.a.* 'Porrum Group' or *A.a.* 'Leek Group' (Hanelt, 1996). The leek is a vegetable, a cultivar of *Allium ampeloprasum* (the broadleaved wild leek) (piazmiaz.com). The name *leek* is also generally used for any of several species of *Allium* that broadly resemble the domesticated plant in appearance in the wild (en.wiktionary.org). Leek (*A.a.* var. *porrum*), native to eastern Mediterranean regions and the Middle East (Britannica, 2021) has a history of several thousand years of human consumption and use (Leek, 2008). More or less bulbous leek was well known and cultivated by ancient Egyptians, Greeks and Romans (Messiaen and Rouamba, 2004). Leeks were eaten in ancient Rome and regarded as superior to garlic and onions (Sanderson and Jane, 2005) and this has been grown in Western Europe since the middle Ages (en.wikipedia.org b).

Leek

By Kowloonese, CC BY-SA 3.0, https://commons.wikimedia.org/w/index.php?curid=2647128

Modern non-bulbous leek is now grown all around the world except the areas having high temperatures or heavy rains. In Africa, leek is produced on a small scale in the Sahel countries during the cool season, and in the East African highlands ~ mostly around the big cities for Western consumers though local cultivars of vegetatively propagated leek can be found in the tropics, *e.g.* in East Africa (Kenya), Indonesia and the West Indies (Messiaen and Rouamba, 2004). Indonesia and Turkey produce the most leeks though France and Belgium are the leading European producers (en.wikipedia.org b).

The edible part of the plant is a bundle of leaf sheaths that is sometimes erroneously called a stem or stalk (piazmiaz.com). They have a delicate, sweet and mild onion-like flavour without the pungency of some other types of onion (Missouri Botanical Garden).

Botany

Leek is a self-compatible outbreeding tetraploid (2n = 4x = 32). It is a robust erect herb that grows up to 1.5m tall (Messiaen and Rouamba, 2004). Leeks are though true perennials, perennating by means of small lateral growths, but are commonly referred to as biennials, forming their tall stocks and flowers in the second year (Leek, 2008). Leek's true stem consists only of a basal plate or disk with little or no bulb formation or with bulbs producing two kinds of cloves ~ small stiped ones and larger sessile ones (Messiaen and Rouamba, 2004). Modern leeks have been bred during more than two centuries for elimination of cloves, leading to strictly seed-propagated plants with perfectly cylindrical pseudostems. However, two vestigial cloves may appear at the base of the scape of seed-producing plants. The leek group is characterized by the development of a broad succulent pseudostem which is actually the edible part of the leek plant where concentric overlapping leaf bases fold over each other to create a hollow stem-like structure though botanically the edible portion of the leek is not a stem (Onions and other allium plants).

Section and root base

By Björn König - schwarzer-knoblauch.org, CC0, commons.wikimedia.org/w/index.php?curid=18336994

Like many other monocots, leeks too have long slender leaves striped with parallel veins (Leek, 2008). Leaves are distichously alternate; sheath tubular, forming a pseudostem up to 50 cm long; blade linear, up to 50 cm × 7 cm and V-shaped in cross-section (Messiaen and Rouamba, 2004). Encircling blue-green leaves form the stalk at the base but branch outward at the top to form fans of flat green leaves and not hollow like onions (Missouri Botanical Garden). Its roots are adventitious and many of its contractile roots are ribbed horizontally like a pushed-up sleeve or a tubular accordion (Botanistinthekitchen.blog, 2013).

Leek types vary in length and thickness of the leaf shanks (pseudo-stem) (Wisconsin Horticulture). In its first season of growth, leek produces long linear leaves arising from a compressed stem or stem plate and have thick overlapping bases arranged concentrically in a nearly cylindrical onion-like bulb (Britannica, 2021). Each new leaf develops at the centre of the cylinder in the form of a tube that surrounds the next little mound of tissue that will develop into a leaf. Innermost leaves can easily be seen by slicing the root end of the leek in half as close to the centre as possible and in mature leaves the top part of the tube opens up into a flat green blade (Leek, 2008). The long cylinder of bundled leaf sheaths is onion-like but milder in taste (en.wikipedia.org b). From the base of the stem plate grows a tuft of fibrous shallow roots.

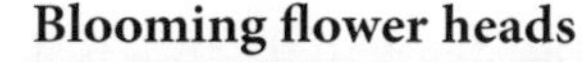

Blooming flower heads

By Derek Ramsey (Ram-Man) - Own work, CC BY-SA 2.5, commons.wikimedia.org/w/index.php?curid=908558

If left unharvested, the second-season leeks produce a large umbel with many flowers (Britannica, 2021) in purple and pink to white colours. Inflorescence is a spherical umbel of 4–12cm in diameter bearing several hundreds of flowers on a solid terete scape up to 1.5m long, umbel subtended by a single long-pointed spathe which sheds at flowering (Messiaen and Rouamba, 2004). Flowers are campanulate and bisexual, pedicel 1–5cm long, tepals 6 in 2 whorls, free, ovate-oblong, 4–6mm long and purple or white, stamens 6, the inner ones tricuspid and the ovary is superior and 3-celled (Messiaen and Rouamba, 2004). When grown as annuals flowers will never be seen (Missouri Botanical Garden). Both self- and cross-pollination occurs in leek and the pollination is entomophilous.

Leek seeds

By Azeite - Own work, CC BY-SA 4.0, https://commons.wikimedia.org/w/index.php?curid=61173042

Leek seedheads develop slowly thus harvested often late during autumn as compared to spring or summer seed maturation of bulb onion, shallot, garlic, and bunching onion. Its up to 6-seeded fruit is a depressed globose to ovoid capsule 2–4mm in diameter bearing small (2-3mm×2mm), black, irregular and angular seeds (Messiaen and Rouamba, 2004). Presence of bulbils as common in the inflorescencees of garlics and tree onions are occasionally seen in some cultivars of leek and kurrat (Onions and other allium plants).

Cultivars

Leek cultivars may be treated as a single cultivar group, e.g., as *A. ampeloprasum* 'Leek Group' (Brewster, 2008; en.wikipedia.org) or *A.a.* 'Porrum Group' (en.wikipedia.org b). Leek cultivars range in size from long, green narrow-leaf varieties with long, thin white stems to long, wide-leaf varieties with thicker, shorter white stems and blue-green leaves. There are no tropical leek cultivars other than vegetatively propagated tropical forms and historical importation from temperate countries/regions (Selvakumar *et al.*, 2021). Though leek cultivars can be subdivided in several ways, the most common types are: i) 'early or summer season leeks' that are intended for harvest in the same season when planted, and 'overwintering or late-season leeks' that are meant to be harvested in the spring of the year following planting (Leek, 2008; Selvakumar *et al.*, 2021).

Compared to summer or early season types, overwintering or late season types are generally strongly flavoured, larger with broader stems, less susceptible to frost and require blanching or hilling for optimum quality (Leek, 2008; Selvakumar *et al.*, 2021). Depending on the cultivars, the early season leeks are ready for harvesting in 50-100 days, whereas late-season leeks mature in 120 to 180 days after sprouting. Moreover, early season varieties do not require blanching so are easy to grow than late-season leeks (Temperate Vegetables, 2012). India does not have its own leek cultivars so what we have in this country all are introductions. The varieties maturing in 80-120 days include **main fall varieties** (August to October) ~ 'American Flag', 'Jolant', 'Kilima', 'King Richard', 'Primor'; **late fall-winter varieties** (October to December) ~ 'Derrick', 'Electra', 'Goldina', 'Goliath', 'Kilima', 'Tivi', 'Wintereuzen'; and some **overwintering varieties** (spring harvest) ~ 'Carina', 'Conqueror' (moderate bulbing), 'Eskimo' and 'Siberia' (Temperate Vegetables, 2012).

Early Season varieties

- **Columbus** (80 days): It is a medium-sized leek with a very little tendency to bulbing and has little of winter-hardiness so requiring mulching. The plant grows tall bearing blue-green leaves.
- **King Richard** (Wisconsin Horticulture) or Richard the Lionheart (Selvakumar *et al.*, 2021) (75 days): It is a summer leek having long and slender stems that stay white, sweet and tender all the season long and is usually densely grown to harvest mini-leeks for use as a garnish or for soups or salads. It has little of frost-resistance and does not overwinter (Wisconsin Horticulture).
- **Lancelot** (70 days): It is a virus tolerant, winter-hardy and bolt-free short leek with a large cylindrical shaft and grey-green leaves.
- **Lincoln** (80 days): It is an early Bulgarian giant kind with uniformly long, white and thick shanks which can be planted during Rabi season.
- **Rival** (80 days): It is up to 92cm tall summer leek bearing 23cm shaft that can have a large 5 cm diameter and green leaves.
- **Varna** (50 days): It is a tall bunching-type variety developed for direct thick seeding to produce clumps of slender plants.

Mid Season

- **Albinstar Baby Leek** (110 days): It is a Dutch variety developed for baby leek market and becomes ready to harvest when attains 1.27cm diameter but usually left in the ground to grow larger. Its leaves are deep green.
- **Dawn Giant** (98 days): It grows up to 38cm tall with 5cm in diameter and is topped with blue-green leaves.

- **Jolant** (100 days): It is an early-blooming medium-sized leek with a 20-23cm shaft and blue-green leaves. It has mild flavour and can be used as mini-vegetable or left as such to grow until August to harvest. Its plants and bulbs have winter-hardiness.
- **Splendid** (105 days): It is a rapidly growing vigorous grower and making good size before fall. It is a frost tender though with winter-protection it can be grown throughout the year. It has medium green 18-20cm non-bulbous stalks that are easily blanched.

Late Season

- **American Flag** (130 days): It is a quite hardy leek and can overwinter in mild climate. The 18-23cm long stems blanch snowy white and are topped with blue-green leaves resembling giant scallions. It has a mild, sweet and onion-like flavour.
- **Bandit** (135 days): It is a short leek growing quite thick with very little bulbing. It is a winter-hardy sweet leek bearing blue-green leaves.
- **Durabel** (125 days): It is a sturdy and thick winter leek with a mild flavour and tender texture.
- **Otina** (120 days): It is a rapid growing quite vigorous French variety bearing blue-green leaves which has mild and delicate flavour.
- **Titan** (120): Though it is quite old but a reliable variety.
- **Winter the Giant:** It is a late variety having thick and heavy stems.

Extra-Late Season

- **Giant Musselburgh** (150 days): It is a long-lived large Scottish heirloom and very cold-hardy variety possessing tender white atalks, medium-dark green tops, broad & lofty leaves, long thick stem and mild flavour.
- **Laura** (180 days): This variety is extremely hardy growing to a medium height with dark blue-green upright leaves.

Some other cultivars/hybrids are

- **Almera:** It has long and thin stems with semi-upright habit bearing mid-green leaves. Its cropping period is mid-July through the end of September, i.e. autumn cropping
- **Atlantic**: It has an excellent tolerance to frost and can be grown as winter crop.
- **Axima**: It has no bulbing and bears long and sturdy shaft with erect dark green foliage.

- **Below Zero (F_1)**: British breeders have blended the vigour of an F_1 hybrid with great cold tolerance to produce high-quality leeks that can resist even the toughest conditions. It has no bulbing, no-bolting, is rust-resistant, stems being pure white, foliage is dark and has long-standing ability.
- **Blanch Mammoth**: It is a great exhibition type leek with a thick, broad flag and extremely long white blanch. It is planted or sown at 15°C from mid-January to early-March but does not tolerate overwatering due to damping off. Starting in early May, the plants may be hardened off and planted out.
- **Carentan**: It has blue-green leaves on thick large stalks. Crops are harvested from late October to early January.
- **Carlton**: It has great consistency in its lengthy shaft nodes. The upright and green leaves of this hybrid species grow quite quickly.
- **Hannibal**: It is a fast-growing, summer and autumn cropping variety.
- **London Flag**: It is the most commonly cultivated cultivar bearing tall and broad leaves.
- **Miner**: It is derived from 'St. George' and is a fall or early winter cultivar bearing long stems and high greens.
- **Oarsman (F_1)**: This bolt-free variety bears a medium to dark flag leaf with second maturation slot occurring from early to mid-season.
- **Parton**:This hybrid has dark green leaves. Its medium growth generates a medium shaft length with outstanding consistency.
- **Pot**: It produces exceptionally giant hefty leeks.
- **Prizetaker/Lyon**: Due to its uniform nature, it produces the thick, long and white stems. It's fruit matures from early October onwards.
- **Renova**:It was created by selecting from a large number of possibilities. It has a longer stem though with poor bulb-forming tendencies and grows more uniform in shape and size.
- **Roven (Large)**: It has dark green leaves that are large and thick, and a short but sturdy stem. It is forcing type because it requires a necessary stem thickness sooner than any other variety.
- **Roxton:** It has a large shaft length and good consistency. This hybrid has a quick growth bearing erect green to dark green leaves.
- **Solaise Bleue:** This variety is a typical French type. It has an excellent flavour and hardiness. Its leaves have a deep blue-purple coloration

- **Tadorna**: It is an exceptionally cold hardy variety, grown from December to March. It has a medium length and upright habit.

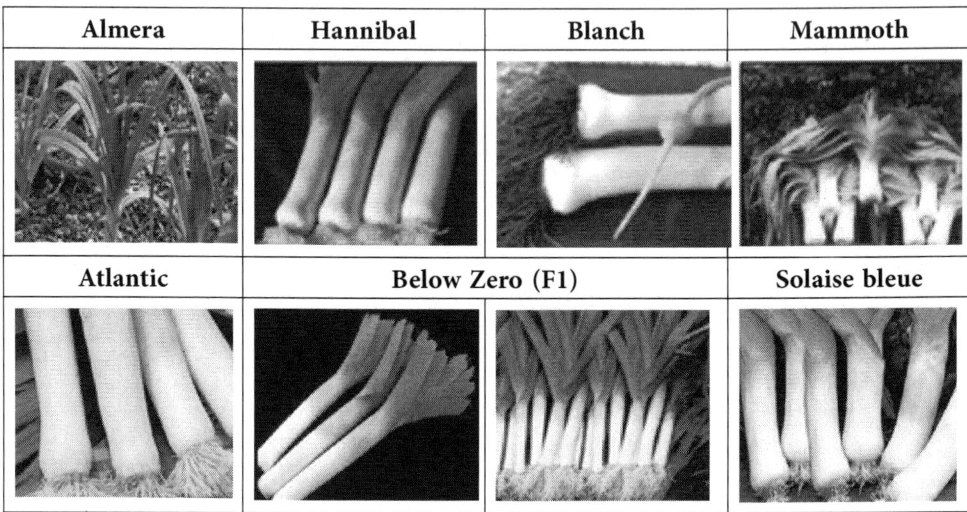

Source: (Selvakumar *et al.*, 2021)

Leek cultivation in India

Leek is not commercially cultivated anywhere in India but grown in kitchen gardens for blanched stems and leaves (agrojiva.com; Selvakumar *et al.*, 2021) or rarely to be sold in local markets. Leeks and green onions (scallions) are locally called *saaga* (greens) in India though scallions are much smaller than leeks and are usually sold in bunches (sweetishhill.com). The variety cultivated in India is PPL-1 becoming readying in 150-160 days which is a selection from exotic germplasm. It has swollen stems, light green leaves and non-bulbing type. In this case whole plant is consumed and is a good replacement for green onion. Palam Paushtik has also resulted through selection and is an alternative of green onion. It matures in 140-150 days with an average yield of 300-350q/ha and is suitable for salad, soup and cooking (Temperate Vegetables, 2012). London Flag and American Flag are also commonly grown in India. Commercially important leek types include 'American Flag', 'Belgium Winter Giant', 'Bluevetia', 'Borde Pearl', 'Copenhagen Market', 'Elephant', 'Gennivillies', 'Large Musselburgh', 'London Flag', 'St. George', and others (Selvakumar *et al.*, 2021).

Nutritional values

Trimmed with outer leaves removed edible white stalk of leak per 100g contains water 90.8g, energy 93 kJ (22 kcal), protein 1.6g, fat 0.5g, carbohydrate with major part being fructans 2.9g, fibre 2.8g, Ca 24mg, P 44mg, Fe 1.1mg, carotene 735µg, thiamine 0.29mg, riboflavin 0.05mg, niacin 0.4mg, folate 56µg, and ascorbic acid 17mg (en.wikipedia.org b). Raw leek (bulb and lower leaves) contains water (83%), carbohydrates (14%), protein (1%) and negligible fat. A 100g reference amount supplies 255 kilojoules (61 kcal) of food energy and is a rich source of 20% or more of the Daily Value (DV) of vitamin K (45% DV) and manganese (23% DV). It is a moderate source (10-19% DV) of vitamin B6, folate, vitamin C, and iron (en.wikipedia.org b).

Table source: en.wikipedia.org b ***Nutritional value per 100g raw leeks, bulb and lower leaves***	
Energy	255 kJ (61 kcal)
Carbohydrates	14.15g
Sugars	3.9g
Dietary fiber	1.8g
Fat	0.3g
Protein	1.5g
Vitamins	**Quantity (%DV†)**
Vitamin A equiv. beta-Carotene lutein zeaxanthin	83μg (10%) 1000μg (9%) 1900μg
Thiamine (B1)	0.06mg (5%)
Riboflavin (B2)	0.03mg (3%)
Niacin (B3)	0.4mg (3%)
Pantothenic acid (B5)	0.14mg (3%)
Vitamin B6	0.233mg (18%)
Folate (B9)	64μg (16%)
Vitamin C	12mg (14%)
Vitamin E	0.92mg (6%)
Vitamin K	47μg (45%)
Minerals	**Quantity (%DV†)**
Calcium	59mg (6%)
Iron	2.1mg (16%)
Magnesium	28mg (8%)
Manganese	0.481mg (23%)
Phosphorus	35mg (5%)
Potassium	180mg (4%)
Other constituents	**Quantity**
Water	83g
Units: μg (micrograms), mg (milligrams), IU (International units)	
†Percentages are roughly approximated using US recommendations for adults. ***Source:*** USDA FoodData Central	

Chemical and nutritional characterization of *Allium ampeloprasum* L. edible part (per 100g)

Proximate Composition	***Units***	***Leek, bulb, raw***
Water	g	86
Energy	Kcal	35
Protein	g	1. 9
Total fat	g	0. 4
Carbohydrate, available	g	5. 9
Dietary fibre(Englyst1988)	g	3. 3
Ash	g	1
Dry matter	g	14
Total nitrogen	g	0. 31
Total saturated fatty acid	g	0. 061
Total MUFA	gv	0. 006
Total PUFA	g	0. 253
Total available sugars	g	5. 9
Glucose	g	2. 4
Fructose	g	2. 4
Sucrose	g	1. 1
Soluble non-starch polysaccharides	g	1. 6
Insoluble non-starch polysaccharides	g	1. 6
Chemical composition		
Oxalic acid composition	(mg/100g)	91. 65
Glutamic acid	(mg/100g)	51. 67
Malic acid	(mg/100g)	132. 86
Citric acid	(mg/100g)	38. 86
Succinic acid	(mg/100g)	2. 14
pH		5. 76
Macro and micronutrient of Allium ampeloprasum (per 100g)		
Sodium	mg	9
Potassium	mg	310
Calcium	mg	63
Magnesium	mg	10
Manganese	µg	188
Iron	mg	1. 1
Zinc	mg	0. 4
Copper	mg	0. 1

Phosphorus	mg	43
Beta-carotene equivalents	µg	49
Total vitamin A equivalents	µg	8
Vitamin E	mg	0. 92
Thiamin	mg	0. 1
Riboflavin	mg	0. 05
Niacin	mg	0. 6
Vitamin B6	mg	0. 25
Total folate	µg	86
Vitamin C	mg	18
Table source: (Dey and Khaled, 2015)		

Phytochemicals

Leek has a high concentration of flavonoids, including kaempferol, as well as a significant level of sulphur. Presence of S-alkyl-cysteine sulphoxides is linked to the leek's distinctive flavour, with propyl- and methyl- cysteine sulphoxides being the most common in modern leeks (Messiaen and Rouamba, 2004; Selvakumar *et al.*, 2021). This precursor is transformed into a number of S-containing volatiles (e.g. thiosulphinates) when raw leek is cut. The amount of quercetin and kaempferol in farmed leek bulbs is very low (below 20mg/kg fresh weight) or traces (below 2mg/kg FW) (Selvakumar *et al.*, 2021). The taste of cooked leek is also linked with S-containing volatiles to which a small amount of saponins may add a little bitterness (Messiaen and Rouamba, 2004).

Five flavonoids with human platelet anti-aggregation activity are also isolated (Fattorusso *et al.*, 2001). The pectic polysaccharides, galacturonic and glucuronic acid isolated from leek have immunostimulating role (Kratchanova *et al.*, 2010). Three dibenzofurans, porric acids A (1), B (2) and C (3) isolated from the bulbs were found to exhibit antifungal activity against *Fusarium culmorum* (Alfonso *et al.*, 1998).

Some of the several isolated steroidal saponins have anti-inflammatory and antiulcerogenic properties (Adao, *et al.*, 2011) while some others have shown to exhibit *in-vitro* antitumour activity (Messiaen and Rouamba, 2004), haemolytic activity, and immunological adjuvant activity on the cellular immune response against ovalbumin antigen (Ayumi *et al.*, 2009). A steroidal saponin from *A. ampeloprasum* named ((3β,5α,6β,25 R)-6-[(β-D- glucopyranosyl)oxy]-spirostan-3-yl-O-β-D-glucopyranosyl-(1→2)-O-[β-D-glucopyranosyl-(1→3)]-β-D-galactopyranoside) was investigated for its anti-inflammatory potential using a carrageen-induced paw edema model in Swiss albino mice. The extract suppressed the edema volume of mice paw prominently, indicating significant anti-inflammatory effect. This was further attributed to the capacity of the compound to modulate the secretion of histamine, serotonin and prostaglandins which are responsible for the generation of biphasic inflammation (Adao *et al.*, 2011). Contrary to the trend observed for other *Allium* vegetables where volatile compounds were identified as the major anti-inflammatory phytoconstituents, *A. ampeloprasum* yielded saponins with prominent anti-inflammatory activity. Saponins are also prevalent among other *Allium* species such as *A. cepa*, *A. sativum*, *A. schoenoprasum*, *A. tuberosum*, *A. Ampeloprasum* and *A. fistulosum* (Alam *et al.*, 2022).

Uses

The edible portions of the leek are the white base of the leaves (above the roots and stem base), the light green parts, and to a lesser extent the dark green parts of the leaves (en.wikipedia.org b). Mostly the white and light green parts are eaten and the dark green portion is usually discarded as it has a tough texture (sweetishhill.com), though the darker green parts have plenty of flavours and can either be cooked longer to tenderize them or used when making homemade soup stock (sweetishhill.com).

Leeks have a mild onion-like taste and can be eaten raw in salads or cooked like vegetables. It is crunchy and firm in its raw state (en.wikipedia.org b). It can be boiled, fried or used as an ingredient in soups. Leeks can be substituted for onions and, unlike onions can be chopped and frozen for use in winter (sweetishhill.com).

The leaves and stem can be cooked or cut into thin slices and added to salads, where these do especially well when they are the prime ingredient. Sliced leek is often dehydrated for use in soups (Messiaen and Rouamba, 2004). The onion-like layers form around a core where tender core may be eaten but as the leek ages the core toughens and becomes fibrous and then it is better to replant these than eaten. The flowers can be consumed raw and used as a garnish on salads though they are rather dry and less desirable (Leek, 2008; en.wikipedia.org b).

Leek is believed to have the same medicinal values as garlic but in a less effective form. These values include as an antiseptic, anthelmintic, diruetic, vadodilator and antispasmodic among other virtues. It is considered to have a beneficial effect on the body, and particularly the circulatory system (Leek, 2008). Leeks are rich in flavonoids, especially kaempferol, and the flavonoids are antioxidants and may have many other health benefits including anti-inflammatory, anti-diabetic and anticancer properties (sweetishhill.com; Selvakumar *et al.*, 2021).

Substitutes

Leek can be substituted by bunching onions with a long thick pseudostem such as some cultivars of *Allium fistulosum* L. or 'Beltsville Bunching Onion' ~ a tetraploid seed-propagated cross between *Allium cepa* and *Allium fistulosum*. In South Africa *Allium dregeanum* (wild leek) is collected from the wild by Hottentot people similar to the wild ecotypes of *Allium ampeloprasum* on both sides of the Mediterranean (Messiaen and Rouamba, 2004).

Precaution

Some sulphuric foods such as garlic, leeks and onions can cause stomach upset or pressure in the abdomen (sweetishhill.com).

Propagation

Leek is propagated either by seed or by lateral shoots. For modern cultivars, about 1000 seeds/m^2 are sown in a nursery roofed with plastic or with palm leaves but roofing is removed immediately after germination. One gramme of packet contains 300–400 leek seeds so for planting one hectare of land some 5-6kg seed is required in the nursery.

Direct seeding is not desired due to lack of registered herbicides, *via-a-vis* 8-12 months or even longer length of time required to harvest the crop. In India, the seeds are generally produced at higher altitude. Similar to other alliums, leek seeds remain viable hardly for two years at ordinary conditions. In this case germination is quite slow (10-12 days) even at optimum conditions.

Climate and soil

Leek is easier to grow at elevations above 1000m but in lowlands the growth gets slowed and may take about 60 days or more in the seedbed and 70 days or more in the field for harvesting leeks more than 1cm in diameter. Leek cultivation in hot and humid lowlands below 500m is rarely successful (Messiaen and Rouamba, 2004) though leeks withstand heat and cold better than onions. These require a cool season of at least 4 months for better growth. Vegetative growth may occur at temperatures of 4–25°C, better being at 13-24°C (sweetishhill.com) though most appropriate being 16–20°C (Messiaen and Rouamba, 2004). Hot weather slows down the growth. For better flavour development, however, 1-2 frosts are required (sweetishhill.com). At temperature above the optimum, *i.e.* 22°C at night and 30°C during the day, leaf senescence can be as rapid as the production of new leaves which results in poor diameter growth.

At a fully sunny location, leeks can easily be grown all the year round in deep and fertile soil with a pH range of 6.5-7, provided the soil is moisture-retentive, well-drained and the climate remains cool to moderate throughout the year

(Temperate Vegetables, 2012). The varieties 'Musselburgh' and 'Autumn Giant' can provide good yield even under shade provided morning sun is available. Flooding in the field is dangerous at any time. Unlike bulb onions, leeks do not have specific photoperiodic requirements for pseudostem formation and can thus be grown in a wide range of latitudes (Onions and other allium plants). However, growth of the floral stem is enhanced by long days but flowering rarely occurs under tropical conditions.

Season

The ideal time of sowing in plains is August–October and seedlings are planted in the garden when they are about 15cm tall though on the hills it is sown in mid-February-April. Temperature is more important than day length in seed stalk development (Temperate Vegetables, 2012). For July-September harvest, the leeks should be sown from mid-February to mid-March, for early winter harvesting it should be sown from mid-March to mid-April, and for spring harvest the sowing should be done from April to mid-May (Temperate Vegetables, 2012), and for each season harvesting, right variety should be chosen.

Field preparation, transplanting and blanching

Seedlings are transplanted when these attain a height of 15-20cm and pencil thickness (agrojiva.com; Messiaen and Rouamba, 2004). Deep ploughing before transplanting is recommended so that a longer shaft can be developed. At transplanting, leek seedlings should be placed vertically into holes 10–12cm deep and at a distance of 10-15cm in the row spaced 30-45cm which will accommodate some 200,000 plants/ha (Messiaen and Rouamba, 2004). For obtaining long pseudostems, planting is done at the bottom of furrows to facilitate earthing-up. When propagated vegetatively, the lateral shoots (tillers) are planted in the same way as seedlings.

For producing larger white shafts that reach almost to the base of the leaves, the lower portion of the stalk should be hilled up so that chlorophyll manufacturing in the covered area is lessened for getting better specimens with long white section of the stalk below the leaves (Britannica, 2021). Care should be taken not to earth up soil too early when the plants are young and also not to allow soil between leaves as it will penetrate the shaft (Temperate Vegetables, 2012).

Alternatively, seedlings can be planted into a trench 15-20cm deep. In either case, the young transplants should be placed up to the first leaf notch into holes. Holes should not be filled with soil immediately after planting as watering these transplants would fill up the hole with soil on its own. In the home garden, the holes should be 15-23 cm apart. For larger plantings, seedlings should be planted at 10-15cm spacing in rows that are 45 to 60cm apart to allow soil to be moved from between the rows around the plants for blanching the stems.

Interplanting

Since they are relatively slow-growing, these can be interplanted with fast maturing crops such as lettuce, spinach, radish, *etc.* Avoid planting beans or other legumes nearby as the leeks can inhibit their growth (Wisconsin Horticulture).

Manure and fertilizers

Harvest of 20t/ha of leek crop removes about 65kg nitrogen, 10kg phosphorus, 75kg potassium, 35kg calcium, 3kg magnesium and 15kg sulphur from the soil so fertilization with 5–20t/ha of organic manure and an adequate quantity of NPK including sulphates is recommended along with liming if necessary (Messiaen and Rouamba, 2004).

However, a soil test is necessary to determine phosphate and potash requirement. Application of 200-250q well-rotten FYM or compost along with 80-100kg N, 60kg P_2O_5 and 80kg K_2O/ha is required to be applied in the field (Temperate Vegetables, 2012). Entire quantity of FYM, P2O5, K2O and half N should be thoroughly mixed in the soil at the time of planting while remaining N is side-dressed one month after.

Use of growth regulators

For proper growth and development, ethephon @ 50mg/l as foliar spray should be applied 20-25 days after sowing; and NAA @ 50mg/l should be applied at 60 and 90 days after planting (Temperate Vegetables, 2012).

Irrigation

Timely irrigation should be given to maintain uniformly vigorous growth with tender stalks. Depending on planting date, seasonal variation and variety a total of 30-38cm of water may be required in one cropping season. Frequency of irrigation depends on the soil type as lighter soils require more frequent though heavy or clayey soils less frequent (Temperate Vegetables, 2012).

Weed control

Similar to onion, the practices for weed removal is the same.

Growth and development

As compared with most other vegetables, the germination in leek is generally slow even at optimal temperatures (16–20°C). First a leafy cotyledon bearing the seed testa at its apex appears about 10-12 days after sowing but afterwards new leaves start appearing every 6 days at optimal temperatures so plants reach the harvest stage when around 15 leaves have appeared, and then these continue to grow in thickness (Messiaen and Rouamba, 2004).

If left unharvested the flowering occurs during second-season, i.e. after about 180 days after sowing only in temperate regions (Britannica, 2021) though it is variety *cum* temperature dependent as low temperature of 5°C may cause earlier bolting and long days enhancing the floral stem growth which is the reason that under tropical regions bolting and flowering rarely occur (Messiaen and Rouamba, 2004).

Harvest

Depending on the variety, leeks require 50 to 180 days to reach harvest stage. Once established in the garden, many of the varieties can be left in the ground as such during winter to be harvested as per requirement. Leeks usually reach maturity during autumn. These can be bunched and harvested early when they are about the size of a finger or pencil or they can be thinned out and allowed to grow to a much larger mature size (Leek, 2008). Leeks do not usually go dormant during fall but tend to grow slowly, so harvest time becomes quite flexible. It can be harvested any time after it reaches 5-6cm in diameter, or can be left in the ground until needed (Wisconsin Horticulture).

Small leeks can be sold from early August and the frost-tolerant varieties can be harvested right from autumn to entire winter months. Machine harvesting has become quite prominent now. Most leeks are lifted or dug by machine and then harvested, cleaned and packed by hand. Single or multiple row harvesters can be used effectively (Temperate Vegetables, 2012).

Harvesting can be continued through a mild winter and into early spring by putting an extra thick layer of mulch around the plants to delay the freezing of the ground and make it easier to dig them up (Wisconsin Horticulture). In a more typical winter, the plants must be pulled before the ground freezes solid. The plants are dug up with a spade and washed, the roots cut at 1–2cm, dried or yellowing leaves are eliminated with their sheaths and the remaining leaves trimmed (Messiaen and Rouamba, 2004). For vegetatively propagated leek the thicker pseudostems are sold, the thinner lateral shoots are retained for own use or for a new planting.

Yield

Leek yield varies depending on choice of varieties, prevailing climatic conditions, planting density and other cultural practices. Different sources mention different yields though it may vary from 250 to 300q/ha (agrojiva.com) to about 925q/ha (Temperate Vegetables, 2012).

Storage

Leek seeds must be stored at 0°C and 95 to 100 per cent relative humidity. Properly harvested leeks can be satisfactorily stored for 2 to 3 months at 0°C to prevent yellowing and decay. Respiration or heat evolution of leeks is about eight times faster at 21.1°C than at 0°C. They should be cooled promptly after harvest to near 0°C through hydro-cooling, crushed ice or vacuum cooling and with high relative humidity throughout storage to prevent wilting. The use of polyethylene film crate-liners and of crushed ice can aid in preventing moisture loss. Good refrigeration will retard the elongation and curvature that develops in leeks at 10°C or 21.1°C (Temperate Vegetables, 2012).

Storage for 4-5 months at 0°C is possible by using a controlled atmosphere (CA) containing from 1-3% oxygen and from 5-10% carbon dioxide though there will be some loss in quality. Atmospheres containing 15-20 per cent carbon dioxide cause tissue injury (Temperate Vegetables, 2012).

Leeks are commonly trimmed to 30cm length, bunched depending on diameter and often placed in polyethylene film bags. They are usually packaged in 5kg cartons or wire bound crates, holding 10 film bags each being 500g. Other crates may be packaged with 18-24 bunches with a net weight up to 15kg (Temperate Vegetables, 2012).

Pests and diseases

Leeks do not have many pests but most insects and diseases that attack onions can also affect leeks. Onion thrips (*Thrips tabaci*) can cause problems mainly under hot and dry conditions. Their feeding produces tiny silvery-white spots on the leaf surface. Though it can be controlled by insecticides, overhead irrigation is usually recommended to decrease thrips infestation (Messiaen and Rouamba, 2004). Onion maggot (*Hylemya antiqua*) feeds on root of the plants causing reduced growth and wilting (Wisconsin Horticulture). In lowlands, roots are invaded by *Pyrenochaeta terrestris* in the same way as for garlic or onion (Messiaen and Rouamba, 2004). The roots and basal plate can be destroyed by the fungus *Sclerotium cepivorum* or the nematode *Ditylenchus dipsaci*, potentially abnoxious in the tropics if introduced in highlands through use of contaminated garlic cloves.

Purple blotch (*Alternaria porri*) is the most important leaf disease, especially in plants grown on calcium-deficient acid soils. Overhead irrigation is also beneficial against purple blotch if applied between 11 a.m. and 3 p.m. Chemical control includes use of dithiocarbamate fungicides or iprodione (Messiaen and Rouamba, 2004). Diseases such as white rot (*Phytophthora porri*), rust (*Puccinia* spp.), pink root, purple blotch, downy mildew, botrytis leaf spot, botrytis neck rot and smudge may occur but are not common in home gardens (Wisconsin Horticulture). Yellow stripe virus may cauuse significant damage to leek.

Elephant garlic/Giant garlic/Great-headed garlic/Levant garlic (*A.a.* var. *ampeloprasum*)

Introduction

Great-headed garlic is grown in the highlands in some countries close to the tropics, *e.g.* Reunion, Saudi Arabia and northern India (Messiaen and Rouamba, 2004). Unlike leeks, elephant garlic has been bred to produce larger edible cloves underground with less of a focus on the green stalks of the plant (Block *et al.,* 2010; Leek, 2008). Its large bulbs and cloves resemble in shape and structure to true garlic but are larger and have a mild garlicky flavour.

By Dave Whitiger davesgarden.com/pfshow image/74930/; CCBy -SA 3.0, commons wikimedia.org/w/index.php?curid=1723518

Botany

Elephant garlic is a hardy cultivated variety of the wild leek with intermediate properties of garlic (*A. sativum*) and onion (*A. cepa*), though milder in flavour (greenharvest.com.au). A fully grown plant has a solid, tall up to 1.8m high flowering stalk and large, broad, blue-green and strap-shaped flat leaves having a central rib (greenharvest.com.au; en.wikipedia.org c). A large leek-like non-seed-producing umbel of pink or purple flowers appears in spring and summer. These plants have large bulbs comprising 3-6 large cloves (up to 60g) with papery skins (en.wikipedia.org c) surrounded by small bulblets that are 8-10mm in diameter and protected by a hard shell (Messiaen and Rouamba, 2004). These bulblets can remain in the soil for more than 1 year before sprouting.

If the plants are grown from small to average size cloves of a multi-cloved bulb or using bulblets, they usually produce a nonflowering plant in their first year (en.wikipedia.org c). Plants that do not flower often form only one large symmetrical single-cloved bulb known as a round, rather than a bulb with multiple cloves (greenharvest.com.au). These rounds are sometimes confused with solo garlic (en.wikipedia.org c). If the plants are grown from these rounds or from very large clove, they produce a regular large bulb with several cloves and a flower stalk (en.wikipedia.org c; greenharvest.com.au). It possesses 2n=4x=32; 2n=6x=48 and 2n=8x=64 chromosome numbers.

By Own work by user Uncaught on the German Wikipedia - Foto, Public Domain, common wikimedia.org/w/index.php?curid=11869346

Nutraceutical parameters

Nutraceutical parameters were measured in elephant garlic bulbs cultivated organically and conventionally in Valdichiana. All values are in mg/g fw except the fresh weight of the cloves which is in grammes. TSI = total sweetness index. * = statistically significant ($p < 0.05$) difference. Source: Loppi *et al.* (2021)

	Organic		***Conventional***	
	M	**SE**	**M**	**SE**
Fresh weight	19.9	8.5	13.6	4.6
Water %	61.4	0.1	61.1	0.1
Soluble proteins	8.4	2.8	10.2	0.2
Polyphenols	2.3	0.3	3.0	0.3
Flavonoids	1.5	0.2	2.0	0.2
ARA%	16.7	2.2	18.1	0.3
Starch	3.9	0.1	3.8	0.1
Sucrose	9.1	3.8	6.5	0.2
Glucose *	5.9	0.2	8.1	0.3
Fructose	13.3	0.2	15.3	0.1
TSI	33.6	5.5	35.7	0.2
Table source: Loppi et al. (2021)				

Range of major and minor element concentrations

Range of major and minor element concentrations (mg/kg dw) in elephant garlic cultivated in Valdichiana, Spain (Garcia-Herrera *et al.*, 2014), and India (Dey and Khaled, 2013; Devi and Brar, 2018).

Major essential elements			
	Valdichiana	***Spain***	***India***
P	2270–015	-	430–1184
K	10635–17194	1466–5332	3100–4576
Mg	834–1153	89–164	100–186
Ca	321–833	302–817	125–630
Minor essential elements			
Fe	41.4–57.9	2.0–9.2	11.0–12.1
Na	15–203	436–671	67–90
Zn	12.9–44.2	0.3–16.7	4.0–6.5
Cu	2.4–6.6	0.5–2.2	1.0–9.6
Table source: *Loppi et al. (2021)*			

Phytochemicals

When crushed and analysed using a DART ion source, elephant garlic produces both allicin, found in garlic, and *syn-propanethial-S-oxide* (onion lachrymatory factor) found in onion and leek but absent in garlic (Block *et al.*, 2010). The onion lachrymatory factor is found to be formed when elephant garlic is crushed.

Of the five saponins (yayoisaponins A-C, dioscin and aginoside) isolated from a new variety of *A. ampeloprasum*, only dioscin exerted significant cytotoxicity against the P388 murine leukemia cell line with an IC50 value of 0.092µg/ml. In contrast, yayoisaponins A-C and aginoside exhibited moderate activity at a concentration of 2.1µg/ml on the same cell line (Sata *et al.*, 1998).

Uses

Elephant garlic (*Allium ampeloprasum*) has very large cloves with a tender, mild and slightly sweet flavour. Some consumers enjoy elephant garlic because it can be eaten raw and used in cooking for a hint of garlic flavour without being overwhelming. Cloves are large, easy to peel and can be eaten raw *cum* sliced into salads (greenharvest.com.au). Fresh elephant garlic contains mostly moisture and foams similar to boiling potatoes, whether on the stove or in a glass dish in the oven. Drying in the basement for a few months reduces the moisture content and brings out a fuller flavour (en.wikipedia.org c). It has in between garlic and onion mild *cum* sweet flavour. Like regular garlic, elephant garlic can be roasted or baked whole to be used as a spread with butter on toast. It can also be steamed or boiled as a vegetable, cooked like onions in a soup or baked in the oven. Young leaves can be sliced and added to salads or used as a garnish. It is high in vitamins A, C and E (greenharvest.com.au).

Propagation

Elephant garlic is not generally propagated by seeds but through cloves or using small bulblets surrounding the cloves. In its first year of cultivation from bulblets or cloves, the plant usually produces a single large round bulb (called rounds) in the first year when the plant doesn't flower, although the plants grown using very large cloves or rounds usually producing a regular large bulb with several cloves and a flower stalk.

For a nonflowering plant, all of the plant's resources go into building up the single bulb which will help it surviving into its second year and sending up flower stalks. In the second year, the single bulb typically divides into multiple separate cloves surrounded by several small bulblets (thespruce, 2022; en.wikipedia.org c; greenharvest.com.au).

Climate and soil

A planting site that is little sheltered from strong and harsh winds is ideal for its cultivation. Elephant garlic tolerates the temperatures of its growing zones fairly well. It can grow successfully from the tropics to temperate regions and is a useful garden vegetable in the warmer and more humid regions where true garlic is difficult to grow (greenharvest.com.au). It grows well in deeply cultivated, rich and well-drained soil with neutral pH in full sun, *i.e.* atleast six hours of direct sunlight on most days but not the waterlogged soil (thespruce, 2022). High humidity can lead to fungal growth and rot on the plants therefore it is necessary to have full access to adequate soil drainage *cum* air circulation.

Season

Elephant garlic in cool climates is planted in autumn or spring whereas in warmer climates in spring, autumn or even in winter for a late-spring harvest. It is planted in the fall to be harvested about eight months later in the following summer or in the spring to get early harvest in about three months (thespruce, 2022). Fall-planted crop usually has enough time to reach maturity and producing flowers and bulbs with several cloves. The plants that do not produce flowers can be left unharvested to finish maturing the next year or the growers may opt to harvest the single bulb. Spring-planted elephant garlics generally develop a large round single-cloved bulb.

Field preparation and planting

After deep ploughing individual cloves or bulblets are separated from the bulb and planted with their pointed end up around 10 to 15cm deep, spaced at 20-30cm apart to allow some air flow in between the plants. After planting, it should be watered lightly (thespruce, 2022).

Irrigation

These plants need regular watering so that soil remains lightly moist but not soggy. Overwatering can rot the bulbs. It's best to water in the mornings, so the foliage can dry before nighttime otherwise damp foliage can lead to fungal growth.

Manures and fertilizer

Before planting, the soil should be fortified with 150q/ha of FYM. Another layer of compost is required to be added in the following year if not harvested in the first year (thespruce, 2022).

Cultural practices

Just like regular garlic, elephant garlic too may produce flower stalks or scapes that will draw energy from the plant which is going towards bulb formation (bulking). Therefore, to save the energy for bulking, scapes are immediately removed as their initiation has been noticed. However, these removed scapes being edible can be used with salad or elsewhere. The weeds should also be brought under control whenever these are noticed as these will compete with the elephant garlic for space, nutrients, sun and water. A layer of mulch around the plants can help to protect the roots from extreme temperatures, to save the moisture and to check growth of weeds.

Harvest, cure and storage

Bulbs should be harvested when the flowers begin to dry out, usually the second year after planting. The whole plant is lifted and bulbs are cleaned by brushing off the soil and by removing dried leaves. For storage, bulbs are dried by spreading on the ground or by hanging for about 3-8 weeks in a cool and dark place but with good air circulation for around 3-8 weeks. Any remaining roots and scape should be cut off leaving only about 5cm of the scape.

Cured bulbs are stored at 7-12^{0}C with less than 50 per cent relative humidity. Under these conditions, these can be retained with for up to 8-10 months (thespruce, 2022).

Unlike many types of garlic, elephant garlic does not have to be harvested or divided each year. It is quite care-free crop so it can be left in the ground as such without any risk of rotting or damage. Once the bulb divides, the plant will spread into a clump with many flowering heads. These are often left in flower gardens as an ornamental and to discourage pests. Once they get overcrowded, the plant growth is stunted and then rotting may also start (en.wikipedia.org c).

Pests and diseases

Elephant garlic generally has relatively few issues with pests and diseases. Slugs might damage young plants, especially during damp seasons. Fungal diseases will often afflict plants growing in wet conditions. In these cases, the diseased plants should be removed and started at better growing conditions.

Kurrat/Egyptian/Salad leek (*A.a.*var. *kurrat*, syn. *A. kurrat* Schweinf. ex K.Krause)

Introduction

Source: *sites.google.com*

Kurrat, grown in the Middle East mainly for its leaves (Messiaen and Rouamba, 2004; plants.ces.ncsu.edu) is closely related to the leek and has similar uses. It has been cultivated in Egypt for at least 2,500 years (sites.google.com; pfaf.org) and is a popular vegetable in Egypt and eastern Mediterranean countries where it is mainly used fresh and for seasoning and is usually grown for successive harvests of green leaves. 'Taree Irani', grown in Iran, is a kurrat-like plant producing in its umbels a mixture of flowers and bulbils (Messiaen and Rouamba, 2004).

Botany

Its chromosome number is 2n=4x=32. Kurrat is a seed-propagated leek-like plant though smaller and growing up to 1m tall (pfaf.org). The hermaphrodite flowers are pollinated by bees and other insects. It has small bulbs. Its cultivation is similar to *A. ampeloprasum* var. *porrum*.

Nutritional Values

Principle	*Nutrient Value*	*Percentage of RDA*
Energy	61 Kcal	2%
Carbohydrates	14.15g	11%
Protein	1.50g	3%
Total Fat	0.30g	1%
Cholesterol	0 mg	0%
Dietary Fiber	1.8g	5%
Vitamins		
Folates	64μg	16%
Niacin	0.4mg	2.50%
Pantothenic acid	0.14mg	3%
Pyridoxine	0.233mg	18%
Riboflavin	0.03mg	2%
Thiamin	0.06mg	5%
Vitamin A	1667 IU	55%
Vitamin C	12mg	20%
Vitamin E	0.92mg	6%
Vitamin K	47μg	39%
Electrolytes		
Sodium	20mg	1%
Potassium	180mg	4%
Minerals		
Calcium	59mg	6%
Copper	0.12mg	13%
Iron	2.1mg	26%
Magnesium	28mg	7%
Manganese	0.481mg	2%
Phosphorus	35mg	5%
Selenium	1μg	2%
Zinc	1.2mg	11%
Phyto-nutrients		
Carotene-ß	1000μg	—
Crypto-xanthin-ß	0 μg	—
Lutein-zeaxanthin	1900μg	—
Source: *Veggiesinfo.com*		

Uses

Whole plant can be cooked and used as the leeks. The leaves and bulbs can be eaten raw or used as flavouring in cooked dishes (Messiaen and Rouamba, 2004). The flowers are normally used as a garnish on salads. These contain sulphur compounds which give them their onion flavor. When it is added to the diet on a regular basis, it helps reducing the blood cholesterol levels, acts as a tonic to the digestive system and also tonify the circulatory system (pfaf.org; sites.google.com).

Cultivation

Its cultivation is similar to leek.

Pearl/Baby/Button onion/Creamers/European leek/Silverskin onion (*A. ampeloprasum* var. *sectivum*/*A.a.* 'Pearl-Onion Group')

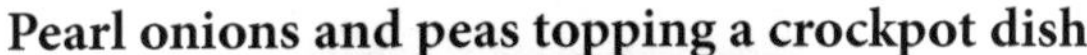

Pearl onions and peas topping a crockpot dish

"By Melissa Doroquez - originally posted to Flickr as Halibut in Creamy Wine Sauce, CC BY-SA 2.0, commons.wikimedia.org/w/index.php?curid=6572908

Pearl onion, a close relative of leek is grown on a small scale in Europe. It may be distinguished from common onions by having only a single storage leaf similar to garlic cloves (en.wikipedia.org d). It is found only in old home gardens in Germany and the Netherlands. This European bulbous leek has sessile cloves covered by a thin silvery-white skin, and sometimes a few shortly stiped cloves or shellless bulblet. It is of unique small size with a taste sweeter than that of a common onion (Messiaen and Rouamba, 2004). When tropical vegetatively propagated forms of leek are experimentally grown in temperate countries, they appear to belong to this category as these produce bulbs usually with a central spherical clove surrounded by small ones. Some of them never flower under long-day conditions, *e.g.* the West Indian leek never reached the flowering stage when grown in southern France for 20 years. On the other hand, the Indonesian 'Anak prei' flowers in the Netherlands, its seeds giving a heterogeneous progeny (Messiaen and Rouamba, 2004). Pearl onion can be used raw as salads, cooked like vegetables, pickled in vinegar or lemon juice, used as a seasoning, or as condiment. It contains chemical compounds that help in improving cardiovascular health and stabilizing blood sugar levels. It is anti-inflammatory and also acts as an antioxidant (en.wikipedia.org d). Known small white varieties of Pearl onion are 'Crystal Wax' or 'White Bermuda' and the red varieties are milder in flavour. Pearl onions are ready to harvest from seed in 90 days and can be stored for up to a month in a cool, dry and dark place (en.wikipedia.org d).

Persian leek or Tareh (*Allium ampeloprasum* ssp. *persicum*)

Persian leek is a cultivated allium native to the Middle East and Iran grown for culinary purposes (en.wikipedia.org a). The linear green leaves have a mild onion flavour and is eaten raw, either alone or in food combinations (Mousavi *et al.*, 2006). Like the leek (*A. ampeloprasun* var. *porrum*), Persian leek too is a tetraploid (2n=4x=32) seed propagated vegetable (Panahandeh, 2016) that does not develop real bulb or produce any cloves.

References

Adao, C.R., B.P. DaSilva and J.P. Parente, 2011. A new steroidal saponin with antiinflammatory and antiulcerogenic properties from the bulbs of *Allium ampeloprasum* var. *porrum*. *Fitoterapia*, **82**(8): 1175-80. (https://www.ijsr.net/archive/v4i7/SUB156374.pdf).

Agrojiva.com/leek-cultivation-varieties-time-of-sowing-yield/

Alam, A., Al Arif Jahan Abdullah, Md. Sazzadul Bari, Labony Khandokar, Md. Hasan Mahmud, Muhammed Junaid, Mohammed Salahuddin Chowdhury, Mohammad Forhad Khan, Veronique Seidel and Md. Areeful Haque, 2022. Allium vegetables: Traditional uses, phytoconstituents, and beneficial effects in inflammation and cancer. *Crit. Rev. Food Sci. Nutrit.*, **43**: 1-32. DOI: 10.1080/10408398.2022.2036094. https://doi.org/10.1080/10408398.2022.2036094.

Alfonso, C., F. Ernesto, L. Virginia and M. Silvana, 1998. Porric Acids A–C – New Antifungal Dibenzofurans from the Bulbs of *Allium porrum* L. *Eur. J. Org. Chem.*, **19**(4): 661–3. https://www.ijsr.net/archive/v4i7/SUB156374.pdf

Ayumi, U., O. Jun, K. Hitomi, A. Makoto, M. Hiroshi, S. Hidetoshi and S. Ichiro, 2009. Mechanisms of Sulfide Components Expression and Structural Determination of Substrate Precursor in Jumbo Leek (*Allium ampeloprasum* L.). *Nippon Shokuhin Kagaku Kogaku Kaishi*, **56**(5): 280-85. https://www.ijsr.net/archive/v4i7/SUB156374.pdf

Block, E., A.J. Dane, S. Thomas and R.B. Cody, 2010. Applications of direct analysis in real time mass spectrometry (DART-MS) in *Allium* chemistry. 2-propenesulfenic and 2- propenesulfinic acids, diallyl trisulfane S-oxide, and other reactive sulfur compounds from crushed garlic and other Alliums, *J Agric Food Chem.*, **58**(8):4617-25.

Botanistinthekitchen.blog/2013/03/03/a-look-at-leeks

Brewster, James L. 2008. Onions and Other Vegetable Alliums (2nd ed.), 30p.. CABI International, Wallingford, UK. CABI International.

Britannica, T. Editors of Encyclopaedia, 2021. Leek. *Encyclopedia Britannica.* (https://www.britannica.com/plant/leek).

Currah, L. 1986. Leek breeding: a review. *J. hort. Sci.*, **61**(4): 407-15.

Devi, P.V.. and D.J.K. Brar, 2018. Comparison of Proximate Composition and Mineral Concentration of *Allium ampeloprasum* (Elephant Garlic) and *Allium sativum* (Garlic). *Chem. Sci. Rev. Lett.*, **7**: 362–7.

Dey, P. and K.L. Khaled, 2915. An Extensive Review on *Allium ampeloprasum,* A Magical Herb. *Intern. J. Sci. Res.,* **4**(7): 371-5. (https://www.ijsr.net/archive/v4i7/SUB156374.pdf).

Dey, P. and K.L. Khaled, 2013. An extensive review on Allium ampeloprasum a magical herb. *Int. J. Sci. Res.*, **4:** 371-7.

En.wikipedia.org a: en.wikipedia.org/wiki/Allium_ampeloprasum

En.wikipedia.org b: en.wikipedia.org/wiki/Leek

En.wikipedia.org c: en.wikipedia.org/wiki/Elephant_garlic

En.wikipedia.org d: en.wikipedia.org/wiki/Pearl_onion

En.wiktionary.org/wiki/leek

Fattorusso, E., V. Lanzotti, O. Taglialatela-Scafati and C. Cicala, 2001. The flavonoids of leek, *Allium porrum*. *Phytochem.*, **57**(4):565-9. (https://www.ijsr.net/archive/v4i7/SUB156374.pdf).

First-Nature. *Allium ampeloprasum* - Wild Leek. https://www.first-nature.com/flowers/allium-ampeloprasum.php

Flora of North America, 1753. FNA, 26:225, 226, 228, 238, 239. *Allium ampeloprasum* Linnaeus, Sp. Pl. 1: 294. http://www.efloras.org/florataxon.aspx?flora_id=1&taxon_id=242101330

Garcia-Herrera, P., P. Morales, V. Fernandez-Ruiz, M.C. Sanchez-Mata, M. Camara, A.M. Carvalho, I.C.F.R. Ferreira, M. Pardo-de-Santayana, M. Molina and J. Tardio, 2014. Nutrients, phytochemicals and antioxidant activity in wild populations of *Allium ampeloprasum* L., a valuable underutilized vegetable. *Int. Food Res. J.*, **62**: 272–9.

Greenharvest.com.au/Plants/Information/ElephantGarlic.html

Hanelt, P. 1996. Proposal to conserve the name *Allium ampeloprasum* against *A. porrum* (Liliaceae). *Taxon*, **45**(4): 691-2.

Kratchanova, M., M. Nikolova, E. Pavlova, I. Yanakieva and V. Kussovski, 2010. Composition and properties of biologically active pectic polysaccharides from leek (*Allium porrum*). *J. Sci. Food Agric.*, **90**(12): 2046-51. (https://www.ijsr.net/archive/v4i7/SUB156374.pdf).

Leek, 2008. New World Encyclopedia. Retrieved 04:27, January 4, 2023 from https://www.newworldencyclopedia.org/entry/Leek

Loppi, S., R. Fedeli, G. Canali, M. Guarnieri, S. Biagiotti and A. Vannini, 2021. Comparison of the mineral and nutraceutical profiles of elephant garlic (*Allium ampeloprasum* L.) grown in organic and conventional fields of Valdichiana, a traditional cultivation area of Tuscany, Italy. *Biology*, **10**(10): 2-10. doi.org/10.3390/biology10101058. pdfs.semanticscholar.org/8672/5412c5310a208b7f8627f274a904b6afafc8.pdf.

Messiaen, C.-M. and A. Rouamba, 2004. *Allium ampeloprasum* L. [Internet] Record from PROTA4U. Grubben, G.J.H. and O.A. Denton (Eds). PROTA (Plant Resources of Tropical Africa / Ressources végétales de l'Afrique tropicale), Wageningen, Netherlands. http://www.prota4u.org/search.asp; https://uses.plantnet-project.org/en/Allium_ampeloprasum_(PROTA).

Missouri Botanical Garden. missouribotanicalgarden.org/PlantFinder/PlantFinderDetails.aspx?taxonid=281717

Mousavi, Amir, Abedolkarim Kashi, Daryoush Davoodi and Mohammad Sanei Shariatpanahi, 2006. Characterization of an *Allium* cultivated in Iran: The Persian Leek. *Belgian J. Bot.*, **139**(1): 115–23.

Onions and other allium plants. https://www.encyclopedia.com/food/encyclopedias-almanacs-transcripts-and-maps/onions-and-other-allium-plants

Panahandeh, J. 2016. Meiosis in Persian leek *Allium ampeloprasum* ssp. *persicum*. *Acta Hortic.*, No. 1143: 23-6. (DOI: 10.17660/ActaHortic.2016.1143.4. https://doi.org/10.17660/ActaHortic.2016.1143.4).

Pandey, A., Pavan Kumar Malav, K.M. Rai and S.P. Ahlawat, 2022. Genus *Allium* L. of the Indian Region: A Field Guide for Germplasm Collection and Identification. ICAR-National Bureau of Plant Genetic Resources, New Delhi, India, 109p + i-vi. http://www.nbpgr.ernet.in/Downloadfile.aspx?EntryId=9197.

Pfaf.org/user/cmspage.aspx?pageid=174

Piazmiaz.com/glossary/leeks-tarreh/

Plants.ces.ncsu.edu/plants/allium-ampeloprasum/

Roper, P. 2018. Edible wild leek bulbs. Ramblings of a Naturalist. http://ramblingsofanaturalist.blogspot.com/2018/09/edible-wild-leek-bulbs.html

Sanderson, H. and R.M. Jane, 2005. The Cultural History of Plants (eds Ghillean Prance and Mark Nesbitt), 121p. Routledge, England (ISBN 0415927463).

Sanyat Misra and R.L. Misra, 2013. *Allium.* In:Commercial Ornamental Bulb Science, pp. 2-10. Westville Pub. House, New Delhi.

Sata, N., S. Matsunaga, N. Fusetani, H. Nishikawa, S. Takamura, and T. Saito. 1998. New antifungal and cytotoxic steroidal saponins from the bulbs of an elephant garlic mutant. *Bioscience, Biotechnology, and Biochemistry,* **62** (10):1904–11. doi: 10.1271/bbb.62.1904.

Selvakumar, R., Praveen Kumar Singh, Jugendra Kumar and D.C. Manjunathagowda, 2021. Varietal Wealth of Gourmet's Onion – Future Vegetable. *Just Agriculture,* pp.30-34. Pub. by ICAR-Directorate of Onion and Garlic Research, Rajgurunagar, Pune. justagriculture.in/files/magazine/2021/september/006%20 Varietal%20Wealth%20of%20Gourmet.pdf

Sites.google.com/site/knowyourvegetables/know-your-onions/know-your-kurrat?pli=1

Sweetishhill.com/does-leek-grow-in-india/

Temperate Vegetables. Leek. https://ecoursesonline.iasri.res.in/mod/page/view.php?id=101636. 21 June 2012. Retrieved 4 January 2023.

thespruce.com/elephant-garlic-1402622. Retrieved December, 2022.

Veggiesinfo.com. Kurrat | Egyptian Leek Properties And Uses. https://veggiesinfo.com/kurrat-egyptian-leek-properties-and-uses/

W3.biosci.utexas.edu/prc/K12/pages/Allium%20ampeloprasum.html

Wisconsin Horticulture. https://hort.extension.wisc.edu/articles/leeks4

Chives (*Allium schoenoprasum* L.)

Subgenus: *Cepa*; **Section:** *Schoenoprasum* (Pandey *et al.*, 2022).

Other common names: Snow mountain garlic, Kashmiri garlic (Alam *et al.*, 2022).

Synonyms: *Allium acutum* Spreng., *A. alpinum* (DC.) Hegetschw., *A. broteri* Kunth, *A. buhseanum* Regel, *A. carneum* Schult. & Schult.f., *A. foliosum* Clairv. ex DC., *A. gredense* Rivas Goday, *A. idzuense* H.Hara, *A. palustre* Chaix, *A. punctulatum* Schltdl., *A. purpurascens* Losa, *A. raddeanum* Regel, *A. reflexum* F.Dietr., *A. riparium* Opiz, *A. sibiricum* L., *A. tenuifolium* Pohl, *A. tenuifolium* Salisb., *A. ubinicum* Kotukhov, *A. udinicum* Antsupova, *Ascalonicum schoenoprasum* (L.) P.Renault , *Cepa schoenoprasa* (L.) Moench, *C. tenuifolia* (Salisb.) Gray, *Porrum schoenoprasum* (L.) Schur, *P. sibiricum* (L.) Schur, *Schoenoprasum vulgare* Fourr (The Plant List, 2017).

Flower with bud	
 **Source:** *GreenMyLife (2017)*	 **By Sporlo - Own work, CC0, https://commons.wikimedia.org/w/index.php?curid=16844696**

Introduction and Origin

The specific epithet *schoenoprasum* has been derived from the Greek *skhoinos* (sedge or rush) and *prason* (leek) (Anderberg, 2023; en.wikipedia.org/wiki/Chives). The English common name 'chives' comes from the French word *cive* and from *cepa* ~ the Latin word for onion (Linford, 2010). It was known as 'rush leek' in the Middle Ages (Linford, 2010). **Chives**, the only species of *Allium* native to both the New and the Old Worlds (Small, 2014; Cullen *et al.*, 2011) is widespread in nature across temperate regions of much of Europe, Asia and North America (FOC, 2017; POWO, 2021; Tardiff and Morisset, 1990). In Europe, chives have been cultivated since the Middle Ages for culinary as well as for medicinal purposes (Mahr, 2023), and as long as 4,000 years ago by the Chinese. Today the leaves are typically used as a culinary herb with a mild onion flavour (Mahr, 2023). It is a highly demanded crop in Manipur (India), where it is grown and produced throughout the year (Salam, 2022). Once planted, it continuously gives high returns up to its 7 years (Salam, 2022). Though, chives have the potential for commercialization to supplement the onion and garlic production in different parts of India, it is still a less known cultivated vegetable in the rest of India. Hence, there is a need to standardize the production technology of chives including post-harvest management for its commercial cultivation. It is related to garlic chives (*A. tuberosum*), a perennial with brighter green, flatter and more angular leaves and white flowers in the fall (Mahr, 2023).

Botany

Chives are herbaceous bulbous perennials growing 30–50 cm tall from the bulbs that are slender, conical, 2–3cm long and 1cm broad, and grow in dense clumps or clusters from the roots, spreading up to 20cm and producing soft-textured hollow and tubular scapes though before floral initiation these may appear stiffer than usual (en.wikipedia.org/wiki/Chives). Its thin grass-like (Linford, 2010) dark green leaves that are shorter than the scapes are also hollow and tubular or terete (round in cross-section) which distinguishes it from garlic chives (*Allium tuberosum*) (en.wikipedia.org/wiki/Chives).

Pale-purple, globular and star-shaped clusters of 10-30 flowers, each with six 1–2cm wide petals are produced in a dense inflorescence. Before opening, the inflorescence is surrounded by a papery bract. The seeds that are produced in a small 3-valved capsule mature in summer. This herb flowers from April to May in the southern parts of its habitat zones though in June in the northern parts (*Allium schoenoprasum* factsheet, 2006).

Genetics

Its diploid chromosome numbers are 2n=16, 24 and 32. A cross between a tetraploid giant form of *Allium Schoenoprasum* (2n = 32) and common diploid chive (2n = 16) gave a triploid F_1 (2n = 24). These all the F_1 plants though were completely self-sterile due to being triploid but in nature through free-flowering habits these produced a rather large number of seeds which gave rise to an F_2 generation of about 700 plants. These arose from crosses within the triploid material where some may possibly have been the result of back-crossings with diploid types of chive flowering in the vicinity (Levan, 1934).

Cultivars

Though disputed but the plants found in North America are sometimes classified as *A. schoenoprasum* var. *sibiricum*. Differences between specimens are significant. It was found growing in northern Maine growing solitarily instead of in clumps producing together with dingy grey flowers (McGary, 2001). There are a number of cultivars of chives in varying heights including dwarf or tall types growing in isolated pockets though may not be readily available (Mahr, 2023).

- 'Album' is a white flowered form
- 'Curly Mauve' has prostrate foliage that curls in different directions and has gray-lavender flowers held erect above the short curled foliage.
- 'Forescate' is larger than the species, growing 45-50cm tall with large and bright flowers.
- 'Marsha' is a very deep purple form.
- 'Pink Giant' is taller than other types, with leaves about twice the size of the species, and large, clear pink blossoms.
- 'Profusion' is a sterile form that flowers over a longer period than the species.
- 'Snowcap' is a dwarf white-flowered cultivar.

Nutritional Values

Chives are composed mainly of carbohydrates, proteins and amino acids, as well as several vitamins and minerals. Being rich in vitamins A & C, Ca and Fe, these also contain sulphur in traces. The unique spicy flavour comes mostly from volatile sulphur and glucosides (Chen, 2006).

Nutritional value per 100g raw chives as given in en.wikipedia.org/wiki/ Chives is shown in the following table:

Chives, raw	
Nutritional value per 100g	
Energy	126 kJ (30 kcal)
Carbohydrates	4.35 g
Sugars	1.85 g
Dietary fiber	2.5 g
Fat	0.73 g
Protein	3.27 g
Vitamins	**%DV† - Quantity**
Vitamin A equiv. beta-Carotene lutein zeaxanthin	27% - 218 μg 24% - 2612 μg 323 μg
Thiamine (B1)	7% - 0.078 mg
Riboflavin (B2)	10% - 0.115 mg
Niacin (B3)	4% - 0.647 mg
Pantothenic acid (B5)	6% - 0.324 mg
Vitamin B6	11% - 0.138 mg
Folate (B9)	26% - 105 μg
Vitamin C	70% - 58.1 mg
Vitamin E	1% - 0.21 mg
Vitamin K	203% - 212.7 μg
Minerals	**%DV† - Quantity**
Calcium	9% - 92 mg
Iron	12% - 1.6 mg
Magnesium	12% - 42 mg
Manganese	18% - 0.373 mg
Phosphorus	8% - 58 mg
Potassium	6% - 296 mg
Zinc	6% - 0.56 mg
Units:- μg = micrograms; mg = milligramsIU = International units	
†Percentages are roughly approximated using US recommendations for adults. ***Source:*** USDA FoodData Centra	

Phytochemicals

The water extract of *A. schoenoprasum* contains flavonoid compounds, glycosides and saponins while the methanol and ethyl acetate extracts contain alkaloids, flavonoids, glycosides, and saponins (Sihombing *et al.*, 2018). The biological activities of flavonoids are mainly due to their antioxidant function. They are also known to inhibit several enzymes, including lipoxgenases and cyclo-oxygenase, *etc.* The green leaves of chive contain mainly kaempferol glucosides (di- and tri-glycosides), dominant as glucose and galactose. The 3-beta-D-glucosides of kaempferol, quercetin and isorthamnetin have been isolated (Chen, 2006).

The flavouring substances of various *Allium* species depend on the quantitative differences in the S-alk(en)yl cysteine sulphoxides (ACSOs) which is hydrolyzed by the enzyme allinase giving rise to the flavour and pungency characteristic. The factors affecting flavour quality and intensity include genetic, ecological and cultivation techniques. Four different ACSOs (2-methyl-2-butenal, 2-methyl-2-pentenal, methyl-propyldisulphide and dipropyldisuphide) that have been identified in *Allium* species have also been found in the green leaves of chives. Propencyl-propyldisulphide is also present in chives but not the allyl disulphide. Of the major thiosulphates found in chives, 75-77% contains the n-propyl group, 10-12% the methyl and 12% contains the 1-propyl group. Total thiosuphates found is 0.19umol/g which is about the middle level of the edible species of *Allium*. In chives, n-propyl group is more abundantly found than methyl, with the methyl/propyl 1:5.8. The pungent and stimulating flavour of both chives and onions is mainly due to the propyl group but the quantitative and qualitative differences in the thiosulphates give each species its own characteristic flavour. The flavour and lachrymatory properties of chives are due to the high n-propyl content while the flavour of onions is due to the high proportion of the 1-propyl content (Chen, 2006).

A. schoenoprasum extracts growth inhibitor (TGI) index values range from 10.2 to 38.4% (Shirshova *et al.*, 2013). Four spirostane-type glycosides and four steroidal saponins isolated from the whole plant of *A. schoenoprasum* were investigated *in vitro* against HCT-116 and HT-29 human colon cancer cell lines. All the tested phytochemicals exerted little to moderate cytotoxic activity against both cell lines. Both diosgenin 3-O-β-D-glucopyranosyl-(1→4)-[α-L-rhamnopyranosyl-(1→2)]-β-D-glucopyranoside, and (25)-furost-5-en-3β,22α,26-triol 26-O-β-D-glucopyranosyl-3- O-α-L-rhamnopyranosyl-(1→2)-[β-D-glucopyranosyl- (1→4)]-β-D-glucopyranoside demonstrated potent cytotoxicity against the HCT 116 cell line (IC_{50} values of 0.40 μM and 1.58 μM, respectively) and the HT-29 cell line (IC_{50} values of 0.75 μM and 1.56 μM, respectively). Only (25 R)-5α-spirostan-3β,11α-diol-3-O-β-D-glucopyranosyl- (1→3)-[β-D-glucopyranosyl-(1→4)]-β-D-galactopyranoside was found to be moderately active (IC_{50} values of 8.45 μM and 8.64 μM) against the HCT 116 and the HT-29 cell line, respectively. Another compound, laxogenin 3-O-α-L- rhamnopyranosyl-(1→2)-β-D-glucopyranoside proved to be ineffective against both cell lines with IC_{50} values greater than 100 μM (Timite *et al.*, 2013; Alam *et al.*, 2022). Bioactive constituents from *A. schoenoprasum* have anti-cancer activity against certain cell lines (Alam *et al.*, 2022).

Uses

In Manipur (India), chives are popularly used as spices and condiments, especially in making popular dishes, fritters and are also consumed raw (Salam, 2022). Chives provide somewhat milder flavour than those of other *Allium* species. Its stems, leaves and florets are used for culinary purposes. The stems and the unopened immature flower buds are used in salads, potatoes, soups, certain other dishes, soft cheese, omelettes or an ingredient for fish (Linford, 2010). The edible flowers with a mild onion flavour can be sprinkled into salads, used to make 'blossom vinegars' or to garnish dishes (Linford, 2010; *Allium schoenoprasum* factsheet, 2006). Chives are one of the *fines herbes* of French cuisine, the others being tarragon, chervil and parsley (en.wikipedia.org/wiki/Chives).

Chives have similar but weaker medicinal properties than garlic though rarely used medicinally. These are mildly stimulant, diuretic and antiseptic properties. It is used for common cold, pulmonary edema, hypertension, digestive problems, colds, flu *cum* lung congestion, pain from sunburn and in sore throat (Singh *et al.*, 2018; Alam *et al.*, 2022). Chives can be sprinkled on food to stimulate appetite and to promote digestion (Singh *et al.*, 2018). Some digestive problems may occur due to over-consumption of chives.

Chives can be used in gardens to control pests because it has insect-repelling properties (Kaufman *et al.*, 1999). The growing plant as well as the leaf juice repel unwanted insect life and together fight mildews, scabs and certain other fungal infections (en.wikipedia.org/wiki/Chives). Although chives are repulsive to insects in general, their flowers attract bees, and they are at times kept to increase desired insect life (en.wikipedia.org/wiki/Chives).

Cultivation

Propagation

Chives, hardy to zone 3, are one of the easiest herbs to grow either from seed or from divisions. Vegetatively the propagation of chives is by division of existing clump of bulbs, keeping some 5 bulblets/clump (Salam, 2022). It forms dense clumps which are easily divided in spring or fall (*Allium schoenoprasum* factsheet, 2006). Every 2-3 years, plants should be divided to prevent overcrowding. Before planting, well decomposed organic matter must be incorporated in the field.

Chives are easily grown from seeds. Seeds mature from early spring to summer. Seeds can be started indoors a few weeks before the average date of the last frost. Once these attain about 10cm height (rhs.org.uk), these are planted outside after the danger of frost is over (Mahr, 2023). These can also be directly seeded once the soil starts to warm up or when atmospheric temperature is 15-20^{0}C but after sowing the soil should be kept moistened. Seeds should be sown about 5mm deep and 10cm apart in the soil of nursery beds (GreenMyLife, 2017). They can also be planted under a cloche or germinated indoors in cooler climates and then planted out later when weather improves. After at least four weeks, the young shoots should be ready to be planted out. They are also easily propagated through division.

Climate and Soil

Chives can easily be grown under partial to fully sunny situation in a moisture-retentive well-drained sandy-loam soil rich in humus, having 6-7 soil pH though flowering may be found reduced under partial shade. Clumps are generally planted 30-50 cm apart (Mahr, 2023). Being a fairly shallow-rooted crop, chives requires special care around it to control weeds so organic mulching would be quite effective. These can also be grown either on their own or in a mixed herb display in large pots having soil-based compost (rhs.org.uk). Small clumps can be potted up and moved indoors in late fall to grow in a bright windowsill through the winter to harvest fresh leaves during winter months (Mahr, 2023). In temperate regions, the chives die back leaving their bulbs underground where new leaves start appearing with improvement in the weather (McGee and Stuckey, 2002).

Irrigation, Fertilization and Other Cultural Practices

Though chives are drought-tolerant once established, but they grow best when given consistent moisture (Mahr, 2023). Adequate nitrogen is important and a pre-plant complete fertilizer or composted manure is recommended with additional nitrogen application after the first harvest every year (Salam, 2022). Weed control is important during the first two months of growth when the plants are growing slowly and cannot compete with weeds. A slow grow-fertilizer with a high nitrogen concentration can liberally be applied to the plants and the crop should regularly be mulched to conserve moisture and to keep weeds under check though when plants are sufficiently grown there is neither any need of fertilization nor the mulching, however, for good bulking and growth the faded leaves and spent flowers are immediately and regularly removed (rhs.org.uk). The plant will self-seed in the garden if spent flowerheads are not promptly dead-headed (*Allium schoenoprasum* factsheet, 2006).

Harvest

During growing season continuously the plants regrow leaves thus allowing for a continuous leaf harvest when these attain 15cm length which is attained in about 30 days period of sowing or planting as taller *cum* older leaves may become fibrous (McGee and Stuckey, 2002; Mahr, 2023). When harvesting, generally its plants after flowering are sheared at 2.5-8cm from soil level so that plants are forced to produce many new tender leaves. As per requirement harvesting of a few leaves per plant will be better though shearing of entire clump will take weeks to regrow. In the first year, harvesting can be done 3-4 times though in subsequent years this can be harvested at monthly intervals. Constant harvesting of leaves is quite necessary to keep the plants healthy, vigorous and in active growth. Every year the clumps should be dug out, divided and planted but after fortifying the soil with sufficient quantity of FYM or any other manure every time. Since continuous growing *cum* harvesting it is difficult to maintain the quality as leaves get thinner and become liable to infection of rust, hence beyond 7th year its growing continuously in the same field is not desirable (Salam, 2022).

Pests and Diseases

Chives have no significant insect or disease problems, *vis-a-vis* not favourite to deer or rabbits (Mahr, 2023), though these may get infested by various pests that attack onion and leek. Aphids are nuisance to this crop which can best be wiped out by thorough washing of the plant with neem or soapy water (rhs.org.uk). These are also susceptible to leek moth larvae (Landry, 2007) which bore into the leaves and bulbs of the plant so while preparing the land for planting, carbofuran granules may thoroughly be mixed in the soil @ 10kg/ha. Rust attack may be a problem in severe cases which can be controlled through the use of sulphur fungicides.

References

Alam, A., Al Arif Jahan Abdullah, Md. Sazzadul Bari, Labony Khandokar, Md. Hasan Mahmud, Muhammed Junaid, Mohammed Salahuddin Chowdhury,

Mohammad Forhad Khan, Veronique Seidel and Md. Areeful Haque, 2022: Allium vegetables: Traditional uses, phytoconstituents, and beneficial effects in inflammation and cancer. *Crit. Rev. Food Sci. Nutr.*, pp. 1-32. DOI: 10.1080/10408398.2022.2036094. https://doi.org/10.1080/10408398.2022.2036094.

Allium schoenoprasum factsheet, from Kemper center for home gardening, retrieved on June 13, 2006, based on the position of the botanical Garden (Missouri). https://www.missouribotanicalgarden.org/PlantFinder/PlantFinderDetails.aspx?taxonid=282062&isprofile=1&basic=Allium%20schoenoprasum.

Anderberg Anna-Lena. 2023. Den virtuella floran: *Allium schoenoprasum* L. - Graslok. nrm.se. en.wikipedia.org/wiki/Chives. Retrieved 15, January, 2023.

Chen, H. 2006. In: Handbook of Herbs and Spices, Vol. 3, pp. 337-46. Woodhead Publishing Series in Food Science, Technology and Nutrition, South Korea.

Cullen James, G. Sabina and H. Knees, 2011. The European Garden Flora Flowering Plants: A Manual for the Identification (ed. Suzanne Cubey), 133p. Google Books, England.

En.wikipedia.org/wiki/Chives.

FOC, 2017. *Allium schoenoprasum*. In: Flora of China @ efloras.org. www.efloras.org. Retrieved 2017-06-18. Vol. 24: 195p.

GreenMyLife. How To Grow Chives. 2017. https://www.greenmylife.in/how-to-grow-chives/

Kaufman, Peter B, J Carlson Thomas, Kaufman B Kaufman, Harry L Brielmann, Sara Warber, J Cseke Leland and A Duke James, 1999. Natural Products from Plants. CRC Press, Boca Raton, 261p. ISBN 9780849331343. 084933134X.

Landry, Jean-François, 2007. Taxonomic review of the leek moth genus *Acrolepiopsis* (Lepidoptera: Acrolepiidae) in North America. *Canad. Ent.*, **139**(3): 319–53. doi:10.4039/n06-098. ISSN 1918-3240. S2CID 86748199.

Levan, A. 1934. Distribution of Chromosome Numbers in a Progeny of Triploid *Allium Schoenoprasum. Nature*, **13**: 254. https://doi.org/10.1038/134254a0; https://www.nature.com/articles/134254a0.

Linford, Jenny, 2010. A Concise Guide to Herbs. Parragon, Bath,. 45p. ISBN 9781405487993.

Mahr, Susan, 2023. Chives, *Allium schoenoprasum*. Wisconsin Horticulture - Division of Extension. University of Wisconsin – Madison. Retrieved 15, January, 2023. https://hort.extension.wisc.edu/articles/chives-allium-schoenoprasum/.

McGary, Mary Jane, 2001. Bulbs of North America. North American Rock Garden Society. Timber Press, Portland, USA, pp. 28–9.

McGee, Rose Marie Nichols and Maggie Stuckey, 2002. The Bountiful Container. Workman Publishing, USA. https://en.wikipedia.org/wiki/Chives.

Pandey Anjula, Pavan Kumar Malav, K.M. Rai and S.P. Ahlawat, 2022. Genus *Allium* L. of the Indian Region: A Field Guide for Germplasm Collection and Identification. *ICAR-National Bureau of Plant Genetic Resources*, New Delhi, India, 109p + i-vi. http://www.nbpgr.ernet.in/Downloadfile.aspx?EntryId=9197.

POWO, 2021 - Plants of the World Online|*Royal Botanic Gardens, Kew. Allium schoenoprasum* L. https://powo.science.kew.org/taxon/urn:lsid:ipni.org:names:528823-1. Retrieved 20 January 2021.

Rhs.org.uk/herbs/chives/grow-your-own

Salam Priyadarshini, 2022. Chives cultivation in Manipur: A lucrative agribusiness option. http://e-pao.net/epSubPageExtractor.asp?src=education.Jobs_Career.Chives_cultivation_in_Manipur_By_Priyadarshini_Salam.

Shirshova, T., I. Beshlei, V. Deryagina, N. Ryzhova and N. Matistov, 2013. Chemical composition of *Allium schoenoprasum* leaves and inhibitory effect of their extract on tumor growth in mice. *Pharmaceutical Chem. J.*, **46**(11)::672–5. doi: 10.1007/ s11094-013-0867-8.

Singh, V., G. Chauhan, P. Krishan, and R. Shri, 2018. *Allium schoenoprasum* L.: A review of phytochemistry, pharmacology and future directions. Natural Product Res., **32**(18): 2202–16. doi: 10.1080/14786419.2017.1367783.

Small, E. 2014. *North American Cornucopia:* Top 100 Indigenous Food Plants, 230p., at Google Books.

Tardiff, B., and P. Morisset, 1990. Clinal morphological variation of *Allium schoenoprasum* in eastern North America. *Taxon.*, **39**(3): 417–29. doi:10.2307/1223088. JSTOR 1223088.

Timite, G., A.-C. Mitaine-Offer, T. Miyamoto, C. Tanaka, J.-F. Mirjolet, O. Duchamp and M.-A. Lacaille-Dubois, 2013. Structure and cytotoxicity of steroidal glycosides from *Allium schoenoprasum. Phytochem.*, **88**:61–6. doi: 10.1016/j.phytochem.2012.12.001.

There are various other *Allium* species which in their common names have though chives but specific names differ. These are described below:

Chinese chives [*Allium ramosum* L. (1753: 296)]

Subgenus: *Buttomissa*; **section**: *Buttomissa* (Wikispecies, 2023).

Common names: Branchy chive (onion), Fragrant-flowered garlic, Fragrant garlic and in Mongolian language: *Gogd*).

Synonyms: *Aglitheis tatarica* (L.f.) Raf.; *Allium beckerianum* Regel; *A. diaphanum* Janka; *A. lancipetalum* Y.P.Hsu; *A. odorum* L.; *A. potaninii* Regel; *A. ramosum* Georgi 1779; illegitimate homonym not L. 1753 nor Jacq. 1781; *A.*

senescens Miq.; *A. tataricum* Dryand. 1811, illegitimate homonym not L. 1782; *A. tataricum* L.f. 1782; *A. umbellatum* Haller f. ex Steud.; *A. weichanicum* Palib.; *Butomissa tatarica* (L.f.) Salisb., not validly published; *Moly odorum* (L.) Moench.

By Michael Wolf - Own work, CC BY-SA 3.0, https://commons.wikimedia.org/w/index.php?curid=7465500

Introduction and origin

Allium ramosum, a wild relative of *A. tuberosum* (Blattner and Friesen, 2006) is sometimes cultivated as a vegetable in NE China (Choi and Oh, 2011). Both these species belong to a small section *A. sect. Butomissa* (Salisb.) Kamelin which is subordinated to a subgenus of the same name, *A.* subgen. *Butomissa* (Salisb.) N.Friesen (Friesen *et al.,* 2006). Chinese chives is native to Kazakhstan, Mongolia, Siberia, the Russian Far East and northern China, as well as naturalized in a few places in eastern Europe (Ballyrobert Gardens, 2023) though in the wild it is found growing at elevations of 500-2100m elevations (FOC, 2023). The plant is often harvested from the wild for local use as a food, and is sometimes also cultivated as a food plant within its native range. This speies is classified as 'Least Concern' in the IUCN Red List (2013).

Botany

Allium ramosum is a perennial herbaceous plant producing a 25-60 cm tall, terete, obscurely angled, flowering scape that is covered with leaf sheaths only at base and a cluster of 3-7 narrowly linear, grass-like, 3-angled, abaxially 1-keeled and fistulose leaves that are 20-50cm long and 2-4mm wide. Clusters of narrow, oblong-ovoid or conical bulbs are produced in groups of 1-3 along a well-developed horizontal root (AgroAtlas, 2023; FOC, 1994). Flowering and fruiting takes place from June to September (FOC, 2023). Umbel consists of many pure white to pale-red flowers having red midveins running down the backs of the tepals (FOC, 2023; Hsu, 1987). As per AgroAtlas (2023), it possesses diploid chromosome number of 2n=32.

Phytochemicals

Compounds found include Skimmianine, Kokusaginine, 4(1H)-Quinazolinone, 1-methyl, Arborine, Dictamnine, Arborinine & 1,5-Dihydroxy-2,3-dimethoxy-10-methyl-9-acridone (NPASS); and 1-Methyl-1,2,3,3a,8,8aalpha-hexahydropyrrolo[2,3-b]indole-3aalpha-ol (LOTUS). A new furostane saponin, ramosaponin (1) and four known furostane saponins, protodioscin (2), dehydrotomatoside (3), *(25 R)-26-O-(β*-D-glucopyranosyl)-furost-5-ene-*3β,22α,26*-triol 3-*O*-*β*-D-glucopyranosyl-(1→4)-*β*-D-galactopyranoside (4), and anguivioside A (5) were isolated from the methanol extract of *Allium ramosum* seeds. All compounds were evaluated for reduction of lipid accumulation in HepG2 cell line. As a result, compound 1 showed significant lipid accumulation inhibitory activity with an IC50 value of 64.32 ± 3.87 µM (Duong *et al.*, 2022).

Uses

Its flavour is somewhat between that of garlic and chives (Facciola, 1990). Leaves and bulbs can be eaten raw or cooked and flowers are used as garnish on salads. From the wild the plant is gathered between May and July in Mongolia and preserved with salt for winter use. The preserved ones then are used to season boiled mutton or stuffed into dumplings (Pieroni, 2005). The flowers, called *soriz*, are gathered in late July & August and salted (Pieroni, 2005).

The leaves and bulbs contain sulphur compounds, saponins and bitter substances (Dan and Nhu, 1989). These possess antibacterial properties and are used in Vietnam in the treatment of haemoptysis, epistaxis, cough, sore throat, asthma, dysentery, dyspepsia, *etc.* (Dan and Nhu, 1989). In Mongolia, it is also used to treat stomach ailments (Pieroni, 2005).

Its nutritional values are almost the same as garlic chives, and cultivation is quite similar to onion.

References

AessenseGrows. https://www.aessensegrows.com/en/fresh-produce-catalog/garlic_chives. Retrieved 21 January, 2023.

AgroAtlas. Interactive Agricultural Ecological Atlas of Russia and Neighboring Countries. http://www.agroatlas.ru/en/content/related/Allium_ramosum/index.html. Retrieved 22 January, 2023.

Ballyrobert Gardens: A Royal Horticultural Society Partner Garden. https://www.ballyrobertgardens.com/products/allium-ramosum. Retrieved 26 January, 2023.

Blattner, F.R. and N. Friesen, 2006. Relationship between Chinese chive (*Allium tuberosum*) and its putative progenitor *A. ramosum* as assessed by random amplified polymorphic DNA (RAPD). Documenting Domestication: New Genetic and Archaeological Paradigms, pp. 134-142 (eds Zeder, M.A., D.G. Bradley, E. Emshwiller and B.D. Smith). California University Press, Berkeley, USA. www.biotaxa.org/Phytotaxa/article/view/phytotaxa.134.1.6.

Choi, H.J. and B.U. Oh, 2011. A partial revision of *Allium* (Amaryllidaceae) in Korea and north-eastern China. *Bot. J. Linnean Soc.*, **167**: 153–211.http://dx.doi.org/10.1111/j.1095-8339.2011.01166.x; www.biotaxa.org/Phytotaxa/article/view/phytotaxa.134.1.6.

Dan, N.V. and D.T. Nhu, 1989. Medicinal Plants in Vietnam. World Health Organisation, Italy. ISBN: 92 9061 101 4. https://temperate.theferns.info/plant/Allium+ramosum

Duong, Thi Hai Yen; Hoang Thi Tuyet Lan, Bui Thi Mai Anh, Phan Van Kiem, Bui Huu Tai, Le Thi Huyen, Pham Hai Yen, Nguyen Xuan Nhiem and Nguyen Thi Mai, 2022. One new furostane saponin from *Allium ramosum* and lipid accumulation inhibitory activity, *J. Asian Nat. Prod. Res.,* DOI: 10.1080/10286020.2022.2098728. www.tandfonline.com/doi/abs/10.1080/10286020.2022.2098728.

Facciola, S. 1990. Cornucopia - A Source Book of Edible Plants. Kampong Publications. ISBN: 0-9628087-0-9. https://temperate.theferns.info/plant/Allium+ramosum

FOC, 1994. Flora of China. Missouri Botanical Garden Press; St. Louis, USA. http://flora.huh.harvard.edu/china/; https://temperate.theferns.info/plant/Allium+ramosum.

FOC, 2023. *Allium ramosum*. Flora of China, vol. v, 180p. www.efloras.org/florataxon.aspx?flora_id=2&taxon_id=200027524. Retrieved 26 January, 2023.

Friesen, N., R.M. Fritsch and F.R. Blattner, 2006. Phylogeny and new intrageneric classification of *Allium* (Alliaceae) based on nuclear ribosomal DNA ITS sequences. *Aliso*, 22: 372–95. www.biotaxa.org/Phytotaxa/article/view/phytotaxa.134.1.6.

Hsu, Y.P. 1987. *Acta Bot. Boreali-Occidentalia Sinica Yangling*, **7**(4): 259. en.wikipedia.org/wiki/Allium_ramosum.

IUCN Red List of Threatened Species, 2013. http://www.iucnredlist.org/; https://temperate.theferns.info/plant/Allium+ramosum. Retrieved 26 January, 2023.

LOTUS - the natural products occurrence database. *Allium ramosum*. https://www.wikidata.org/wiki/Q7219554; https://pubchem.ncbi.nlm.nih.gov/taxonomy/105309

NPASS (Natural Product Activity and Species Source). *Allium ramosum*. https://bidd.group/NPASS/organism.php?org_id=NPO1568;https://pubchem.ncbi.nlm.nih.gov/taxonomy/105309.

Pieroni, Andrea; 2005. The Cultural History of Plants (eds Prance, Ghillean and Mark Nesbitt, 35.p), Routledge. ISBN 0415927463.

Wikispecies, 2023. Allium ramosum. https://species.wikimedia.org/wiki/Allium_ramosum. Retrieved on 12,February, 2023.

Garlic chives [*Allium tuberosum* Rottler ex Sprengel (1825: 38)]

Subgenus: *Buttomissa*; **section**: *Buttomissa* (Pandey *et al.*, 2022).

Other common names: Asian chives, Chinese chives, Chinese leek, Fragrant-flowered garlic and Oriental garlic. Bunga, Dume and Gandana are its vernacular names (Pandey *et al.*, 2022).

Regional names: *Maroi nakuppi* in Manipuri, *Nira* in Japan, *Buchu* in Korea, *Dunduko sag* in Nepal and *Gui chai* in Thailand.

Synonyms: *Allium angulosum* Lour. in Fl. Cochinch.: 203 (1790), nom. illeg.; *A. argyi* H.Lev. in Mem. Pontif. Accad. Romana Nuovi Lincei 24: 350 (1906); *A. chinense* Maxim. in Mem. Acad. Imp. Sci. St.-Petersbourg Divers Savans 9: 284 (1859), nom. illeg.; *A. clarkei* Hook.f. in Fl. Brit. India 6: 344 (1892); *A. roxburghii* Kunth in Enum. Pl. 4: 454 (1843); *A. sulvia* Buch.-Ham. ex D.Don in Prodr. Fl. Nepal.: 53 (1825); *A. tricoccum* Blanco in Fl. Filip.: 239 (1837), sensu auct.; *A. tuberosum* Roxb. in Fl. Ind. ed. 1832. 2: 141 (1832), nom. illeg.; *A. tuberosum* f. yezoense (Nakai) M.Hiroe in Pl. Basho's & Buson's Hokku Lit. 8(1): 275 (1971); *A. uliginosum* G.Don in Mem. Wern. Nat. Hist. Soc. 6: 60 (1827); *A. yesoense* Nakai in Bot. Mag. (Tokyo) 36: 117 (1922); *Nothoscordum sulvia* (Buch.-Ham. ex D.Don) Kunth in Enum. Pl. 4: 462 (1843) (POWO, 2023).

Allium tuberosum

Plant

By KENPEI - KENPEI' photo, CC BY-SA 3.0, commons.wikimedia.org/w/index.php?curid=1181701

Introduction and origin

Garlic chive is native to Assam, Nepal, West Himalaya and SW China, especially Shanxi province but is introduced into USA in the states of Alabama, Iowa, Kansas, Massachusetts, Nebraska, New York and Wisconsin; Austria, Bangladesh, Belarus, Cambodia, Caroline Island, Comoros, Cuba, Czechoslovakia, Great Britain, Japan, Korea, Myanmar, Nauru, Philippines, Romania, Seychelles, Thailand and Tibet under Chinese occupation (POWO, 2023). It has been found growing wild in scattered locations in the United States (Illinois, Michigan, Ohio, Nebraska, Alabama, Iowa, Arkansas and Wisconsin) (USDA, 2015; Hilty, 2015; NAPA, 2015). However, it is believed to be more widespread in North America. *A. tuberosum*, often grown as an ornamental plant in gardens, is one of several *Allium* species known as wild onion or wild garlic that are listed as abnoxious weeds in various parts of the world as in Australia (USDA, 2015; Randall, 2007).

It is easy to differentiate between onion chives and garlic chives. Garlic chives have a flat, grass-like leaves and not the hollow one as in case of onion chives (AEssenseGrows, 2023). Moreover, garlic chives have mild garlic-like flavour whereas onion chives have a mild onion-like flavour.

Plants in garden	Inflorescence	Seeds
By WingkLEE - Own work, Public Domain, commons.wikimedia.org/w/index.php?curid=10504249	By WingkLEE - Own work, Public Domain, commons.wikimedia.org/w/index.php?curid=10572260	By Taraxacumseeds - Own work, CC BY-SA 3.0, commons.wikimedia.org/w/index.php?curid=22984074

Botany

This vigorous, stout, clump-forming, perennial herb bears tough and fibrous cylindrical bulbs (about 1cm across) and has mild onion-garlic flavour throughout including tuberous roots. It bears a clump of some 4-9 sheathing green, flat *cum* keeled, solid, long but narrow edible leaves that are strap-shaped having triangular bases some 1.5-8mm wide (McGee and Stuckey, 2002). Garlic chive's sweet-scented and star-shaped white umbellate flowers appear in late summer to autumn (RHS, 2015) on 25-60cm long stalks. Blooming occurs later than most native or naturalized species of *Allium* (Hilty, 2015). Its hard capsule bears black and triangular seeds (Susan, 2010; Shah, 2014). Its chromosome number is 2n=4x=32.

Cultivars

Several of varieties of *A. tuberosum* have been developed either for improved leaves (*e.g.* 'Shiva') or for production of floral stem (*e.g.* 'Nien Hua') (Larkcom and Douglass, 2008). Its giant cultivar is 'Monstrosum' (Staudengartnerei, 2015).

Nutritional values

Garlic chives are rich in vitamin C which prevents humans from common cold and fever. It is also rich in vit. A, riboflavin, thiamine, beta carotene, potassium and iron. These elements help in blood count increase, maintaining blood pressure and in increasing immunity power (AEssenseGrows, 2023).

Nutritional value per 100g		
	Source: AEssense Grows (2023)	Source: SuperHerb (2023)
Energy	23 kcal	30 kcal
Dietary fiber	2.5g	2.5g
Carbohydrates	4.35g	44 g
Sugars	N/A	1.9 g
Fat	0.73g	Total fat 0.7g Saturated Fat 0.1g Monounsaturated Fat 0.1g Polyunsaturated 0.3g
Protein	3.27g	3.3g
Minerals		
Calcium	92mg	92mg
Copper	0mg	0.2mg
Iron	1.6mg	1.6mg
Magnesium	42mg	42mg
Manganese	0.373mg	0.4mg
Phosphorus	58mg	58mg
Potassium	296mg	296mg
Selenium		0.9mcg
Sodium	3mg	3mg
Zinc	0.56mg	0.6mg
Vitamins		
Vitamin A IU	4353mg	1305mcg
B1 - Thiamine	0.078mg	0.1mg

Nutritional value per 100g		
B2 - Riboflavin	0.115mg	0.1mg
B3 - Niacin	0.647mg	0.6mg
B5 - Pantothenic acid	0.324mg	0.1mg
Vitamin B-6	-	0.1mg
B9 - Folate	105mg	105mcg
Vitamin C	58.1mg	58.1mg
Vitamin E	0.21mg	0.2mg
Vitamin K	212.7mg	213mcg

Phytochemicals

Data source: Jannat *et al.* (2019).

Allium tuberosum consists of a substantial number of phyochemicals. Sulphides, linalool and flavonoid glycosides were reported to be found in leaves. Methanolic extract of aerial parts of the plant gave two new phenylpropane glycosides specified by tuberonoid A (1) and B (2). With these two, four common flavonoids named [kaempferol 3-*O*-b-sophoroside], [3-*O*-b-D-(2-*O*-feruloyl)-glucosyl-7, 40-di-*O*-b-D-glucosylkaempferol], [3-*O*-b-sophorosyl-7-*O*-b-D-(2-*O*-feruloyl)glucosylkaempferol], and [kaempferol 3,4-*O*-di-*O*-b-D-glucoside] were found (Sang *et al.*, 2015). Alcoholic extract of garlic chives seeds has been reported to contain several chemical constituents. The compounds were identified as [nicotianoside C], [(22S)-cholest-5-ene-1 beta, 3 beta, 16 beta, 22-tetrol-1-O-alpha-L-rhamnopyranosyl-16-O-beta-D-glucopyranoside], daucosterol, adenosine and thymidine which were isolated for the first time from seed of this plant (Sang *et al.*, 2000).

Up to 2003, 21 steroid saponins [tuberoside (A-U)] were identified from seeds of *Allium tuberosum*. Of the last eight steroidal saponins [tuberoside (N-U)] that were identified in 2003, four are spirostanols ($C_{45}H_{74}O_{18}$, $C_{33}H_{54}O_{10}$, $C_{39}H_{64}O_{14}$, $C_{39}H_{64}O_{15}$), three are furostanols ($C_{39}H_{64}O_{15}$, $C_{51}H_{84}O_{22}$, $C_{51}H_{84}O_{21}$) and one is cholesterol saponin ($C_{51}H_{86}O_{22}$) (Sang *et al.*, 2003). Methanolic seed extract gave three oligoglycosides where two were furostane-type steroidal oligoglycosides [26-*O*-b-D-glucopyranosyl-(25R)-3b,22x,26-trihydroxyl-5a-furostane3-*O*-b-chacotrioside] and [26-*O*-b-D-glucopyranosyl-(25S)-3b,5b,6a,22x,26-pentahydroxyl-5b-furostane 3-*O*-a-L-rhamnopyranosyl-(1→4)-b-D-glucopyranoside], and another one was pregnane-type oligoglycoside [3-*O*-a-L-rhamnopyranosyl-(1→4)-b-D-glucopyranosyl 3b,5b,6a,16b-tetrahydroxypregnane 16-(5-*O*-b-D-glucopyranoyl-4(S)-methyl-5-hydroxypentanoic acid) ester] (Tsuyoshi *et al.*, 2014).

Along with three known compounds, three new spirostanol saponins ~ tuberosine A [(25S)-5b-spirostan-2b,3b-diol 3-*O*-b-D-glucopyranoside], tuberosine B [(25S)-5b-spirostan-2b,3b,19-triol 3-*O*-b-Dglucopyranoside] and tuberosine C [(25S)-5b-spirostan-2b,3b-diol 3-*O*-a-L-rhamnopyranoyl-(1-4)-*O*-b-D-glucopyranoside] have been found from the roots of garlic chives (Fang *et al.*, 2015). Molecular formulae of the three tuberosines were $C_{33}H_{54}O_9$, $C_{33}H_{54}O_{10}$, and $C_{39}H_{64}O_{13}$, respectively (Fang *et al.*, 2015). A new phenylpropanoidglucoside along with a new chain compound was isolated from the methanolic extract of garlic chives roots: the glucoside compound was named tuberosine D ($C_{19}H_{28}O_{12}$) and the second chain compound was (*Z*)-11*R*, 12*S*, 13*S*-trihydroxy-9-octadecenoate ($C_{18}H_{34}O_5$) (Fang *et al.*, 2017). Some novel compounds were isolated from the ethanolic extract of *A. tuberosum* among which four were new pyrazines with unique flavour, one lignan and three new flavonoids. From this study, a total of 40 compounds were identified including 32 known compounds. The four pyrazine compounds were colourless oil with distinguishing roasting odour and their structures were established to be (R)-2-(1-ethoxyl)-5-(2-ethoxyl)-pyrazine ($C_8H_{12}N_2O_2$), 2-methyl-3-methylol-5-ethyl-pyrazine ($C_8H_{12}N_2O$), (R)-2-(1-ethoxyl)-6-ethyl-pyridazine ($C_8H_{12}N_2O_2$) and (R)-2, 3-di-methyl-6-(1-ethoxyl)-pyridazine ($C_8H_{12}N_2O$). Three new flavonoid compounds were found from garlic chives and their structures were elucidated to be quercetin-3-*O*-(6-trans-feruloyl)-b-D-glucopyranosyl-(1/2)-bDglucopyranoside-7-*O*-b-D-glucopyranoside ($C_{43}H_{48}O_{25}$), kaempferol-3-*O*-(6-trans-feruloyl)-b-D-glucopyranosyl-(1/2)-b-D-glucopyranoside-7-*O*-b-D-glucopyranoside ($C_{43}H_{48}O_{24}$), and quercetin-3-*O*-(6-*trans*-p-coumaroyl)-b-D-glucopyranosyl-(1/2)-b-D-glucopyranoside-7-*O*-b-D-glucopyranoside ($C_{42}H_{46}O_{24}$). A new lignan compound isolated from the plant was identified as (7R, 8S)-dihydrodehydrodiconiferyl alcohol-di-9, 90-*O*-b-Dglucopyranoside ($C_{32}H_{44}O_{16}$). Some of the known compounds identified reportedly are uridine, thymine, deoxythymidine, thymidine, guanosine, adenine, adenosine, deoxyadenosine, 1-methyl-l,2,3,4-tetrahydro-carboline-3-carboxylic acid, D-tryptophan, D-phenylalanine, 2-[(1R,2R)-2-hydroxy-2-(4-hydroxy-3-methoxyphenyl)-1-(hydroxymethyl)-ethoxy]-5-(3-hydroxypropyl)-phenyl-b-D-glucopyranoside, hydroxybenzoic acid, *etc.* (Amalesh *et al.*, 2011; Panche *et al.*, 2016; Quan *et al.*, 2018).

Plants may contain several volatile compounds that may serve as perfumes, flavours and as insect repellents (Ian, 2010). Different plant parts of *A. tuberosum* have been analyzed which showed that methyl and 2-propenyl (allyl) radicals are present as volatile disulphides (Ian and David, 1977).

Thirty seven essential oils have been found from *Allium tuberosum* and they comprise mainly of dimethyl disulphide (19.58%), allyl methyl disulphide

(14.37%), dimethyl trisulphide (14.34%), allyl methyl trisulphide (7.24%), methyl 1-propenyl disulphide (6.07%) and diallyl disulphide (5.14%) (Dima *et al.*, 2014).

Nucleosides and nucleotides were also identified from different solvent (petroleum ether, ethyl acetate, n-butanol and water) extract of the plant. Uracil was identified as 4-pyrimidinediol, a white powder holding the molecular formula $C_4H_4N_2O_2$. The second compound was thymine, determined as 2,4-hydroxy, 5-methylpyrimidine whose molecular formula was decided to be $C_5H_6N_2O_2$. Compound 3 was thymidine or thymidine deoxyriboside, determined to be 1-(2-deoxy-β-D-ribofuranosyl)-5-methyluracil and its molecular formula was established to be $C_{10}H_{14}N_2O_5$. The last one was $C_{10}H_{13}N_5O_4$ which was named adenosine or adenine riboside and was identified as 9-(beta-D-ribofuranosyl)-adenine (Shuang *et al.*, 2018).

Sulphur rich compounds

Twenty seven volatile sulphur rich compounds were extracted from *A. tuberosum* including sulphide, disulphides, trisulphides and tetrasulphides with ethyl, butyl and pentyl groups (Borlinghaus *et al.*, 2014; Krest *et al.*, 2000).

Uses

The flavour of garlic chive is more like garlic than chives (McGee and Stuckey, 2002). Although in Asia the garlic chives are primarily used as culinary herbs and as traditional medicine, in North America the interest has been more as an ornamental (Mahr, 2010). It has been widely cultivated for centuries in East Asia for its culinary values. The leaves, the stalks and unopened immature flower buds are used as flavouring (Larkcom and Douglass, 2008) in a similar way to chives, scallions or garlic and are included as a stir-fry ingredient. In the NE states of India especially in Manipur it is known as *maroi nakuppi* where it is grown and used as a substitute for garlic and onion in cooking; in Japan, it is known as *nira* and used for both garlic and sweet flavours, in soups and salads; and in Japanese-Chinese dishes such as *gyoza dumplings* with a combination of egg, shrimp and pork (en.wikipedia.org/wiki/Allium_tuberosum).

Though mainly used as a culinary herb, garlic chives have medicinal values as well. It is helpful in urinary incontinence, kidney and bladder weakness and stomach chills with vomiting (seeds). In traditional Chinese medicine, the seeds are commonly used to treat impotence and nocturnal emissions (Hu *et al.*, 2006). They are also used as tonic and aphrodisiac (Alam *et al.*, 2022). In contrast, the leaves are usually employed to cure abdominal pain, diarrhoea, hematemesis, diabetes, snake-bite and asthma; the root, bulb, leaves and cloves in Arunachal Pradesh (India) against common cold and coughs, as well as a hair tonic and antilipidemic; in another region of India, the Meitei tribe of Manipur uses a decoction of the whole plant to relieve various gastrointestinal ailments; a poultice prepared from the whole plant is used commonly in India to heal spermatorrhea; the plant juice is used to stem severe bleeding; the Subanen tribe in the Philippines uses a poultice of the whole plant to alleviate fever; this is also used in the country to treat asthma; and in Thailand and Indo-China, the seeds are used as a mouthwash to alleviate toothache through their antiseptic effect (Jannat *et al.*, 2019).

Cultivation

Garlic chives can be propagated by seeds or by dividing the clumps. These are easy to grow under various conditions in slowly expanding perennial clumps though also sprout readily from seeds (Soule, 2016). Division is usually done in the spring (Naturalmedicinalherbs, 2023; Temperate Plants Database, 2023). The plants divide successfully at any time in the growing season and the divisions can be planted straight out into their permanent positions if required. Seeds are usually sown in spring in a cold frame and then the seedlings can be pricked out into individual pots when they are large enough to handle. They must be grown in the greenhouse for at least their first winter and then can be planted out into their permanent positions in spring once they are growing vigorously and are large enough.

The plant remains evergreen in warmer areas but in cold areas leaves and stalks completely die back to the ground and resprout from roots or rhizomes in the spring when weather improves (Soule, 2016). Garlic chive is cold-hardy to USDA zones 4–10 (−34 to 2 °C) (). This likes full sun and does well in many soil types though these are most compatible with damp soil having a pH of 6-7 and high organic content (HeirloomOrganics, 2023).

Chives are related to onions and share many of the same companion plants such as beets, carrots, tomatoes, strawberries, potatoes, rhubarb, kohlrabi, parsley, broccoli, cabbage, eggplant, mustard and peppers, and are thought to enhance their flavours and growth intensity (MDidea, 2012). Garlic chives discourage the spread of Japanese beetles, slugs, aphids and cabbage worms so these should be grown in areas where plants in full sun have a problem with these insect infestations. Growing garlic chives near asparagus, peas, spinach and beans should be avoided as they will compete for similar soil nutrients (MDidea, 2012).

Seeds can cause uncontrolled spread so flowers are sheared off immediately as soon as these appear (Susan, 2010; Shah, 2014). It takes about 50 days to harvest chives for best flavour (HeritageGarden, 2023). Regular harvesting of the leaves will promote more vigorous growth and spreading behaviour though allowing its white flowers to grow may create dormancy in the plant during hot summer temperatures (HeirloomOrganics, 2023). Plant should be rotated every 3 to 5 years to ensure soil fertility.

Pests and diseases

Garlic chives is susceptible to slugs, snails, allium leaf mining fly, onion fly, onion white rot and downy mildew but usually trouble free in the border (RHS, 2015). All these are controlled as defined under onion and garlic chapters.

References

AessenseGrows. https://www.aessensegrows.com/en/fresh-produce-catalog/garlic_chives. Retrieved 21 January, 2023.

Alam, A., Al Arif Jahan Abdullah, Md. Sazzadul Bari, Labony Khandokar, Md. Hasan Mahmud, Muhammed Junaid, Mohammed Salahuddin Chowdhury, Mohammad Forhad Khan, Veronique Seidel and Md. Areeful Haque, 2022. Allium vegetables: Traditional uses, phytoconstituents and beneficial effects in inflammation and cancer. *Crit. Rev. Food Sci. Nutr.,* pp. 1-32. DOI: 10.1080/10408398.2022.2036094. https://doi.org/10.1080/10408398.2022.2036094.

Amalesh, S., D. Gouranga and K.D. Sanjoy, 2011. Roles of flavonoids in plants. *Int. J. Pharm. Sci. Res.,* **6**:12-35.

Anjula, P., K. Pradheep and G. Rita, 2014. Chinese chives (*Allium tuberosum* Rottler ex Sprengel): A home garden species or a commercial crop in India. *Gen. Res. Crop Evol.,* **61**:1433-40. Doi: 10.1007/s10722-014-0144-z.

Borlinghaus, J., F. Albrecht, M.C. Gruhlke, I..D. Nwachukwu and A.J. Slusarenko, 2014. Allicin: Chemistry and biological properties. *Molecule.,* **19**(8): 12591-618. (https://www.sciencedirect.com/science/article/pii/S2667142522000811).

Dima, M., A.S. Fabiano-Tixier, P. Emmanuel, H. Tayssir, N. Nancy and F. Christine, 2014. Chemical composition, antibacterial and antioxidant activities of six essentials oils from the *Alliaceae* family. *Molecules,* **19**:20034-53. Doi: 10.3390/molecules191220034.

en.wikipedia.org/wiki/Allium_tuberosum

Fang, Y.S., C. Le, L. Ying, J.P. Wang, X. Huai and Z.T. Ding, 2015. Spirostenol steroids from the roots of *Allium tuberosum. Steroids,* **100**:1-4. http://dx.doi.org/10.1016/j.steroids.2015.03.015.

Fang, Y.S., S.X. Liu, Y.C. Ma, J.W. Dong, C. Le and Z.T. Ding, 2017. A new phenylpropanoid glucoside and a chain compound from the roots of *Allium tuberosum. Nat. Prod. Res.,* **31**:70-76. http://dx.doi.org/10.1080/14786419.2016.1212034.

HeirloomOrganics, 2023. How to Grow Garlic Chives: Guide to Growing Garlic Chives. http://www.heirloom-organics.com/guide/va/guidetogrowinggarlicchives.

html; http://heritagegarden.uic.edu/garlic-chives-allium-tuberosum. Retrieved 21 January, 2023.

HeritageGarden, 2023. http://heritagegarden.uic.edu/garlic-chives-allium-tuberosum. Retrieved 21 January, 2023.

Hilty, John, 2015. Garlic chives. Illinois Wildflowers. Retrieved 14 October 2015. https://en.wikipedia.org/wiki/Allium_tuberosum.

Hu, G., Y. Lu and D. Wei, 2006. Chemical characterization of Chinese chive seed (*Allium tuberosum Rottl.*). *Food Chem.,* **99**(4):693–7. doi: 10.1016/j.foodchem.2005.08.045.

Ian, A.M. and A.F. David, 1977. The composition of volatiles from different parts of *Allium tuberosum* plants. *Phytochem.*, **16**:763-4.

Ian, T.B. 2010. Plant Volatiles. *Curr Biol.* ,**20**:392-7.

Jannat, K., Taufiq Rahman and Mohammed Rahmatullah, 2019. Traditional uses, phytochemicals and pharmacological properties of *Allium tuberosum* Rottler ex spreng. *J. medic. Pl. Stud.,* **7**(2): 214-20. https://www.plantsjournal.com/archives/2019/vol7issue2/PartC/7-2-33-247.pdf.

Krest, I., J. Glodek and M. Keusgen, 2000. Cysteine sulfoxides and alliinase activity of some *Allium species. J. Agri. Food Chem.,* **48**(8): 3753-60. (https://www.sciencedirect.com/science/article/pii/S2667142522000811).

Larkcom, Joy and Elizabeth Douglass, 2008. Oriental Vegetables : The Complete Guide for the Gardening cook (2nd ed.), pp. 75-80. Kodansha International, New York. USA. ISBN 978-1-56836-370-7. Retrieved 17 October 2015.

Mahr, Susan, 2010. Garlic chives, *Allium tuberosum.* University of Wisconsin Extension Master Gardener Program. Retrieved 19 October 2015. https://en.wikipedia.org/wiki/Allium_tuberosum

McGee, Rose Marie Nichols and Stuckey Maggie, 2002. The Bountiful Container. Workman Publishing. ISBN 978-0-7611-1623-3. https://en.wikipedia.org/wiki/Allium_tuberosum.

MDidea, 2012. What is Chinese Leek-Chinese Chives and Its Quotes. *Modern Research,* November 26, 2012. http://mdidea.com/products/proper/proper03007.html; http://heritagegarden.uic.edu/garlic-chives-allium-tuberosum.

NAPA, 2015. *Allium tuberosum*. County-level distribution map from the North American Plant Atlas (NAPA). Biota of North America Program (BONAP). 2014. Retrieved 15 October 2015. https://en.wikipedia.org/wiki/Allium_tuberosum.

Naturalmedicinalherbs, 2023. Chinese Chives, *Allium ramosum*. http://www.naturalmedicinalherbs.net/herbs/a/allium-ramosum=chinese-chives.php. Retrieved 22 January, 2023.

Panche, A.N., A.D. Diwan and S.R. Chandra, 2016. Flavonoids: an overview. *J. Nutr. Sci.,* **5**: 1-15. Doi: 10.1017/jns.2016.41.

Pandey, Anjula; Pavan Kumar Malav, K.M. Rai and S.P. Ahlawat, 2022. Genus *Allium* L. of the Indian Region: A Field Guide for Germplasm Collection and Identification. *ICAR-National Bureau of Plant Genetic Resources*, New Delhi, India, 109p + i-vi. http://www.nbpgr.ernet.in/Downloadfile.aspx?EntryId=9197.

POWO, 2023. *Allium hookeri* Thwaites. Plants of the World Online. Royal Botanic Gardens, Kew. https https://powo.science.kew.org/taxon/urn:lsid:ipni.org:names:528166-1. Retrieved 16 February, 2023.

Quan, G., X.B. Li, S. Jia, E.D. Xia, T. Feng and H.Q. Cao, 2018. Isolation and identification of new chemical constituents from Chinese chive (*Allium tuberosum*) and toxicological evaluation of raw and cooked Chinese chives. *Food Chem. Toxicol.,* **112**: 400-11. Doi: 10.1016/j.fct.2017.02.011.

Randall, R.P. 2007. The introduced flora of Australia and its weed status (PDF). Australian Weed Management. University of Adelaide. ISBN 978-1-920932-60-2. Retrieved 15 October 2015. https://en.wikipedia.org/wiki/Allium_tuberosum.

RHS (Royal Horticultural Society), 2015. Allium tuberosum (Chinese chives). Retrieved 14 October 2015. https://www.rhs.org.uk/plants/877/i-allium-tuberosum-i/details?returnurl=%2Fplants%2Fbulbs%3Faliaspath%3D%252fplants%252fbulbs.

Sang, S.M., Z.H. Xi, S.L. Mao, A.N. Lao, Z.L. Chen, 2000. Study on chemical constitutents in seeds of *Allium tuberosum* Rottl. *ZhongguoZhong Yao Za Zhi - China J. Chinese Mat. Med.,* **25**: 286-8.

Sang, S., S. Mao, A. Lao, Z. Chen and C.T. Ho, 2003. New steroid saponins from the seeds of *Allium tuberosum. Food Chem.,* **83**:499-506.

Sang, H.H., W.S. Suh, J.P. Kyoung, K.H. Kim and R.L. Kang, 2015. Two new phenylpropane glycosides from *Allium tuberosum* Rottler. *Arch. Pharm. Res.,* **38**:1312-6. Doi: 10.1007/s12272-015-0603-x.

Shah, N.C. 2014. Status of cultivated & wild *Allium* species in India: A Review. *The SciTech J.,* **1**: 28-36. https://www.plantsjournal.com/archives/2019/vol7issue2/PartC/7-2-33-247.pdf.

Shuang, G., W. Xiangyang, Z. Rong and H. Jiaying, 2018. Isolation and identification of nucleosides/nucleotides raising testosterone and NO levels of mice serum from Chinese chive (*Allium tuberosum*) leaves. *Andrologia,* e13191. Doi: 10.1111/and.13191.

Soule, J.A. 2016. Month by Month Guide to Gardening in the Southwest. Cool Springs Press. https://en.wikipedia.org/wiki/Allium_tuberosum.

Staudengartnerei Gaißmayer, 2015. Allium tuberosum 'Monstrosum'. Retrieved 19 October 2015. https://en.wikipedia.org/wiki/Allium_tuberosum.

SuperbHerb, 2023. https://superbherb.co.nz/product/garlic-chives/. Retrieved 21 January, 2023.

Susan, M.A.. 2010. Garlic chives, *Allium tuberosum.* Master Gardener Program. University of Wisconsin, Madison, USA. Accessed 30 Aug, 2010. https://www.plantsjournal.com/archives/2019/vol7issue2/PartC/7-2-33-247.pdf.

Temperate Plants Database, 2023. Ken Fern. temperate.theferns.info. 2023-01-21. https://temperate.theferns.info/plant/Allium+ramosum

USDA PLANTS database. 2015. Allium tuberosum. https://en.wikipedia.org/wiki/Allium_tuberosum.

Hooker chives (*Allium hookeri* Thwaites)

Subgenus: *Amerallium.*

Section: *Bromatorrhiza.*

Common names: Hooker Chives and Garlic chives (en.wikipedia.org, 2023).

Indian names [Manipuri: Maroi napakpi; Tangkhul: Namrei; and Mizo: Runphek (FOI, 2023)]. In Nepal it is kinown as Dung Dung and Dudu Sag (Anjula Pandey *et al.*, 2022).

Introduction and origin

Hooker Chive is a small ornamental plant *cum* spicy herb, having thick fibrous root system and compact underground stem. It is found growing gregariously in the wild form on hilly slopes at altitudes of 1400-4200m in Eastern Himalayas (Tibet, and Sichuan & Yunnan provinces of China; FOI, 2023). In India it is found in Manipur, Arunachal Pradesh, Meghalaya, West Bengal and Sikkim (POWO, 2023a; Anjula Pandey *et al.*, 2022). Apart from there, it is also recorded from Sri Lanka, Nepal, Myanmar and Bhutan.

Hooker Chive

By Uleli - Own work, CC BY-SA 3.0, https://commons.wikimedia.org/w/index.php?curid=15860387

The plant is widely cultivated outside its native range and valued as a food item in much of South and SE Asia (en.wikipedia.org, 2023). The species includes 2 accepted intraspecifics: *Allium hookeri* var. *hookeri* and *A.h.* var. *muliense* Airy Shaw (POWO, 2023b). Underground stems produce side shoots which bear long, simple, smooth, thick and alternate leaves with prominent midrib though leaf bases are scaly and fleshy. During summer and rainy seasons (April to September), this produces floral shoots bearing clustered white flowers on the top, followed by globular capsular fruits. For daily use as a spice or condiment, *vis-à-vis* in salad and soups; people grow and use it since a long in their kitchen garden or as pot-herb for daily use otherwise it is collected from the wild daily as per requirement. Its thick fibrous roots are used basically for temering and while cooking the fish, and its addition causes superb taste of the fish curry. In folklore medicines it is used as diuretic and as wormicide. Its leaves are also used in salads and for preparing soup.

Botany

Hooker Chive plants attain the height of up to 90cm bearing thick, elongate & fleshy roots and linear, 0.5-1cm wide leaves with distinct mid-vein, being slightly shorter than or equal to scape that is 20-60cm long, arising laterally from bulb-base without leaf sheaths. Its bulbs are poorly formed or absent, clustered and cylindric and the tunic is membranous and entire. The plant flowers from July to October. The hemispheric-globose umbellate inflorescence is around 4.5cm in diameter comprising of many flowers. Spathe is ovate, 3x0.8cm, 2-valved and usually deciduous. Flowers are campanulate though stellate on maturity exposing the ovary (FOI, 2023; Anjula Pandey *et al.,* 2022).

Pedicels are 2-3cm, equalling perianth and ebracteolate. Perianth is white to greenish-yellow (yellow in var. *muliense*; a Chinese taxon); segments similar, lanceolate, 4-7.5x1-1.5mm and apex acuminate, but sometimes unequally 2-lobed. Filaments are subulate, slightly shorter than to sub-equalling perianth segments, connate at base and adnate to perianth segments. Stamen without tooth and anthers are oblong with 1.5-2x0.7-1mm dimension. Ovary is obovoid, constricted at base stipe, 1.5-3x1.5-2mm and stipitate. Ovule is 1 per locule. The style is longer than ovary, inconspicuous, and the stigma is punctiform. Capsules are globose-subglobose, 5x5mm (immature) and brown (FOI, 2023; Anjula Pandey *et al.,* 2022; FOC, 2023b). It has diploid (2n=22) and tetraploid (2n=44) chromosome numbers.

Uses

In places of its origin, apart from being used as an ornamental pot plant, it is grown in the kitchen garden or as a pot-herb for daily use since a long as a spice or condiment, *vis-à-vis* in salad and soups for daily use. Its roots are used for preparing superb fish curries. In Manipur, it is a fairly common ingredient in preparing pakoras or omelette; *cum* as an essential ingredient in *paaknam* ~ a delicious dish by mixing gram flour, *Maroi napakpi cum* fermented fish followed by baking them after wrapping the mixture in turmeric leaf (FOI, 2023). In folklore medicines, it is being used as diuretic and as wormicide.

Its cultivation is similar to onion.

References

Anjula Pandey, Pavan Kumar Malav, K.M. Rai and S.P. Ahlawat, 2022. Genus *Allium* L. of the Indian Region: A Field Guide for Germplasm Collection and Identification. ICAR-National Bureau of Plant Genetic Resources, New Delhi, India, 109p + i-vi. http://www.nbpgr.ernet.in/Downloadfile.aspx?EntryId=9197.

3en.wikipedia.org/wiki/Allium_hookeri. Retrieved on 16 February, 2023.

FOC, 2023. Flora of China FOC, vol. 24, page 174. Allium hookeri Thwaites, Enum. *Pl. Zeyl.*,No. 339. http://www.efloras.org/florataxon.aspx?flora_id=2&taxon_id=200027488. Retrieved 16 Feb 2023.

FOI, 2023. Flowers of India. http://www.flowersofindia.net/catalog/slides/Hooker%20Chives.html. (Retrieved on 16 February, 2023).

POWO, 2023a. Plants of the World Online|Kew Science. *Allium tuberosum* Rottler ex Spreng. *First published in Syst. Veg.*, 2: 38 (1825). https://powo.science.kew.org/taxon/urn:lsid:ipni.org:names:529062-1. Retrieved 20 January 2023.

POWO, 2023b. *Allium hookeri* Thwaites. Plants of the World Online. Royal Botanic Gardens, Kew. https https://powo.science.kew.org/taxon/urn:lsid:ipni.org:names:528166-1. Retrieved 16 February, 2023.

5

Welsh Onion or Japanese Bunching Onion (*Allium fistulosum* L.)

Subgenus: *Cepa*; **section**: *Cepa* (Anjula Pandey *et al.*, 2022).

Other common names: Bunching onion, Long green onion, Japanese bunching onion (en.wikipedia.org/wiki/Allium_fistulosum) and Stone leek (Singh and Ramakrishna, 2017).

Indian local names: Vilayati lahsun (in Hindi); Khorat (in Marathi); Vellulli (in Malayalam); Mizo-purun (Mizo); and Ja-uat (in Khashi) (Singh and Ramakrishna, 2017).

Synonyms: *Allium bouddae* Debeaux, *A. kashgaricum* Prokh., *Cepa fissilis* Garsault, *C. fistulosa* (L.) Gray, *C. ventricosa* Moench, *Kepa fistulosa* (L.) Raf., *Phyllodolon fistulosum* Salisb. (L.) , *Porrum fistulosum* (L.) Schur (POWO, 2023).

Introduction and origin

The species epithet *fistulosum* is Latin for *fistular* which refers to the hollow tube structure of the leaves (National Park, 2023). Welsh onion is neither indigenous to Wales nor common in Welsh cuisine (green *Allium* common to Wales is leek, the national vegetable of Wales). The word 'Welsh' preserves the original meaning of the old English word 'welisc' or old German 'welsche' meaning 'foreign'. This was probably applied when it was first introduced into

Germany near the end of the Middle Ages. The other common names commonly applied to *A. fistulosum* include green onion, salad onion, scallion and spring onion but these are ambiguous as these also refer to any young green *Allium* stalks whether grown from Welsh onions, common onions or other similar members of the genus *Allium* (Singh and Ramakrishna, 2017).

Allium fistulosum was introduced to Europe in the 17th century (Christman, 2003). It is not known in the wild though its nearest wild relative *A. altaicum* Pall (Wild bunching onion or Butun in Siberia) seems to be progenitor which is widespread in the parts of Mongolia and Siberia where it was occasionally collected as a vegetable for local use or for export to China (Padula *et al.,* 2022; Oyen and Messiaen, 2004). It has been domesticated in the neighbourhood of Lake Baikal in Siberia near the Altai Mountains in Russia and in Western or NW China (Borna, 1982; Malachowski, 1982; Chen *et al.,* 2001; Tang *et al.,* 2011; Peng *et al.,* 2017; Singh and Ramakrishna, 2017). The crop is still popular in Russia, especially in its central part and in Siberia (Ivanova *et al.,* 2015; Shishkina *et al.,* 2019). From Siberia, the crop has spread to Japan during the 8th century AD (Purseglove, 1978).

Growth and development

Allium fistulosum is perennial but commercially it is mostly grown as an annual though in home gardens usually as a perennial. Welsh onion does not have a long-day dormant stage similar to onion (*Allium cepa*) so it continues its vegetative growth without developing a real bulb. However, some cultivars which originated from cold temperate areas show short-day dormancy. These stop growing and their leaves dry out under short days even when the temperature would permit normal growth. The lateral buds in the leaf axils elongate and develop as tillers to form a clump. This tillering characteristic is more pronounced in cultivars grown for the green leaves than in those grown for the long blanched pseudostems (Oyen and Messiaen, 2004).

Flower induction is controlled by temperature and daylength. Low temperatures and short days induce flowering though requirements vary strongly with the origin of cultivars. Flowering is generally induced by temperatures below 13°C when seedlings have formed a certain number of leaves or a pseudostem of a certain thickness. In the tropics where conditions are more favourable for vegetative rather than generative growth, only some well-adapted cultivars will flower (Oyen and Messiaen, 2004). Roots are readily colonized by arbuscular mycorrhizal fungi which enhances P uptake and stimulates growth (Oyen and Messiaen, 2004).

Botany

Similar to *Allium cepa*, *A. fistulosum* is also diploid with 2n=16 chromosomes with similar chromosome morphology though *A. fistulosum* chromosomes are considerably smaller (Jones and Rees, 1968). Welsh onions multiply by forming perennial evergreen clumps (Thompson, 1995) up to 30cm in diameter (Christman, 2003) and some 50-90cm tall (Singh and Ramakrishna, 2017). Among the *Allium* species, *A. fistulosum* roots are much spreading *cum* longer which changes entire environment in the soil in the manner that growth of fungus *Aspergillus niger* becomes quite unsuitable inside the soil (Das, 2010).

These form a poorly developed bulb (1-2cm in diameter) hardly exceeding its pseudostem diameter (Singh and Ramakrishna, 2017). Its bulbs usually have white tunic, leaves numbering 5-13 are tube-like hollow, tip-pointed, glaucous, distichously alternate and inflated to entire length, blades being D-shaped in cross-section and 30-75cm long with tubular sheath (Singh and Ramakrishna, 2017; Christman, 2003).

After a cold spell, the plants initiate the hollow stalks topped with dense and more or less greenish-white centrifugal blooms in almost spherical umbels but without bracts and having 2.5-7.5cm diameter. It has campanulate perianth, unequal (0.4-2cm long) flower pedicel, filaments being 8–12mm long and pistils lower than stamens (Singh and Ramakrishna, 2017; Christman, 2003). Its fruits being 4-6mm in diameter are round possessing minute black seeds.

lsh onion	**Multi-stemmed *Allium fistulosum***
By Ping an Chang - Own work, CC BY-SA 4.0, https://commons.wikimedia.org/w/index.php?curid=80247562	**Singh and Ramakrishna (2017)**

Comparision with onion, leek and chives

Allium fistulosum is often considered to be a kind of scallion having taste and odour almost similar to *Allium cepa*. Also there are hybrids, between the two (Christman, 2003; Thompson, 1995). The first hybrid between *A. cepa* and *A. fistulosum* were reported by Emsweller and Jones (1935). The leaves of *A. fistulosum* have a fairly intermediate smell of onion and leek (Singh and Ramakrishna, 2017). Morphologically, there is a strong resemblance between *A. fistulosum* and *A. cepa*, but *A. fistulosum* does not have a long-day dormant stage like *Allium cepa* so it continues its vegetative growth without developing a real bulb (Oyen and Messiaen, 2004; Singh and Ramakrishna, 2017) though possesses hollow leaves and scapes (Christman, 2003; Thompson, 1995; Singh and Ramakrishna, 2017). However, some slight thickening of the base of the pseudostem may occur. The foliage leaves of welsh onions are not flattened adaxially but somewhat rounder in cross-section, flowering in *A. fistulosum* progressing from the top of the umbel downwards, and the flowers are campanulate to urceolate and larger but lack bracteoles while in case of *A. cepa* the flowers have bracteoles and are smaller *cum* stellate (Oyen and Messiaen, 2004). Its larger varieties resemble the leek as to that of Japanese negi while smaller ones the chives (Christman, 2003; Thompson, 1995).

Classification

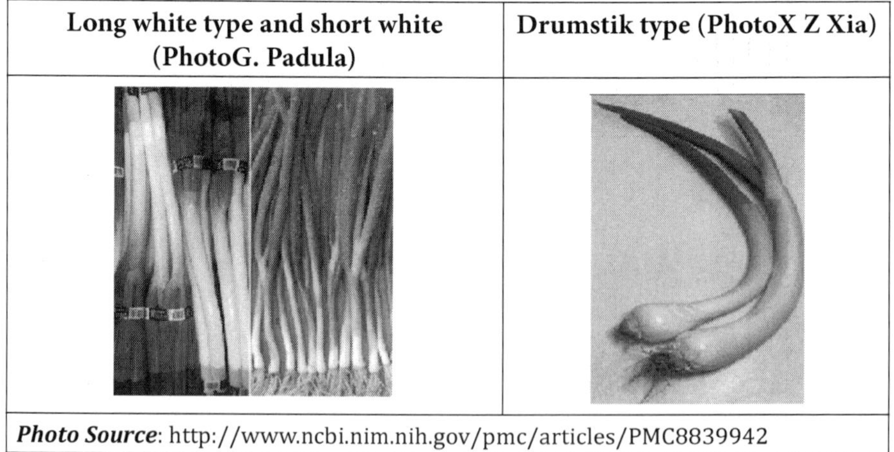

Long white type and short white (PhotoG. Padula)	Drumstik type (PhotoX Z Xia)
Photo Source: http://www.ncbi.nim.nih.gov/pmc/articles/PMC8839942	

Allium fistulosum cultivars are classified in several ways, depending on their traits or area of cultivation.

In Europe, *A. fistulosum* cultivar-groups are broadly categorized into two types (sometimes taxonomically treated as two subspecies) (Kazakova, 1978; Oyen and Messiaen, 2004; Singh and Ramakrishna, 2017) such as:

- **Japanese bunching onion group** (subsp. *fistulosum*) are single-stemmed cultivars for cool climates which are grown for their thickened and blanched pseudostems. They are eaten as a potherb, in sukiyaki and in chicken dishes.
- **Welsh onion group** (subsp. *caespitosum*) being most common in Africa are multi-stemmed cultivars grown in warm climates for their green leaves in salads and for flavouring soups & other dishes.

Japanese commonly classify it into three groups (Oyen and Messiaen, 2004):

- **Kaga Group** grown in the coolest parts of Japan for its pseudostems
- **Kujyo Group** grown in the warmest parts of Japan mainly for its leaves
- **Senju Group**, intermediate in use and ecological requirements.

As Chen *et al.* (2001) and Ma *et al.* (2009) mention that based on the plant's shape, sometimes Welsh onion is also divided into:

- **Rod-shaped common type** with narrow base
- **Tillering type** with a wide base

Sometimes *Allium fistulosum* cultivars are categorized into three groups: Chinese, Japanese and Russian cultivars as per different morphological traits [number, size, leaf colour & taste, earliness and overwintering potential (Padula *et al.*, 2022).

As per Chen *et al.* (2001) and Ma *et al.* (2009), in Asia, Welsh onion cultivars are more commonly divided into three types based on their pseudostem length:

- **Long white type**: Its white part is over 40cm long with pseudostems exhibiting 10:1 length-to-thickness ratio. This type contains high yield accompanied with light pungency, sweet taste, high water content and a short shelf life. Representative cultivars of this type include 'Zhangqiu', 'Chishui' and 'Beijing Gaojiaobai'.
- **Short white type**: It has a thick but short pseudostem with narrower length-to-thickness ratio. Its cultivars do not require deep soils for cultivation and are resistant to lodging. The pseudostem yield in this case is quite high along with the longer shelf life. Representative cultivars of this type include 'Shouguang Bayeqi' and 'Tonghua Xiaohong'.
- **Drumstick type**: Its pseudostem length is similar to that of the short white type though inverted drumstick- or garlic-shaped pseudostem base is significantly widely enlarged though upper part is thinner. It has stronger flavour than other two types though with poor yield. This type has shown a long shelf life. Representative cultivars of this type include 'Longyao' and 'Yinchuan Datou'.

Cultivars and hybrids

There are many cultivars including ones with red skins ('Santa Clause', 'Red Beard'); some with shorter but thicker stalks ('Shimonita'); some with larger bulb size ('Yoshima'); and certain that are exceptionally cold-hardy, such as 'White Lisbon', 'Evergreen White Bunching' and 'Winter Over' (Christman, 2003). It is not unusual to harvest these under the snow. Due to problems in introgression of 'Japanese Bunching Onion' traits into onion, till now only four new varieties are developed and these are tetraploid 'Beltsville Bunching', triploid 'Delta Giant', 'Wakegi Onion' and 'Top Onion' ('Tree Onion', 'Egyptian Onion' or 'Walking Onion') as this produces topsets or bulbils instead of flowers as onions produce. The 'top onion' has been the home garden crop of Europe and NE Asia and is termed as *A. proliferum* which in fact is a cross between *A. cepa* and *A. fistulosum* (Frietsch and Friesen, 2002). 'Wakegi Onion' used in Japan and China is a diploid bred through *A. fistulosum* and shallot type *A. cepa*. The tetraploid 'Beltsville Bunching' being favourite of Americans acquires two sets of chromosomes from *A. cepa* and the same from *A. fistulosum*, however, 'Delta Giant' a triploid is also grown in USA as a minor crop (Kik, 2002). Havey (1991) stated about the hybrid 'Pran' (*Allium × cornutum*) being grown especially in Jammu region of Kashmir (India), Europe, Canada and elsewhere whose seed parent is said to be *A. cepa* but male parent is not known.

Allium fistulosum hybridizes easily with its wild relative *A. altaicum* and also with *A. cepa*. Hybrids with *A. altaicum* have high pollen fertility and form ample seeds whereas those with *A. cepa* usually do not produce flowers but in case if flower during spring is produced also, usually those turn out to be sterile (specialtyproduce.com). Dole *et al.* (1980) have suggested use of embryo rescue to increase the success rate. However, certain tetraploids such as 'Beltsville Bunching' may produce even seeds and the plants produced from these turn out to be tolerant to hot and dry weather (Christman, 2003). Its hybrids with *A. cepa* become perennial plants forming small bulbs (Oyen and Messiaen, 2004).

Several old hybrids between *Allium fistulosum* as female parent and tropical shallots (*A. cepa* Aggregatum Group) exist. These are known as *Allium × proliferum*, *Allium × wakegi* or *Allium fistulosum* var. *caespitosum*, such as Indonesian shallot 'Sumenep' which is well adapted to tropical conditions but never flowers, and the top onion in which the inflorescence is composed of bulbils. In the vegetative stage these plants resemble *A. fistulosum*. Recent commercial hybrids between *A. fistulosum* as female parent and *A. cepa* grown mainly in the United States and Europe for their green tops include 'Beltsville Bunching' (a seed-producing tetraploid), 'Louisiana Evergreen' (a sterile diploid) and 'Delta Giant' (a sterile triploid originating from a backcross of the hybrid with *A. cepa*).

Another group of hybrids grown in Japan is 'Yakura Negi Group'. Plants of this group tiller abundantly during spring to summer but remain dormant during winter. These are propagated through inflorescence bulbils or by division of clumps (Oyen and Messiaen, 2004).

Cultivation areas

Though Welsh onion is grown throughout the world from temperate to tropical conditions but China and Japan are the leading countries for its commercial cultivation. It is cultivated in Siberia and Ciscaucasia (Russia), China, Japan, Korea, Taiwan, Thailand, Indonesia, Malaysia, Philippines, Vietnam, Northeast India, Norway, Germany, Sudan, Kenya, Cameroon, Congo, Zaire, Sierra Leone, Zambia and Zimbabwe (Singh and Ramakrishna, 2017).

In India, it is traditionally grown in the NE states of Mizoram, Manipur, Nagaland, Arunachal Pradesh, Manipur, Assam and Tripura; though most commonly in Mizoram (called Mizo-purun) where it is one of the most important spice crops grown frequently in the home gardens and *Jhum* lands. It is grown sparsely in Odisha, West Bengal, Andaman & Nicobar Islands, and only in few parts of Tamil Nadu, Kerala and Maharashtra; but is unknown to other parts of India (Singh and Ramakrishna, 2017).

Nutritional values

The nutritional value of green tops and blanched pseudostems differs, green tops being more nutritious.

Welsh onions, raw	
Nutritional value per 100g	
Energy	142 kJ (34 kcal)
Carbohydrates	6.5 g
Sugars	2.18 g
Dietary fiber	2.4 g
Fat	0.4 g
Protein	1.9 g
Vitamins	**%DV† (Quantity)**
Thiamine (B1)	4% (0.05 mg)
Riboflavin (B2)	8% (0.09 mg)
Niacin (B3)	3% (0.4 mg)
Pantothenic acid (B5)	3% (0.169 mg)

Welsh onions, raw	
Vitamin B6	6% (0.072 mg)
Folate (B9)	4% (16 μg)
Vitamin C	33% (27 mg)
Vitamin E	3% (0.51 mg)
Vitamin K	184% (193.4 μg)
Minerals	**%DV† (Quantity)**
Calcium	5% (52 mg)
Iron	9% (1.22 mg)
Magnesium	6% (23 mg)
Manganese	7% (0.137 mg)
Phosphorus	7% (49 mg)
Potassium	5% (212 mg)
Sodium	1% (17 mg)
Zinc	5% (0.52 mg)
Units: μg = micrograms; mg = milligrams; IU = International units †Percentages are roughly approximated using US recommendations for adults. Source:Singh and Ramakrishna (2017)	
Table source: *https://en.wikipedia.org/wiki/Allium_fistulosum*	

The nutritional composition of the blanched *cum* thickened pseudostems and green leaf blades differs; the green leaf types contain higher levels of β-carotene, vitamin B1 (thiamine), vitamin B2 (riboflavin) and vitamin C (ascorbic acid) though contents of each vitamin in the green leaf of Welsh onion are lower than those of other vitamin-rich vegetables (Singh and Ramakrishna, 2017). The raw green leaves are reported to contain water 90.5g, energy 34kcal, protein 1.9g, fat 0.4g, carbohydrate 6.5g, Ca 18mg, Mg 23mg, P 49mg, Fe 1.2mg, Zn 0.52mg, vitamin-A 1160 IU, thiamine 0.05mg, riboflavin 0.09mg, niacin (vitamin B3) 0.40mg, folic acid (vitamin B9) 16μg and ascorbic acid 27mg in each 100g edible portions (Singh and Ramakrishna, 2017). A large proportion of the storage carbohydrates are sugars and oligosaccharides (Darbyshire and Steer, 1990). Besides glucose, fructose and sucrose; they consist of maltose, rhamnose, galactose, arabinose, mannose and xylose and the sugar and protein contents were found increased in plants grown under low temperatures which improves eating quality (Messiaen and Rouamba, 2004).

Phytochemicals

Welsh onion leaves contain high levels of quercetin ~ a flavonol compound with potential benefit to human health, *i.e.* as protective effects in reducing the risk of cardiovascular disease, acts as anticancer due to its antiprostanoid and anti-inflammatory responses, and decreases the rate of DNA degradation (Crystal *et al.*, 2003; Feng and Liu, 2011). The aroma of Welsh onion is not very strong as in onion and garlic (Singh and Ramakrishna, 2017). Chemically, its leaves have a fairly intermediate smell of onion and leek which it derives from volatile propyl cysteine sulphoxide (leek aroma) and propenyl cysteine sulphoxide (onion aroma) (Singh and Ramakrishna, 2017). The antioxidant activity of *Allium* species is due to a variety of sulphur-containing compounds and their precursors, polyphenols, dietary fibre and microelements (Nencini *et al.*, 2007). Stajner and Varga (2003) estimated higher antioxidant activities in the leaves of *A. sativum*, *A. cepa*, *A. vineale*, *A. fistulosum*, *A. nutans*, *A. flavum* and *A. ursinum*. Consumption of onion and related *Allium* is associated with a reduction in blood lipids, cholesterol and platelet activity, contributing to decrease risk of cardiovascular disease (Augusti, 1990; Block, 1992).

The leaves contain carotenoids that have antioxidant activity (e.g. beta-carotene, lutein). Lutein is responsible to improve the vision of patients with age-related macular degeneration, a disease that causes vision loss.

Also a novel antifungal compound fistulosin (octadecyl 3-hydroxyindole) has been isolated from the roots of Welsh onion that showed high activity against *Fusarium oxysporum* primarily inhibiting protein synthesis (Phay *et al.*, 1999). Antifungal constituents have also been isolated from the seeds (Oyen and Messiaen, 2004).

Uses

Welsh onions can be used raw, in stir fries or cooked like vegetables. They are used raw in salads, as a garnish, and as a substitute for chives (*A. schoenoprasum*) or leeks (*A. ampeloprasum*, Porrum Group) (Christman, 2003). Young inflorescence is sometimes deep-fried and eaten as a snack. In some places, the dehydrated welsh onion is available that can be used as an additive to preprocessed food such as instant noodles (Oyen and Messiaen, 2004). Whole plants are consumed by Mizo tribes to flavour soups, steamed-boils, salads, vegetables, dals and certain other culinaries (Singh *et al.*, 2012).

It is also grown in a bunch as an ornamental plant (en.wikipedia.org/wiki/Allium_fistulosum). Locally, it is said that planting of Mizo-purun reduces termite infestations in the home gardens as well as in *Jhum* lands (Singh and Ramakrishna, 2017).

Cultivation

Propagation

Welsh onion is propagated through division of clumps and seeds (Singh and Ramakrishna, 2017). In temperate areas where seed production is more successful propagation is mainly by seed @ 8-16kg per hectare through direct sowing though only 2-4kg in case of transplanting. Seed is either sown directly into the field or first in nurseries which requires roughly 1/10th of the field area. The weight of 1000 seeds is about 2.2–2.5g. In nursery beds the seeds are usually sown in rows or in 5–6cm wide bands and the seedlings are ready for transplanting when these attain a pencil thickness and 25–30cm length (Oyen and Messiaen, 2004).

While seed production is possible at elevations above 1000m, *vis-a-vis* imported seeds of several cultivars are also available in the tropics but under tropical conditions Welsh onions are raised mainly by basal tillers or side shoots because they are more difficult and time consuming when raised from seeds. Moreover, through tillers it can be planted the whole year round. As in other alliums the seeds of *A. fistulosum* are also short-lived and their vitality tends to be lost easily during storage (Oyen and Messiaen, 2004).

Once the Welsh onion has been established in the field it can be taken out by loosening the soil around the clump and after separating the clumps these can be transplanted in field at the spacing of about 20×10cm (Singh and Ramakrishna, 2017) or at 20×25cm (Oyen and Messiaen, 2004). The separated side shoots (tillers) should be planted deep so that more of the lower stem is found blanched (Christman, 2003).

Climate

Welsh onion is adapted to a wide range of climates, and being quite tolerant to cold weather can overwinter in most regions (Oyen and Messiaen, 2004). These grow easily and are fast growing in nature. Welsh onion is though a perennial vegetable being harvested throughout the year, in the colder climates it is grown as an annual (Christman, 2003). It is tolerant to hot humid conditions and grows well above an altitude of 200m though feels more comfortable above 500m; most of the local cultivars are well adapted to variations in rainfall and are quite tolerant to heavy rainfall than other *Allium* spp., *vis-a-vis* well established plants are quite tolerant to moisture stress and drought (Oyen and Messiaen, 2004).

Welsh onion is cultivated during May-November in Mizoram-like climatic conditions where mean monthly temperature range is between 12 to 29°C, monthly relative humidity is 52-91%, monthly rainfall is around 35-705 mm and monthly number of rainy days is 5-26 days with the altitudes ranging from 900-1100 m m.s.l. (Singh and Ramakrishna, 2017).

These easy to grow and fast growing Welsh onions are hardy to USDA Zones 6 - 9 (Christman, 2003) under partial shade or full sun and can be grown in a wide range of climatic conditions though thriving best under long-day *cum* humid conditions though tending to bolt profusely under short days during winter which hampers quality, leaf growth and aroma (Singh and Ramakrishna, 2017). However, the genotypes markedly differ in the number of leaves developed before reaching the post-juvenile age, the cold and photoperiod requirement for scape elongation and floral induction (Kamenetsky and Rabinowitch, 2006).

Soil

These prefer well-drained heavy soils rich in organic matter, with a pH of 5.5-6.5 though can tolerate higher pH of 8-10, however, it does not relish waterlogging as this condition kills the active roots quickly (Singh and Ramakrishna, 2017). Most of the local genotypes or landraces of *A. fistulosum* are well adapted to wide range of soil moisture so they are more tolerant to excess moisture and high humidity than the other *Allium* species. Besides, it is also a potential donor of resistance to biotic stresses, high soluble solids and folic acid content (Peffley, 1986; Kumar and Banerjee, 2001).

Field preparation and transplanting

In Mizoram, the best time to transplant these is after the onset of rains, *i.e.* May-June (Singh and Ramakrishna, 2017). For blanched pseudostem production fields are deeply cultivated. Furrows of 10–20cm deep are made, the soil being thrown to one side forming a ridge which will support the young plant and facilitate earthing up later. Tillers or seedlings are transplanted into raised beds or ridges alternated with furrows for irrigation and drainage. Planting distances are about 20 cm × 25 cm (200,000 plants/ha). About one-third of the top part of the tiller is usually trimmed to reduce transpiration (Oyen and Messiaen, 2004).

Manures and fertilizers

Planting holes are filled with 50–100g of manure (10–20t/ha) and the shoots inserted slanting to stimulate tillering (Oyen and Messiaen, 2004). Urea or ammonium sulphate at the rate of 3g/plant (about 600kg/ha) is applied 3 weeks after planting and again at 6 weeks after if soil fertility is low (Oyen and Messiaen, 2004).

Irrigation

Allium fistulosum needs plenty of water. At lower elevations, it is usually grown during rainy season. During dry season the crop requires to be irrigated daily (Oyen and Messiaen, 2004).

Weeding and earthing up

Weeding and earthing up are usually practised 6–7 weeks after planting. Earthing up is essential for blanching of pseudostem and for softening the leaf-sheath cylinder. Earthing up also affects aeration of the roots and thus checks weed growth. Earthing up should be done at a proper stage and not too early (Oyen and Messiaen, 2004).

Mixed cropping

Mixed cropping with white cabbage, carrot and potato is common in the highlands (Oyen and Messiaen, 2004).

Other factors effecting growth and quality

Though Welsh onions can tolerate drought but best growth can be obtained through regular watering especially during summer (Christman, 2003). Early nipping (removal of apical buds) promotes vigorous shoot growth and thereby increasing yield by 10- 20% (Singh *et al.*, 2014).

Delay in harvesting may result in a substantial increment of yield but may cause depletion of the contents of vitamin C, carotenoids, chlorophyll 'a', chlorophyll 'b', sugars, volatile oils, nitrates, total N, K and Ca contents (Kolota *et al.*, 2012).

The priming of seeds with 2% potassium nitrate (KNO_3) increases the activity of antioxidant enzymes (dehydrogenase, superoxide dismutase, catalase) and lipid peroxidation, thereby increasing the seed vigour (Dong *et al.*, 2014). Higher doses of nitrogen (>70 kg/ha) adversely affects the quality of the crop, *i.e.* more accumulation of nitrates (Kolota *et al.*, 2013), *vis-a-vis* lowering down pungency level in the plant tissues (Singh and Ramakrishna, 2017). The levels of light and CO_2 ~ the basic requirements for photosynthesis, have positive effect on pungency. Increasing light and CO_2 levels increases the net accumulation of flavour precursors and/or alliinase activity (Levine *et al.*, 2005).

Postharvest durability of *A. fistulosum* is very short mainly due to its high metabolic rate and high water content coupled with poor techniques of management and transport to the consumer (Singh and Ramakrishna, 2017). To manage this, Freddo *et al.* (2013) suggests application of salicylic acid as it decreases the weight loss as well as shoot and leaf rot.

Harvest, yield and postharvest

The growing period is only 2.5–3 months in Japan, Korea, Taiwan and Indonesia compared to 9 months in East Asian countries. Welsh onion can be harvested year-round in the tropics. These are pulled out about 2.5 months after planting the tillers and the part used as planting material for the next crop is left in the field until it is needed. Harvesting is a labour-intensive operation, especially for pseudostems which have to be dug up, cleaned and bundled. Mechanized harvesting equipments are also now developed in Japan (Oyen and Messiaen, 2004). After harvesting, the harvest (pseudostem and leaves) is thoroughly cleaned of adhering soil, dirts, damaged leaves, *etc.*, surface-dried, bunched and packed in boxes for transport to the market (Oyen and Messiaen, 2004). Average yield in Japan and Korea is about 25t/ha though in Taiwan it is only 10–15t/ha. Yield in Indonesia is comparatively 7-15t/ha, usually only 7-8 tonnes (Oyen and Messiaen, 2004).

Pests and diseases

Though *Allium fistulosum* is tolerant to various biotic stresses such as pink root disease (*Phoma terrestris* or *Pyrenochaeta terrestris*), onion smut (*Urocystis cepulae*), neck rot (*Botrytis aclada, Botrytis allii*), leaf blight (*Botrytis squamosa*), anthracnose (*Colletotrichum gloeosporioides*), downy mildew (*Peronospora destructor*), *Fusarium* basal rot (*Fusarium oxysporum*), thrips (*Thrips tabaci*), onion fly (*Delia antiqua*) and OYDV-onion yellow dwarf virus (Emsweller and Jones, 1938; Khrustaleva and Kik, 1998, 2000; Kik, 2002) though it may be affected by a number of diseases most of which being common to other *Allium* crops. These diseases include **purple blotch** (*Alternaria porri*) causing characteristic concentric spots on the leaves and **white rot** (*Sclerotium cepivorum*) causing serious losses under successive or repeated croppings as the pathogen is very persistent in the soil. *Allium fistulosum* is resistant to onion yellow dwarf virus (OYDV) but susceptible to Welsh onion yellow stripe virus (WoYSV) which is most prevalent in Japan and Indonesia. These cause similar mosaic-type symptoms including chlorotic mottling, streaking and stunting and distorted flattening of the leaves. Relative tolerance is found in cultivars of 'Kujyo Group' (Oyen and Messiaen, 2004).

Poor, unbalanced nutrition and heavy rains stimulate the development of the diseases. The practice of vegetative propagation is conducive to virus infection but many landraces seem to be relatively tolerant. Diseased plants should be removed by rigorous visual inspection of the planting material (Oyen and Messiaen, 2004) and destroyed.

The **beet army worm** (*Spodoptera exigua*) and the **American bollworm** (*Helicoverpa armigera*) are the most serious pests. They are difficult to control because the larvae hide deep into the soil during daytime and inside the hollow leaves, as well as waxy leaf layer also hinder the wetting (Oyen and Messiaen, 2004).

References

Anjula Pandey, Pavan Kumar Malav, K.M. Rai and S.P. Ahlawat, 2022. Genus *Allium* L. of the Indian Region: A Field Guide for Germplasm Collection and Identification. ICAR-National Bureau of Plant Genetic Resources, New Delhi, India, 109p + i-vi. http://www.nbpgr.ernet.in/Downloadfile.aspx?EntryId=9197.

Augusti, K.T. 1990. Therapeutic and medicinal values of onion and garlic. In: Onion and allied crops, vol III (eds Rabinowitch, H. and J. Brewster). CRC Press, Boca Raton, FL, USA, pp 94–108. http://www.kiran.nic.in/pdf/IJHF/Vol_30-2/8.pdf.

Block, A, 1992. The organosulfur chemistry of the genus *Allium*: implications for the organic chemistry of sulfur. *Angewandte Chemie International Edition*, 31: 1135–78. http://www.kiran.nic.in/pdf/IJHF/Vol_30-2/8.pdf.

Borna, Z. 1982. Welsh onion. Vegetable Production. PWRiL; Warsaw, Poland, pp.198–200. (In Polish).

Chen, H., Q.D. Kong, D.Z. Fu, H. Liu, P.Y. Liu, W.F. Du, G.S. Wu, S.L. Zhang, L.L. Rao and S.N. Xu, 2001. Atlas of the Traditional Vegetables in China (Chinese). Zhejiang Science and Technology Publishing House; Hangzhou, China, 61p.

Christman, S. 2003. *Allium fistulosum*. A Floridata Plant Profile. https://floridata.com/plant/627. Retrieved 25 January, 2023.

Crystal, S., K.A. Lombard, E.B. Peffley and W.X. Liu, 2003. Genetic analysis of quercetin in onion (*Allium cepa* L.) —Lady Raider⍰. *Texas J. Agric. Nat. Resour.*, **16**: 24–8. http://www.kiran.nic.in/pdf/IJHF/Vol_30-2/8.pdf.

Darbyshire, B. and B.T. Steer, 1990. Carbohydrate biochemistry. In: Onions and Allied Crops, vol. III. - Botany, Physiology and Genetics (eds Rabinowitch, H.D. and J.L. Brewster), pp. 1-16. CRC Press Inc, Boca Raton, Florida, USA. http://www.kiran.nic.in/pdf/IJHF/Vol_30-2/8.pdf.

Das, R. 2010. Conservation of medicinal plant *Hibiscus rosasinensis* through *Allium fistulosum* plantation. *Der Pharmacia Sinica*, **1**(1): 166–72. http://www.kiran.nic.in/pdf/IJHF/Vol_30-2/8.pdf.

Dole, el J., E.J. Nowak and J. Luny, 1980. Embryo development and *in vitro* culture of *Allium cepa* and its interspecific hybrids. *Z. Pflanzenzüchtg.,* **85**: 177-84.

Dong, L., Z. Hao, Z. Li, J. Zhu and Q. Wang, 2014.. Enhancement of Welsh onion (*Allium fistulosum* L.) seed vigor by KNO3 priming. *J. Agric. Sci. Technol.*, 16: 1345–53. http://www.kiran.nic.in/pdf/IJHF/Vol_30-2/8.pdf.

Emsweller, S.L. and H.A. Jones, 1935. An interspecific hybrid in *Allium*. *Hilgardia*, **9**: 265-73.

Emsweller, S.L. and H.A. Jones, 1938. Crossing-over, fragmentation and formation of new chromosomes in an *Allium* species hybrid. *Bot. Gaz.*, **99**(4): 729–72.

en.wikipedia.org/wiki/Allium_fistulosum.

Feng, X. and W. Liu, 2011. Variation of quercetin content in different tissues of Welsh onion (*Allium fistulosum* L.). *African J. agric. Res.*, **6**: 5675–79. http://www.kiran.nic.in/pdf/IJHF/Vol_30-2/8.pdf.

Freddo, A.R., F.E. Cechim and S.M. Mazaro, 2013. Conservation of post-harvest leaves of green onion (*Allium fistulosum* L.) with the use of salicylic acid solution. *Pesquisa Aplicada Agrotecnologia* (Impresso), **6**: 87–92. http://www.kiran.nic.in/pdf/IJHF/Vol_30-2/8.pdf.

Frietsch, R.M. and N. Friesen, 2002. Evolution, domestication and taxonomy. In: *Allium* Crop Science: Recent Advances (eds Rabinnowitch, H.D. and L. Currah), pp. 5-30. CABI Publishing, Wallingford, Oxfordshire, UK.

Havey, M.J. 1999. Seed yield, floral morphology and lack of male fertility restoration of malke sterile onion (*Allium cepa* L.) population possessing the cytoplasm of Allium galanthum. *J. Am. Soc. Hort. Sci.,* **124**:626-9.

Ivanova, M.I., A.F.Bukharov, A.I. Kashleva and D.N. Baleev, 2015. Complex of traits of Welsh onion in the annual culture. *Russ. Veg.,* **27**:36–9. doi: 10.18619/2072-9146-2015-2-36-39 (In Russian).

Jones, R.N. and H. Rees, 1968. Nuclear DNA variation in *Allium*. *Hered.*, **23**: 591–605.

Kamenetsky, R. and H.D. Rabinowitch, 2006. The genus *Allium*: a developmental and horticultural analysis. *Hort. Rev.,* **32**: 329–78. http://www.kiran.nic.in/pdf/IJHF/Vol_30-2/8.pdf.

Kazakova, A.A. 1978. Luktatarka (*Allium fistulosum* L.) (Russian). In: Kulturnaja Flora (ed. Brezzhnev, D.D.), pp. 54-61. 10 Leningrad, Koloz, USSR.

Khrustaleva, L.I. and C. Kik, 1998. Cytogenetical studies in the bridge cross *Allium cepa* × (*A. fistulosum* × *A. roylei*). *Theoret. and Appl. Gen.,* **96**: 8–14.

Khrustaleva, .L.I. and C. Kik, 2000. Introgression of *Allium fistulosum* into *A. cepa* mediated by *A. roylei. Theoret. and Appl. Gen.*, **100**: 17–26.

Kik, C. 2002. Exploitation of wild relatives for the breeding of cultivated *Allium* species. In: *Allium* Crop Science: Recent Advances (eds .Rabinowitch, H.D. and L. Currah), pp. 81-100. CABI Publishing, Oxon, Wallingford, U.K.

Kolota, E., K. Adamczewska-Sowinska and C. Uklanska-Pusz, 2012. Yield and nutritional value of Japanese bunching onion (*Allium fistulosum* L.) depending on the growing season and plant maturation stage. *J. Element.*, **17**(4): 587–96. http://www.kiran.nic.in/pdf/IJHF/Vol_30-2/8.pdf.

Kolota, E., K. Adamczewska-Sowinska and C Uklanska-Pusz, 2013. Response of Japanese bunching onion (*Allium fistulosum* L.) to nitrogen fertilization. *Acta Scientia. Polonorum Hort.*, **12**(2): 51–61.

Kumar, S. and M.K. Banerjee, 2001. Evidences for three paracentric inversions between Welsh and common onion. *J. Genet. Cytol.*, **2**: 149–52. http://www.kiran.nic.in/pdf/IJHF/Vol_30-2/8.pdf.

Levine, L.H., J. Bauer, S. Edney, J. Richards, N. Yorio, K. Li, P. Pare and R. Wheeler, 2005.. Scallion (*Allium fistulosum* L.) chemistry affected by variety and environmental conditions (light and CO2). In: *Proc. 35th intern. Conf. on environ Syst. and 8th Europ. Symp. on Space environ. Control Syst.*, 11, held on July 14, 2005 in Rome (Italy). http://www.kiran.nic.in/pdf/IJHF/Vol_30-2/8.pdf.

Ma, D..X., Y.S. Ma, C.W. Ma, N. Wang, S. Wang, Z.G. Wang, S.T. Wang, Z.S. Wang, B.Y. Wang and G.C. Wang, 2009. Cultivation of Chinese Vegetables (2nd ed), pp. 384-94 (in Chinese). China Agriculture Press, Beijing, China.

Malachowski, A. 1982. Onion-like Vegetables (2nd ed), pp. 149-160 (In polish).. Panstwowe Wydawnictwo Rolnicze i Lesne; Warsaw, Poland.

Messiaen, C.M. and A. Rouamba, 2004. *Allium cepa* L.-PROTA (Plant Resources of Tropical Africa), Wageningen, the Netherlands. http://database.prota.org/search.htm (accessed in October 2012). http://www.kiran.nic.in/pdf/IJHF/Vol_30-2/8.pdf.

National Park. Flora & Fauna Web. *Allium fistulosum* L. https://www.nparks.gov.sg/florafaunaweb/flora/4/9/4944. Retrieved 25 January, 2023.

Nencini, C., F. Cavallo, A. Capasso, G.G. Franchi, G. Giorgio and L. Micheli, 2007.. Evaluation of antioxidative properties of *Allium* species growing wild in Italy. *Phytother. Res.*, **21**: 874–8. http://www.kiran.nic.in/pdf/IJHF/Vol_30-2/8.pdf.

Oyen, L.P.A. and C.-M. Messiaen, 2004. *Allium fistulosum* L. [Internet] Record from PROTA4U. Grubben, G.J.H. & Denton, O.A. (Editors). PROTA (Plant Resources of Tropical Africa / Ressources vegetales de l'Afrique tropicale), Wageningen, Netherlands. <http://www.prota4u.org/search.asp>.

Padula, G., X. Xia and R. Hołubowicz, 2022. Welsh Onion (*Allium fistulosum* L.) Seed Physiology, Breeding, Production and Trade. *Planta*, **11**(3):343. https://doi.org/10.3390/plants11030343; https://www.mdpi.com/2223-7747/11/3/343.

Peffley, E.B. 1986. Evidence for chromosomal differentiation of *Allium fistulosum* and *A. cepa. J. Amer. Soc. Hort. Sci.*, **111**: 126–9. http://www.kiran.nic.in/pdf/IJHF/Vol_30-2/8.pdf.

Peng, S., Y. Yang, L. Chen aqnd Z.P. Jiang, 2017. Analysis on planting and mechanized harvesting of Welsh onion. *J. Chin. Agric. Mech.*, **38**:30–35 (In Chinese).

Phay, N., T. Higashiyama, M. Tsuji, H. Matsuura, Y. Fukushi, A. Yokota and F. Tomita, 1999. An antifungal compound from roots of Welsh onion. *Photochem.*,**52**(2): 271–4. http://www.kiran.nic.in/pdf/IJHF/Vol_30-2/8.pdf.

POWO, 2023. *Allium fistulosum* L. Plants of the World Online|Kew Science. https://powo.science.kew.org/taxon/urn:lsid:ipni.org:names:528046-1. Retrieved 20 January 2023.

Purseglove, J.W. 1978. Monocotyledons. *Allium fistulosum* L. Tropical Crops, 52p. The English Language Book Society, Longman, London, UK.

Shishkina, Y.V., S.U. Zharkova and O.U. Malykina, 2019. Welsh onion (*Allium fistulosum*) variety 'Premyera' for conditions of the south of western Siberia. *Russ. Veg.*, **45**:65–67 (Russian). doi: 10.18619/2072-9146-2019-1-65-67.

Singh, B.K. and Y. Ramakrishna, 2017. Welsh onion (*Allium fistulosum* L.): A promising spicing-culinary herb of Mizoram. *Indian J. Hill Farm.*, **30**:201–8. http://www.kiran.nic.in/pdf/IJHF/Vol_30-2/8.pdf.

Singh, B.K., V.K. Verma and Y. Ramakrishna, 2014. Leaf vegetables of Mizoram. In: Climate Resilient Crops for the Future (ed. Peter, K.V.), pp. 383-404.. NIPA, New Delhi. http://www.kiran.nic.in/pdf/IJHF/Vol_30-2/8.pdf.

specialtyproduce.com/produce/Walking_Tree_Onions_12607.php

Stajner, D and I.S. Varga, 2003. An evaluation of antioxidant abilities of Allium species. *Acta Biologica Szegdiensis*, **47**: 103–6. http://www.kiran.nic.in/pdf/IJHF/Vol_30-2/8.pdf.

Tang, P., L.W. Cui, Y.L. Zhang, C.Y. Hua and Y.L. Zhang, 2011. Research of storage processing techniques in *Allium fistulosum* L. var. *giganteum* Makino seed (Chinese). *Hortic.Seed*, **3**:28–9.

Thompson, Sylvia; 1995. The Kitchen Garden. Bantam Books. ISBN 9780553081381. https://archive.org/details/isbn_9780553081381.

French Grey Shallot (*Allium oschaninii O.Fedtsch.*)

Subgenus: *Cepa*; **Section**: *cepa* (Wikispecies, 2023).

Other common names: Dutch red shallot, Eschalote grise, French gray shallot, Griselle or True shallot.

Synonyms: *Allium cepa* var. *sylvestre* Regel (Heterotypic) (POWO, 1906).

Introduction and origin

Shallots - *Allium oschaninii*
By David Monniaux –Own work, CC BY-SA 3.0, https://commons.wikimedia.

French grey shallot is native to NE Iran and Central Asia (POWO, 1906). It is a specific form of shallots long cultivated in France and Italy (masterclass.com,

2022). It is widely cultivated in parts of southern France for culinary use (Aliza, 2004). It has greyish hue in its outer skin and is the ultimate culinary variety (masterclass.com, 2022). These shallots are believed to be the true shallots as these do not set seeds so are only propagated through bulbs (hudsonvalleyseed.com). Some people consider shallots the tastiest member of the onion family (greyduckgarlic.com.).

All shallots including multiplier onions, French grey shallots, Persian shallots, *etc.*, respond to day length, *i.e.* they adapt to the latitude where they are planted. As a result, a particular variety being grown for years in one ecosystem may not do well in different ecosystem as a shallot variety from Israel is not likely to do well north of Pennsylvania and likewise a shallot from northern France may wither in the scorching heat of Texas (Weaver, 2023).

***Source*:** Weaver, 2023

Comparision with French red shallot

The French grey shallots differ from other shallots and are found to be more closely related either to *A. oschaninii* or *A. vavilovii* than to *A. cepa* and other shallots (Maass, 1996). French red shallots (a subgroup of *Allium cepa* var. *aggregatum*) are more widely grown, easier to peel, and may be started from both bulb and seed whereas the French grey shallots are almost always propagated through bulbs but are not easy to peel (sonomamg.ucanr.edu; kitchengardenseeds.com) though have sweeter and milder taste than the 'shallot onion' cross (Sweetser, 2022). French grey shallots are not easy-peeling though these elevate the delicacy of almost any recipe with their distinctive rich creamy flavour and texture (kitchengardenseeds.com). French grey and French red shallots can be stored only up to about 3-4 months (hudsonvalleyseed.com). The clusters of bulbs of *A.c.* var. *aggregatum* are usually larger and rounder compared to the French grey shallot (*A. oschaninii*) that are elongated. French grey shallot is an easy-to-grow allium that is planted like garlic in fall for early summer harvest unlike French red shallots which can also be planted in spring (hudsonvalleyseed.com).

Botany

French grey shallot is a bulbous perennial that after 1-2 years reaches to a height of around 40cm and a spread of 20cm with upright, blue-green, strap-like leaves, and a cluster of small (2.5-5cm), elongated, teardrop-shaped underground bulbs much like the French red shallots though covered by a layer of pale brown-grey skin (Shoot, 2023) concealing creamy-purple or pink-purple flesh. Similar to garlic, the shallots multiply in the ground but individual bulbs have concentric layers similar to onions so these are used also in place of onions in cuisines (sonomamg.ucanr.edu). Moreover, shallots are generally smaller than garlic bulbs and have a mild onion flavour (Iannotti, 2022). In the spring, the shallot bulbs will multiply and sprout several green stalks per planted bulb producing 6-12 bulbs per plant (hudsonvalleyseed.com).

Like most other alliums, shallots too develop a bulb in one year and then bloom the next year. However, shallots, unlike onions, have lost the ability to produce flowers because they have been developed from clones for a long time, and if these produce the blossom too, the flowers are sterile. That's why shallots must be reproduced by replanting bulbs harvested the previous season (Weaver, 2023).

Uses

Shallots having sweeter and milder flavour can be used like onions in most recipes. These can be added to salads and to flavour food and drinks (Shoot, 2023). Shallots add distinctive flavour to vinaigrette dressings, elegant sauces and savory winter soups (kitchengardenseeds.com).

Cultivation

Propagation

French grey shallots are almost always grown by bulb sets and not like multiplier onions which are also propagated through seeds (Shoot, 2023). Shallots grow in clusters that can be gently broken apart like garlic cloves and planted (hudsonvalleyseed.com).

Climate and season of planting

French grey shallot grows primarily in the temperate biome (POWO, 1906). Vernalization (exposure to winter freezing) usually results in larger and better-flavoured shallots (Weaver, 2023). Fall planting of grey shallot bulb sets work for gardens in USDA Hardiness Zones 5 to 8. Shallots are sensitive to severe winters especially if they remain unprotected by organic mulch (Weaver, 2023), therefore the bulb sets are not likely to survive in zones 4 or below where the planting can be done in the spring instead, however, spring planted shallots will be smaller and are not quite as flavourful as the fall-planted ones (Sweetser, 2022).

In warmer regions these shallots will not get the required day length to produce big bulbs unless a short day shallot variety is grown which will start producing bulbs with only 10 to 12 hours of daylight, therefore, it is recommended that gardeners in Zones 9 to 11 should grow multiplier or potato onions instead. Being quite similar to shallots producing clusters of onions from a single planted clove, these are good for hot climates (Sweetser, 2022).

Soil

French red shallot grows well under full sun in loose, alkaline to neutral and well-drained sandy-loam soil (Shoot, 2023) with plenty of organic matter and good fertility (Sweetser, 2022). soil should be amended with compost to maintain fertility and to provide excellent drainage as in heavy soil drainage is a major problem (sonomamg.ucanr.edu).

Field preparation and planting

Field beds are prepared in the same way as for garlic (Sweetser, 2022). The bulbs during winter should not be allowed to stay in soggy soil as they will rot, so to avoid soil being soggy the planting should be done on raised beds or on ridges prepared in the field (Sweetser, 2022). Growing beds should be prepared with plenty of compost by mixing some 10-15t/ha of FYM or compost in the soil (hudsonvalleyseed.com).

After the first frost individual bulb-sets should be planted at a distance of about 15-20cm and a row distance of about 25cm (sonomamg.ucanr.edu). Bulbs should be planted 2.5 to 5cm deep with root end down (hudsonvalleyseed.com) which will send up green shoots in 10-14 days.

Mulching

Like most onions, shallots too are shallow-rooted and easily heaved out of the ground during alternate freeze-thaw cycles that occur over the winter (Sweetser, 2022). To protect the crop from this problem the growers should mulch sparingly until tops appear and then augment depth to hold in moisture and control weeds (sonomamg.ucanr.edu). Straw or shredded leaves work well and won't oppress the early spring growth. It is up to growers whether they want to remove the mulch in the spring or not. If mulch is removed, growers will have to keep up with weeding because their shallow roots don't compete well. If kept in place, it should be slightly thinned so that the new shoots can push through the mulch as these grow (Sweetser, 2022).

Manures and fertilizers

Before planting, the soil should be well composted. High nitrogen fertilizer promotes leafy growth at the expense of bulb size so it should be avoided. In spring and before bulbs begin to develop, side dressing with additional compost or a very light but balanced organic fertilizer is recommended to promote large bulb size (sonomamg.ucanr.edu).

Irrigation

Like all alliums, shallots respond quite positively to rainwater or irrigation (hudsonvalleyseed.com). Watering should be done at least 2.5cm deep per week regularly during dry periods but stopped when onions are swollen, however, before a few weeks of bulb harvesting the watering is stopped (Shoot, 2023).

Clipping

Flower heads may appear in some shallots, especially in June due to the day-length sensitivity of the plants (Weaver, 2023) so these should be sheared off from the base so that entire energy is diverted towards bulb development (Shoot, 2023).

Harvest and postharvest

A few baby shallots can be harvested in early spring to use as scallions similar to green onions (sonomamg.ucanr.edu). The bulbs can be harvested when the leaves become yellow and fall over (Shoot, 2023) in late spring or early summer, usually in July (hudsonvalleyseed.com). They are gently dug, cleaned by shaking or brushing off the dirt, and then placed on screens or wire racks in a warm and dry shady place to cure for about a week or so (Sweetser, 2022; hudsonvalleyseed.com). Once the green tops have dried up to become crispy these are removed along with the soft and rotten bulbs (hudsonvalleyseed.com) and then these are stored in a cool, dry and dark place but not under refrigeration system for whole of winter. Thick-necked ones since do not store well so these are used first (Sweetser, 2022; sonomamg.ucanr.edu). French grey shallots can be stored only for about 3-4 months (hudsonvalleyseed.com).

Pests and diseases

Pests and diseases are rarely a problem. Eelworm (stem and bulb), leek moth and onion fly may infest the crop though it has good tolerance to mildew, however, the crop faces only a few problems such as bolting, bullneck, neck rot, rust, saddleback, shanking, smut, white rot, white tip (Shoot, 2023), *etc.* which are controlled as with other alliums.

References

Aliza, G. 2004. Field Guide to Produce: How to Identify, Select, and Prepare Virtually Every Fruit and Vegetable at the Market, 256p. Quirk Books. ISBN 978-1-931686-80-8.

Greyduckgarlic.com/How-to-Grow-Shallots.html

Hudsonvalleyseed.com/products/french-grey-shallot?variant=31280912334893

Iannotti, M. 2022. How to Grow Shallots. Reviewed by Julie Thompson-Adolf. Fact checked by SARAH SCOTT. Updated on 09/05/22. The Spruce. https://www.thespruce.com/growing-shallots-in-the-vegetable-garden-1403464

Kitchengardenseeds.com/seed-index/tubers-sets-slips/french-gray-shallot-bulb-sets.html

Maass, H.I. 1996. About the origin of the French gray shallot. *Genet. Res. and Crop Evol.*, **43**: 291–2.

Masterclass.com, 2022/articles/what-is-a-shallot. Retrieved on 29 November, 2022.

POWO - Plants of the World Online|Kew Science, 1906.. *Allium oschaninii* O.Fedtsch. *Progr. Sadov. Ogorad.* 3: 332. https://powo.science.kew.org/taxon/urn:lsid:ipni.org:names:528547-1.

Shoot. *Allium oschaninii* (French grey shallot). Retrieved January 10, 2023. https://www.shootgardening.co.uk/plant/allium-oschaninii

Sonomamg.ucanr.edu/Vegetable_of_the_Month/Shallots/

Sweetser, R. 2022. Growing Shallots: How to Plant Shallots in Fall. ALMANAC. https://www.almanac.com/growing-shallots-how-plant-shallots-fall

Weaver, W.W. 2023. How to Grow Shallots. Fine Gardening. Retrieved January 11, 2023. https://www.finegardening.com/project-guides/fruits-and-vegetables/how-to-grow-shallots

Wikispecies, 2023. *Allium oschaninii.* https://species.wikimedia.org/wiki/Allium_oschaninii. Retrieved on 12,February, 2023.

Persian Shallot (*Allium stipitatum* Regel)

Other common names: Mooseer in Iran (Moradi *et al.*, 2013).

Synonyms: *Allium hirtifolium* Boissier and *A. atropurpureum* var. *hirtulum* Regel.

Introduction and origin

***Allium stipitatum* 'Mt Everest'**	***A. altissimum***
Source: By I, KENPEI, CC BY-SA 3.0, commons.wikimedia.org/w/index.php?curid=19916461	By Rillke - Own work, CC BY-SA 3.0, commons.wikimedia.org/w/index.php?curid=15801006

The specific epithet *stipitatum* means 'with a little stalk'. Persian shallot is native to SE Turkey to Central Asia and Pakistan, and grows as a wild plant in Afghanistan, Iran, Iraq, Kazakhstan, Kirgizstan, Pakistan, Tadzhikistan, Turkey, Turkmenistan and Uzbekistan (POWO, 1881). It grows on rocky slopes and in fields at elevations of 1,500–2,500m (Davies, 1992; Mathew, 1978). It is one of the important edible alliums in Iran, where it grows as a wild plant across the Zagross Mountains (Ebrahimia *et al.*, 2009; Ghahreman, 1984) at high elevations of different provinces from NW to south of Iran (Sanandaj, Kangavar, Siakhdarengoon, Sahneh, Ashtian, Dashtearzhan, Koohrang, Sepidan, Divandareh, Boroujerd, Khomein, Yasuj, Nahavand, Khansar, Harsin, Arak, Doshmanziare and Koohmaresorkhi) with very cold to moderate cold climate (Ghahreman, 1984), thus exhibiting a great diversity in colour, shape, pungency and dry matter content (Moradi *et al.*, 2013). Like other shallots, Persian shallots are also able to successfully adapt to a very wide range of environments (Moradi *et al.*, 2013).

Persian shallot

Source: *http://www.edimentals.com/blog/?page_id=893*

Some sources regard *A. hirtifolium* as synonym of *A. stipitatum* (POWO, 2022), while others treat *A. stipitatum* and *A. hirtifolium* as distinct species, the latter being the Iranian species (Davies, 1992).

Allium stipitatum is one of at least two botanical species mostly wild collected in West Asia as Persian shallots; the other species is *A. altissimum* Regel (tall onion) having a more slender stature, narrower and only slightly toothed or even smooth leaves, a smaller umbel, and more intensely-coloured tepals that in late anthesis are spirally enrolled (Fritsch *et al.*, 2007). *A. altissimum* is native to Central Asia, Iran and Afghanistan (POWO, 2023) and its best known cultivar is 'Goliath' (RHS, 2021).

The Persian shallot cultivars 'Mount Everest' and 'Violet Beauty' (RHS, 2017a, b) have gained the Royal Horticultural Society's Award of Garden Merit.

Botany

Persian shallot bulbs		Seeds
		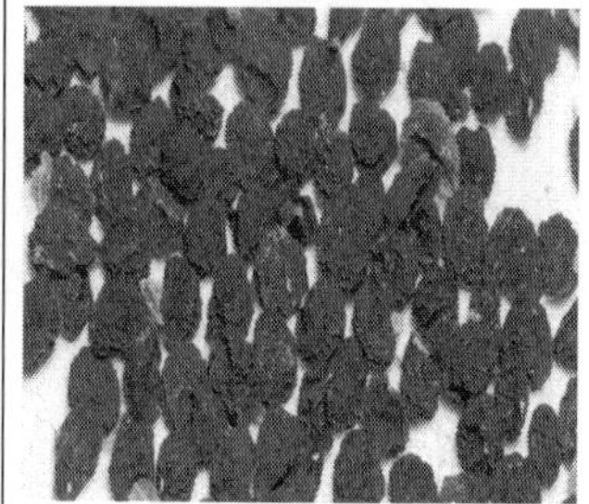
Source: *www.edimentals.com/blog/?page_id=893*	***Source:*** *By Chaojoker - Own work, CC BY-SA 3.0, https://commons.wikimedia.org/w/index.php?curid=18158497*	***Source:*** *www.edimentals.com/blog/?page_id=893*

Persian shallot is a wild perennial herb (Moradi *et al.*, 2013) having 4–6 broad, green to grayish-green and variably hairy basal leaves (en.wikipedia.org) that are 20 to 30cm long, linear and lanceolate (Moradi *et al.*, 2013). The leaves are usually withered by the time the plant flowers (en.wikipedia.org). The plant produces 0.8-1.2m tall, naked and erect scape (Moradi *et al.*, 2013) on which several flowers are arranged in a more-or-less spherical umbel about 8-12cm in diameter (en.wikipedia.org). Each flower has a superior ovary and six, lilac to purple or sometimes white tepals that are 2.5-5cm long (en.wikipedia.org). The underground bulbs are yellow, oval (Moradi *et al.*, 2013), 3-6cm in diameter and have blackish, paper-like tunics (en.wikipedia.org) or white skin (Moradi *et al.*, 2013). Usually the plant consists of a 1-2 main bulbs, each bulb weighing 8–15 times of garlic clove (Moradi *et al.*, 2013).

Comparatively, Persian shallot produces more seeds than multiplier onion shallot but its seeds have low germination percentage and velocity as well as slow growth of the subsequent seeding (Moradi *et al.*, 2013). Therefore bulbs are more convenient way of propagation than seeds (Etemadi *et al.*, 2011).

Nutritional values

When compared to other alliums, dry matter in Persian shallot is higher except garlic, as well as Persian shallot is rich in Cu, Zn and Mn elements

(Moradi *et al.*, 2013). For human consumption, Persian shallot landraces are more preferred due to their higher contents of linolenic and linoleic acids than common shallot and onion. Landraces of Persian shallot are important in mineral elements and essential fatty acid contents therefore these are recommended for human nutrition (Ebrahimia *et al.*, 2009). Persian shallots are high in vitamin C, folic acid, fibre, iron, potassium and calcium, *vis-a-vis* a good source of protein so its vegetables to vegetarians it quite preferable (Moghaddasi, 2011).

Phytochemicals

Like other alliums, Persian shallot too comprises of sulphur- as well as non-sulphur containing compounds. As in case of garlic, Persian shallot flavour is also largly due to allicin formed by the cleavage of S-alk(en)yl- L-cysteine sulphoxides (alliin) by alliin alkylsulphenate-lyase (alliinase) enzyme (Moradi *et al.*, 2013). Both alliin and the enzyme alliinase are stored in separate cellular compartments in intact tissue of fresh bulbs (GhodratiAzadi *et al.*, 2008), so when the bulb is cut or crushed, both the compartments get disrupted resulting in hydrolysis of alliin by allinase to give allicin (Jansen *et al.*, 1989).

Other components such as saponins, sapogenins, ajoene, sulphuric compounds (thiosulfinates) and flavonoids (including quercetin and kaempferol) as found in several species of *Allium* may also be present in Persian shallot for which further investigations are required (Moradi *et al.*, 2013; Rose *et al.*, 2005).

Uses

In Iran, the bulbs of Persian shallots are harvested from the wild and then sliced, dried and sold. Buyers will often soak the shallots for a number of days then boil them to obtain a milder flavour or often crushed and mixed with yogurt (en.wikipedia.org). Its powder transmits good taste to foods therefore it is used as a tasty additive or spice in foods, especially the sauces (Moradi *et al.*, 2013). Its dried bulb-slices are used as an additive to yogurt and pickling mixtures (Ebrahimia *et al.*, 2009).

Allium stipitatum is a medicinal plant in Central Asia (en.wikipedia.org). In traditional medicine Persian shallot is recommended for treatment of rheumatic and inflammatory disorders, gout, arthritis, diarrhoea, stomach pain, psoriasis and hemorrhoids (Etemadi *et al.*, 2011). Its bulb extracts as contain several pyridine-*N*-oxide derivatives therefore exhibit activity against *Mycobacterium tuberculosis*. Analysis of the cut plant using a mass spectrometer equipped with a DART ion source showed the presence of *N*-hydroxypyridine-2-thione (pyrithione) ~ a compound which is moderately cytotoxic towards human tumour cell lines, and highly active against fungi and gram-negative bacteria. Plant extracts containing pyrithione- are used in herbal medicine for treatment of malaria. Zinc pyrithione is used in commercial anti-dandruff shampoos (O'Donnell et al, 2009; Kusterer *et al.,* 2010; Block *et al.,* 2011).

Cultivation

Persian shallot being an invasive plant so its growing on the borders should be avoided. It can be propagated through bulbs, offset divisions and through seeds. Propagation through bulbs is by placing these during autumn in the soil to a depth of up to three times their diameter (candide.com). Since species produces numerous offsets so these prove to be a potential source for further multiplication. One year it may take a little growth of bulbs but second year attaining full-sized bulbs as then it is a 2-year crop, planting in 1st autumn and bulb harvesting during 3rd autumn, *i.e.* full 2-year term (Davies, 1992; candide.com). Seeds are not meant for direct cropping as these are sown thinly during spring at least 5cm apart in rows spaced 30cm apart in the first year to produce onion sets (bulblets) and then planted the second year for taking for further bulb growth, and then again the third year for harvesting full-sized bulbs (candide.com), however, from seeds it takes more than three years to bloom (Mathew, 1978).

Persian shallot is primarily a crop of temperate regions (POWO, 2022). It is able to survive under low moisture conditions but for good yield it requires an adequate supply of water (Moradi *et al.,* 2013), full sun and adequate provision of drainage. It can also be grown in large containers for home consumption but container size should match the plants when in bloom. However, sunny situation, well-drained fertile soil and raised beds are most suitable for its successful growing (candide.com).

While preparing the field, the soil is thoroughly ploughed at least thrice 15-20cm deep, perennial rootstocks of weeds are removed along with other constraining materials and at third ploughing some 10-15t/ha of well-rotten dung or farmyard manure in the field should thoroughly be mixed with the soil, then beds of convenient sizes are prepared and crop is sown (seeds) or planted (bulbs or bulblets/offsets). Since in this case we consume only the bulbs similar to onion so we will have to wait up to next autumn for harvest though in case of bulblet planting only after completion of two autumns and in case of seed sowing only after completion of three years in the field. First irrigation is given only when sprouts are visible.

The bulbs remain dormant at both, *i.e.* low and high temperatures. Plants are lifted at maturity and dried to keep well (Moradi *et al.*, 2013). Its fresh or dried bulbs are exported to Persian Gulf (Khalij-e-Fars) countries (Moradi *et al.*, 2013; Mozaffarian, 1996).

Among all the tall alliums, Persian shallots are easiest to grow. Its cultural practices are almost similar to other alliums.

Pests and diseases

Like other alliums, common problems with Persian shallot are onion thrips (*Thrips tabaci*), onion fly (*Delia antiqua*) and maggot, allium leafminer (*Phytomyza gymnostoma*), stem and bulb nematode (*Ditylenchus dipsaci*), *etc.* Some diseases include onion white rot (*Stromatinia cepivora*), downy mildew (*Peronospora cepae*), *etc.* (candide.com). Their full details are furnished under the chapter 'Onion'.

References

Block, E., A.J. Dane and R.B. Cody, 2011. Crushing Garlic and Slicing Onions: Detection of Sulfenic Acids and Other Reactive Organosulfur Intermediates from Garlic and Other Alliums Using Direct Analysis in Real Time-Mass Spectrometry (DART-MS). *Phosphorus, Sulfur, Silicon and the Related Elements*, **186** (5): 1085–93. doi:10.1080/10426507.2010.507728, S2CID 98520689.

Candide.com/IE/plants/6bbe8487-70dd-4c72-ae3c-4e0a8f0c1f5b

Davies, Dilys; 1992. Alliums: the ornamental onions. Pub. by B.T. Batsford (in association with the Hardy Plant Society), London, pp. 101 & 137. ISBN 978-0-7134-7030-7.

Ebrahimia, R., Z. Zamani and A. Kash, 2009. Genetic diversity evaluation of wild Persian shallot (*Allium hirtifolium* Boiss.) using morphological and RAPD markers *Sci. Hort.*, **119** (4): 345–51. doi:10.1016/j.scienta.2008.08.032.

En.wikipedia.org/wiki/Allium_stipitatum

Etemadi, N., M. Haghighi and N. Zamani, 2011. *Afr. J. agric. Res.*, 6: 5133. https://www.primescholars.com/articles/medicinal-properties-of-persian-shallot-90731.html.

Fritsch, Reinhard; Maia Gurushidze; Jarmila Jedelska and Keusgen Michael, 2007. Ornamental Drumstick Onions of *Allium* subgenus *Melanocrommyum* Used as Medicinal Plants in Southwest and Central Asia. *Planta Medica*, **73**:10.1055/s-986919.

Ghahreman, A. 1984. Color Atlas of Iranian Plants. Botany Division, Institute of Forestries & Grasslands, Iran, pp. 512.

GhodratiAzadi, H., S.M. Ghaffari, G. Riazi, S. Ahmadian and F. Vahedi, 2008. *Cytotechn.*, 56: 179. https://www.primescholars.com/articles/medicinal-properties-of-persian-shallot-90731.html.

https://www.edimentals.com/blog/?page_id=893

Jansen, H., B. Muller and K. Knobloch, 1989. *Planta Med.*, 5:434. https://www.primescholars.com/articles/medicinal-properties-of-persian-shallot-90731.html.

Kusterer, J., A. Vogt and M. Keusgen, 2010. Isolation and Identification of a New Cysteine Sulfoxide and Volatile Sulfur Compounds from *Allium* Subgenus Melanocrommyum. *J. Agric. Food Chem.*, **58**(1): 520–26. doi:10.1021/jf902294c, PMID 19919098.

Mathew, Brian; 1978. The Larger Bulbs. Pub. by B.T. Batsford (in association with the Royal Horticultural Society), London, 32p. ISBN 978-0-7134-1246-8.

Moghaddasi, M. 2011. *Adv. environ. Biol.*, 5: 1965. https://www.primescholars.com/articles/medicinal-properties-of-persian-shallot-90731.html.

Moradi Younes, Moradi-Sardareh Hemen, Ghasemi Hasan, Mohamadi Nejad, Mohammad-Nabi Moradi and Seyed-Mostafa Hosseini-Zijoud, 2013. Medicinal properties of Persian shallot. *Eur. J. exper. Biol.*, **3**(1). https://www.primescholars.com/articles/medicinal-properties-of-persian-shallot-90731.html.

Mozaffarian, V. 1996. Dictionary of Iranian Plant Names (Latin–English–Persian). Farhange Moaser, Tehran. https://www.primescholars.com/articles/medicinal-properties-of-persian-shallot-90731.html.

O'Donnell, G., R. Poeschl, O. Zimhony, M. Gunaratnam, J.B.C. Moreira, S. Neidle, D. Evangelopoulos, S. Bhakta, J.P. Malkinson, H.I. Boshoff., A. Lenaerts and S. Gibbons, 2009. Bioactive pyridine-N-oxide disulfides from *Allium stipitatum*. *J. Nat. Prod.*, **72**(3): 360–65. doi:10.1021/np800572r, PMC 2765505, PMID 19093848.

POWO, 2022. Plants of the World Online|Kew Science. *Allium stipitatum* Regel. First published in *Trudy Imp. S.-Peterburgsk. Bot. Sada*, **7**: 546 (1881). https://powo.science.kew.org/taxon/urn:lsid:ipni.org:names:528920-1. Retrieved 17 December, 2022.

POWO, 2023. Plants of the World Online|Kew Science. *Allium altissimum* Regel. https://powo.science.kew.org/taxon/urn:lsid:ipni.org:names:527568-1. Retrieved 17 February, 2023.

RHS, 2017a. *Allium stipitatum* 'Mount Everest'. Retrieved 2018-01-05. https://www.rhs.org.uk/plants/243759/i-allium-stipitatum-i-mount-everest/details.

RHS, 2017b. *Allium stipitatum* 'Violet Beauty'. Retrieved 2018-01-05. https://www.rhs.org.uk/plants/195244/i-allium-stipitatum-i-violet-beauty/details.

RHS, 2021. *Allium altissimum* 'Goliath'. Archived from the original on 3 August 2021. Retrieved 3 August 2021. https://www.rhs.org.uk/plants/138663/i-allium-altissimum-i-goliath/details.

Rose, P., M. Whiteman, P.K. Moore and Y.Z. Zhu, 2005. *Nat. Prod. Rep.*, **22**: 351. https://www.primescholars.com/articles/medicinal-properties-of-persian-shallot-90731.html.

8

Ramps/Wild Garlics and Ramsons(*Allium tricoccum* and *A. ursinum*)

Introduction

Ramps (*Allium tricoccum*) of eastern North America, and **Ramsons** (*A. ursinum*) of Eurasia, the two closely related wild alliums grow wild in moist deciduous woodlands, being most common in areas with slightly acidic soils (Keith, 2023). Both these wild onions are harvested from the wild and also grown for their greens and bulbs that have the pleasant taste of sweet spring onions with a strong garlic-like aroma (Davis and Greenfield, 2023). Though both these species have some similar common names and grow in similar habitats but they have different floral characteristics (Keith, 2023).

In *Allium tricoccum*, flowering occurs after the leaves have died back, though in *A. ursinum* leaves and flowers both are seen together (en.wikipedia.org/wiki/Allium_tricoccum).

The Anglo-Saxon ancestor of ramson was *hramsa* meaning 'garlic'. The name ramps (usually plural) is one of the many dialectical variants of the English word ramson, a common name of *Allium ursinum*, a broad-leaved species of garlic much cultivated and eaten in salads (Core, 1973).

Uses

Excepting roots, all the plant parts of ramps and ramsons are edible. Flowers can be used to garnish salads; leaves as cooked vegetables, in raw salads or to replace garlic and certain other herbs in many recipes; and the bulbs in a similar way to garlic cloves. The leaves and bulbs can also be pickled or used in soups and other foods in place of onions and garlic. The white lower leaf stalks and the broad green leaves are regarded as an early spring vegetable with a strong garlic-like odour and a pronounced onion flavour (en.wikipedia.org/wiki/Allium_ursinum; en.wikipedia.org/wiki/Allium_tricoccum; Seidemann, 2005).

In North American system of medicines, *Allium tricoccum* is used as a spring tonic to treat cold and sore throat, and externally against the earache by Cherokee people of SE Woodlands of US (Hamel Paul and Chiltosky Mary, 1975) and against insect stings by some native Americans (Niering and Olmstead Nancy, 1985). Its decoction is used as a quick-acting emetic by the Ojibwe ~ the indigenous people of the Subarctic and NE Woodlands (Densmore, 1928). First Nations peoples ~ the Iroquois in northeast North America/Turtle Island use decoction of the root to treat worms in children (William Herrick James, 1977).

Allium ursinum is a dietary and medicinal plant species with a long tradition of use. It has many medicinal properties and is a popular homoeopathic ingredient. It is often used for treating digestive, respiratory and cardiovascular problems as well as for sterilisation of wounds (Danuta Sobolewska *et al.*, 2015). *Allium tricoccum* has also been used in traditional medicine for its health-promoting roles in lowering blood pressure and cholesterol (Dabeek *et al.*, 2019).

Phytochemicals

Allium ursinum

Sulphur compounds

The distinct onion- or garlic-like smell of *Allium ursinum* and various *Allium* species is associated with various sulphur compounds, of which glutamyl peptides and sulphoxides are considered as primary (Danuta Sobolewska *et al.*, 2015). These plants usually contain a high concentration of odourless S-alk(en) yl-l-cysteine-sulphoxides, secondary non-volatile sulphur metabolites which through subsequent hydrolysis by specific enzymes (C,S-lyases) give rise to many volatile compounds including thiosulphinates and (poly)sulphides responsible for specific *Allium* flavour and odour (Boscher *et al.*, 1995; Yoo and Pike, 1998). In the intact cell, cysteine sulphoxides are localized in the cytoplasm while the hydrolytic enzyme is found in vacuoles. Cellular compartmentalization damage results in its release and subsequent hydrolysis of sulphoxides (Boscher *et al.*, 1995; Sendl, 1995). This reaction occurs upon tissue damage, *e.g.* when organ is crushed, minced, otherwise processed or in case of pathogenic infection (Ankri and Mirelman, 1999).

Ramson is one of the methiin/alliin-type allium containing mainly a mixture of (+)-S-methyl-l-cysteine-sulphoxide (methiin) and (+)-S-allyl-l-cysteine-sulphoxide = (+)-S-2-propenyl-l-cysteine-sulphoxide (alliin) (Schmitt *et al.*, 2002; Kubec *et al.*, 2000), however, isoalliin—(+)-S-(1-propenyl)-l-cysteine-sulphoxide and propiin—(+)-S-propyl-l-cysteine-sulphoxide are also present (Schmitt *et al.*, 2002) though ethiin (S-ethyl-cysteine-sulfoxide) is reported in the sample of fresh leaves collected in April in Czech Republic (Kubec *et al.*, 2000).

In contrast to similar enzymes isolated from *A. sativum*, *A. cepa* or *A. porrum*, ramson's C,S-lyase is not a glycoprotein (Sobolewska *et al.*, 2015). The C,S-lyase in ramson catalyses S-alk(en)yl-cysteine-sulphoxides hydrolysis with alliin as the most preferred substrate (Landshuter *et al.*, 1994). Its comparison with common garlic's alliinase shows that though it is less specific to alliin but still shows higher relative activity towards other cysteine sulphoxides, especially methiin and isoalliin (Schmitt *et al.*, 2005). Primary products formed as a result of C,S-lyases action are thiosulphinates, pyruvic acid and ammonia (Sobolewska *et al.*, 2015).

Thiosulphinates primarily responsible for odour and flavour of freshly prepared ramson macerates are unstable reactive compounds that easily decompose to (poly)sulphides, dithiins, ajoenes and other volatile and non-volatile compounds (Sobolewska *et al.*, 2015). This occurs during storage, processing as in the presence of organic solvents, and also when heat-treated. Allicin is very unstable even at room temperature. The major thiosulphinates found in ramson extracts are allicin (diallyl thiosulphinate = di-2-propenyl thiosulphinate), and methyl-allyl- or dimethyl thiosulphinates (Sendl and Wagner, 1991). Comparative analysis of *A. sativum* and *A. ursinum* extracts showed that ajoene dominated in garlic while its methyl- and dimethyl homologues were the main components in ramson extract (Wagner and Sendl, 1990). Vinyldithiins are cyclic compounds which are another group of degradation products of allicin (Sendl, 1995). Various other sulphur compounds have also been detected as constituents of ramson's essential oil such as alkyl sulphides, alkyl polysulphides and trace amounts of mercaptan (Sendl, 1995).

Phenolics

Allium ursinum is also a good source of phenolic compounds, predominantly kaempferol derivatives. Phytochemical investigation of ethanol extract of leaves has led to the isolation of kaempferol 3-O-beta-neohesperidoside-7-O-[2-O-(trans-p-coumaroyl)]-beta-D-glucopyranoside, kaempferol 3-O-beta-neohesperi-doside-7-O-[2-O-(trans-feruloyl)]-beta-D glucopyranoside, kaempferol 3-O-beta-neohesperidoside-7-O-[2-O-(trans-p-coumaroyl)-3-O-b eta-D- glucopyranosyl]-beta-D-glu-copyranoside, kaempferol 3-O-beta-glucopyranoside and kaempferol 3-O-beta-neohesperidoside (Carotenuto *et al.*, 1996). Furthermore, extract prepared from freeze-dried fresh leaves of *A. ursinum* contained 3-O-β-d-glucopyranoside, 3-O-β-d-glucopyranosyl-7-O-β-d-glucopyranoside, 3-O-α-l-rhamnopyranosyl-(1→2)-[3-acetyl]-β-d-glucopyranoside and 3-O-α-l-rhamnopyranosyl-(1→2)-[6-acetyl]-β-d-glucopyranoside in addition to several of the previously mentioned compounds (Wu *et al.*, 2009). Ivanova *et al.* (2009) reported that fresh flowers of wild garlic contain thymidine, adenosine and astragalin (kaempferol-3-O-beta-D-glucopyranoside, kaempferol-3-O-beta-D-glucopyranosyl-7-O-beta-D-glucopyranoside, kaempferol-3-O-beta-D-neohesperoside and kaempferol-3-O-beta-D-neohesperoside-7-O-beta-D-glucopyranoside).

Steroidal glycosides

Steroidal glycosides also occur in *A. ursinum*. Danuta Sobolewska *et al.* (2006) reported diosgenin 3-O-α-l-rhamnopyranosyl-(1→4)-α-l-rhamnopyranosyl-(1→4)-[α-l-rhamnopyranosyl-(1→2)]-β-d-glucopyranoside and (25R)-spirost-5,25(27)-dien-3β-ol 3-O-α-l-rhamnopyranosyl-(1→4)-α-l-rhamnopyranosyl-(1→4)-[α-l-rhamnopyranosyl-(1→2)]-β-d-glucopyranoside together with 3-hydroxy-pregna-5,16-dien-20-on 3-O-α-l-rhamnopyranosyl-(1→4)-α-l-rhamnopyranosyl-(1→4)-[α-l-rhamnopyranosyl-(1→2)]-β-d-glucopyranoside from the underground parts of wild garlic.

Other compounds

The major predominant biologically active substances in *A. ursinum* are undoubtedly sulphur and phenolic compounds. However, their bulbs also contain lectins, polysaccharides, fatty acids (palmitic, linoleic, oleic, palmitoleic, stearic, α-linolenic, and myristic acid), many amino acids (asparagine, glutamine, aspartic acid, glutamic acid, arginine, alanine, glycine, threonine), pigments, *etc.* (Danuta Sobolewska *et al.*, 2015; Smeets *et al.*, 1997; Stajner and Szollosi, 2003).

Allium tricoccum

Though phytochemicals in both these species are more or less same but have little difference in their concentrations. Major flavonol glycosides in *Allium tricoccum* are kaempferol sophoroside glucuronide, quercetin sophoroside glucuronide, kaempferol rutinoside glucuronide, quercetin hexoside glucuronide, quercetin sophoroside and kaempferol sophoroside. All conjugates were detected in leaves though quercetin and kaempferol sophoroside glucuronide conjugates were also detected in the stem but no flavonol glycosides were detected in the bulb (Dabeek *et al.*, 2019).

Cultivation

Propagation

Both the species can be propagated through seeds and bulblets both. The seeds can be difficult to start with and if started from seed, they will take 2-3 years before they reach harvestable size whereas through bulbs planted in the fall the crop can be harvested within a year (Keith, 2023).

Climate and soil

Both these alliums grow naturally beneath deciduous trees preferring a well-drained, moist and a little acidic soil (pH 5.5) rich in organic matter (Keith, 2023). They further mention that soil rich in calcium and magnesium are preferred by these crops but their addition should be done only after soil testing. Throughout cropping period, there should be sufficient moisture in the soil.

Season

Ramp seeds can be sown anytime from late summer to early fall but not in the frozen soil. Seeds break the roots in warm and moist period though for breaking shoot dormancy a subsequent cold period is required. In case there is no adequate warm period after sowing, the seed will not germinate until the second spring. If the crop is being raised from bulbs, the best time to transplant these is February to mid-March.

Field selection and preparation

When planting under a forested canopy, weeds, tree sprouts and roots should be removed after raking back the leaves on the forest floor and if the soil is not naturally high in organic matter organic materials such as composted leaves and decaying plant material from the forest should be incorporated. In a field site under artificial shade, the field should be ploughed properly and organic matter should be added 10 tonnes per hectare at the time of field preparation.

Direct seeding

A fine seed bed should be prepared by loosening the soil and then the seeds should be sown thinly on top of the ground pressing them gently into the soil. After sowing the seeds are quite thickly (atleast 5cm) pressed with composted leaves or leafmould to retain moisture in the soil and to protect the seeds from wildlife. Though not waterlogged but always there should be sufficient moisture in the soil (Davis and Greenfield, 2023).

Bulb transplanting

For direct bulb planting, the field should be well-prepared after addition of some 10t/ha of organic matter, beds prepared of the convenient sizes and leafed-out bulbs are then planted around 7.5cm deep in the rows 10-15cm apart, allowing all the roots to be buried and keeping just the very tip of the bulb peeping out. Clumps of 4 or 5 plants can be grouped together if the growing area is limited (Davis and Greenfield, 2023).

Mulching

After the planting is over the beds are mulched with dry or rotting leaves some 5-7.5cm deep which will protect the germinating seeds and bulbs from wildlife, will suppress the growth of weeds, will retain moisture in the soil which is very necessary for this crop, under sub-zero temperature this will act as insulator and protecting the plants, and through decay these leaves will improve soil texture *cum* structure, *vis-à-vis* provide nutrients in the long run (Davis and Greenfield, 2023).

Harvest

Harvesting of ramps can be done in several ways including the young leaves. For human consumption the dense stand of leaves from the large and vigorous plants should be harvested in the manner of thinning out. Bulbs should be harvesting as per requirement patch by patch. Bulbs are harvested only when these have attained large size and have once flowered. If entire plots are harvested at one time, there should be enough plants to allow for a 5 to 7 year rotation. Harvesting just the foliage and leaving the bulb to regrow year after year is increasing in popularity among consumers and harvesters. Harvests should be limited only up to 5 to 10% of the plants in each plot. When harvesting bulbs from a portion of a plot, no more than 15% of the ramps should be removed (Davis and Greenfield, 2023).

Postharvest

While harvesting bulbs or leaves, the harvested ramps should be kept cool and moist. After harvest, the leaves are gently rinsed and allowed to dry. Bulbs are washed, rootlets trimmed off, surface-dried and then these are packed in waxed but perforated cardboard boxes and stored in a cool place as airtight containers may cause rotting of the material (Davis and Greenfield, 2023).

Pests

In North Carolina and Tennessee, *Septoria* leaf spot has been observed in the wild and cultivated ramps. Although the spot is unsightly on the foliage but in reported cases it does not appear to cause lasting damage, *vis-a-vis* not affecting adversely the plant yields. The allium leaf miner, *Phytomyza gymnostoma* Loew has been reported on ramps particularly in Pennsylvania and elsewhere in the Northeast. Its damage and control measures are still under way. Various larval stages of leek moth (*Acrolepiopsis assectella* Zeller) prefer eating cultivated leeks than the wildly growing ones, so its control is the same as described in the chaper on onion (Davis and Greenfield, 2023).

Individual Description of Both *Allium* species

Ramps (*Allium tricoccum* Ait. 1789)

Subgenus: *Anguinum* (Wikispecies, 2023).

Common names: **Ramps**, Ramson, Wild garlic, Wild leek and Wood leek (GRIN, 2012; Core, 1973).

Synonyms: *Aglitheis tricoccum* (Aiton) Raf., *Allium burdickii* (Hanes) A.G.Jones, *A. pictum* Moldenke, *A. tricoccum* f. *pictum* Moldenke, *A. triflorum* Raf., *Ophioscorodon tricoccon* (Aiton) Wallr., *Validallium tricoccum* (Aiton) Small (en.wikipedia.org/wiki/Allium_tricoccum; POWO, 2023a).

Allium tricoccum	

By Hardyplants at English Wikipedia – Own work by the original uploader, Public Domain, https://commons.wikimedia.org/w/index.php?curid=19248780	**By Fritzflohrreynolds Own work, CC BY-SA 3.0, https://commons.wikimedia.org/w/index.php?curid=95963387**

Introduction and origin

The specific epithet *tricoccum* refers to its 3-seeded capsular fruits (Aiton, 1789). It is native from eastern Canada to NC&E USA (POWO, 2023a), growing naturally under a forest canopy of beech, birch, sugar maple, poplar and other temperate deciduous trees (Davis and Greenfield, 2023). Naturally these are found growing under deciduous trees where they can get some sun until the trees leaf out. Ramps are one of the first plants to emerge in the spring so 'Annual Spring Ramp Festivals' are held everywhere throughout eastern US including many western North Carolina counties. One of the most popular is the North Carolina Ramp Festival, first started in the early 1930s and is celebrated each May in Waynesville, North Carolina.

Growth cycle

In the SE United States the plants produce and develop new leaves from March to April before the full development of tree leaves. The delicate greenish-brown flower stalks appear by May or June when the leaves begin to die back (Keith, 2023) as the days lengthen and temperatures rise (Davis and Greenfield, 2023). Flowering most often occurs in June though sometimes in July-August after the leaves have completely died back (en.wikipedia.org/wiki/Allium_tricoccum). The seeds mature atop a leafless stalk and eventually fall to the ground to germinate near the mother plant (Davis and Greenfield, 2023). The timing of these events is usually delayed at high elevations and locations north of North Carolina and Tennessee.

Botany

Ramps are true forest spring ephemerals as these appear in early spring before new leaves in the forest trees spring out but disappear by summer by the time the trees provide full shade (Keith, 2023), a natural arrangement of growing conditions. Ramps grow in strongly-rooted close groups just beneath soil surface (en.wikipedia.org/wiki/Allium_tricoccum). This hardy perennial plant usually produces a cluster of 2-6 ovoid-conical white bulbs (McNeal *et al.*, 2002) that give rise to light green, smooth, broad, flat and 10–30cm long leaves with narrow petioles (Gleason Henry, 1963) often patterned with deep purple or burgundy tints on the lower stems. The 2-6cm long bulbs (Gleason Henry, 1963) are surrounded by brownish to greyish sheathing and each cluster of bulbs give rise only to one flowering stem (McNeal*et al.*, 2002). The inflorescence has two ovate bracts that enclose the flowers before they open and fall away at anthesis (Gleason Henry, 1963).

Blooming occurs in early summer and the seeds develop in late summer (Jackie Davis and Greenfield, 2023). The flowers with 4-7mm long white, cream or yellowish tepals are arranged into an umbel that has an erect and 10-40cm long scape (FNA, 2021). The stamens are about as long as the tepals and the filaments of the stamens with widened bases are inserted on the corolla (en.wikipedia.org/wiki/Allium_tricoccum). After flowering *cum* fertilization, 3-lobed green fruits are produced that open by way of three valves (Gleason Henry, 1963). The seeds are round, black and shiny (McNeal and Jacobsen, 2002).

Varieties

var. *tricoccum*	var. *burdickii*

By Nathan Martineau – https://www.inaturalist.org/observations/12337426, CC BY 4.0, https://commons.wikimedia.org/w/index.php?curid=74197776	**By Jacob Meismer –https://www.inaturalist.org/observations/5503124, CC0, https://commons.wikimedia.org/w/index.php?curid=74197733**

Kew's Plants of the World Online accepts two varieties *Allium tricoccum* var. *tricoccum* and *A. tricoccum* var. *burdickii* which is commonly known as 'Hanes narrow-leaf ramps', White ramps, Chicago leek and Burdick's leek (POWO, 2023a) which several other sources also accept (McNeal *et al.,* 2002), though some literature sources treat these two taxa as two different species, *Allium tricoccum* Ait. and *A. burdickii* Hanes (Michael Hough, 2004). The epithet *burdickii* is in honour of Justin Herbert Burdick (1851–1939), a Midwestern physician who first described the differences between these two species (Michael Hough, 2004).

A.t. var. *tricoccum* occurs among woods in the moist grounds where soil is rich on colluvial slopes and along streamside bluffs whereas *A.t.* var. *burdickii* is found growing in dry soils of upland woods. *A.t.* var. *tricoccum* has in the soil shallowly buried bulbs usually with bulb diameter larger in size (4-6cm), leaf width 3-9cm, the spathe bracts measuring 2-3cm, umbels bearing 15-50 flowers and petioles *cum* leaf sheaths being red or purplish while in case of *A.t.* var. *burdicki* the bulb size remains smaller (2-5cm) but buried deeply, leaf width usually 1.5- 4.5cm, the spathe bracts measuring 1-2cm, per bract floral number being 5-24, and the leasf stalks and leaf sheaths remain white (Michael Hough, 2004; FNA, 2021; en.wikipedia.org/wiki/Allium_tricoccum).

Ramsons (*Allium ursinum* L.)

Subgenus: *Amerallium.*

Section: *Arctoprasum* (Wikispecies, 2023).

Common names: Bear leek, Bear's garlic, Broad-leaved garlic, Buckrams, Cow's leek, **Ramsons**, Wild cowleek, Wild garlic and Wood garlic.

Synonyms: *Aglitheis ursina* (L.) Raf., *Allium latifolium* Gilib., *A. longipetiolatum* St.-Lag., *A. nemorale* Salisb., *A. petiolatum* Lam., *A. ucrainicum* (Oksner & Kleopow) Bordz., *A. ursinoides* G.Don ex Sweet, *A. ursinum* subsp. *ucrainicum* Kleop. & Oxner, *A. ursinum* subsp. *ucrainicum* Oksner & Kleopow, *A. ursinum* var. *ucrainicum* (Oksner & Kleopow) Soo, *A. vincetoxicum* Pall. ex Ledeb., *Cepa ursina* (L.) Bernh., *Geboscon ursinum* (L.) Raf., *Hylogeton ursinum* (L.) Salisb., *Moly latifolium* (Gilib.) Gray, *Ophioscorodon ursinum* (L.) Wallr. (POWO, 2023b)

Introduction

Allium ursinum	Fruit with seeds	Inflorescence	Small flowers and flower buds on a slender stem
By Zonki, CC BY-SA 3.0, https://commons.wikimedia.org/w/index.php?curid=2031844	By Stefan.lefnaer – Own work, CC BY-SA 4.0, https://commons.wikimedia.org/w/index.php?curid=53309481	By Dominicus Johannes Bergsma - Own work, CC BY-SA 4.0, https://commons.wikimedia.org/w/index.php?curid=68661741	By Dominicus Johannes Bergsma - Own work, CC BY-SA 4.0, https://commons.wikimedia.org/w/index.php?curid=68661738

Latin specific epithet *ursinum* translates as 'bear', referring to the supposed fondness of the brown bear for the bulbs (Danuta Sobolewska *et al.*, 2015). *Allium ursinum*, native to northern Eurasia (Keith, 2023), Asia Minor, the Caucasus and

Siberia up to Kamchatka Peninsula (Rola, 2012; Oborny *et al.*, 2011; Djurdjevic *et al.*, 2004) is a wild allium occurring in deciduous woodlands with moist soils that are little acidic. Though it does not grow in the far North and at altitudes above 1900m, it can be found on natural stands from the Mediterranean region to Scandinavia (Oborny *et al.*, 2011).

Various minerals in much higher amounts than in clove garlic are present in *Allium ursinum*. Its leaves have high levels of magnesium ~ the anti-stress mineral which protects the circulatory system, especially the heart (en.wikipedia.org/wiki/Allium_ursinum). It is a popular homoeopathic ingredient and its fresh leaves are used against high blood pressure and hardening of arteries and for treating cardiovascular, respiratory and digestive problems as well as for sterilisation of wounds (Danuta Sobolewska *et al.*, 2015).

Subspecies

The two recognized subspecies of *Allium ursinum* are *A.u.* ssp. *ursinum* from W&C Europe and *A.u.* ssp. *ucrainicum* from E&SE Europe (Rola, 2012; Oborny *et al.*, 2011). These both the subspecies differ in their smoothness of pedicel surface (Karpaviciene, 2006). Pedicels of ssp. *ursinum* are scabrid *cum* rough with numerous papillae while the ssp. *ucrainicum* has smooth pedicels without papillae (Rola, 2012).

Growth cycle

Period of active growth of *A. ursinum* lasts 3.5–4 months starting in early spring, *i.e.* late February to early March when the trees under whom these are growing are leafless (Jandl *et al.*, 1997). This provides enough of light in the first stage of the plant growth and helps to avoid competition for light with the canopy (Shmanova and Krichfalushii, 1995; Oborny *et al.*, 2011). The canopy cover protects the plant from direct sunlight during late spring and helps to keep the appropriate humidity of the topsoil air. Unlike ramps, the showy clusters of white flowers appear when plants are in full foliage and this makes the ramsons even better candidates for ornamental woodland gardens (Keith, 2023). The above-ground parts wither as summer arrives.

Botany

Ramsons from different ecotypes may differ significantly in the width of leaf blades, the length of leaf stalks and flowering stems, and number of flowers in inflorescence (Błazewicz-Wozniak and Michowska, 2011). These hardy perennial herbs grow up to 50cm high bearing 2-6.4cm wide 2-3 bright green leaves which are shorter than scape. These elliptic-lanceolate, flat and smooth leaves with a distinct well-developed blade are sharpened at the apex and gradually narrowed into petiole at the base (Danuta Sobolewska *et al.*, 2015; Szafer *et al.*, 1988; Błazewicz-Wozniak and Michowska, 2011).

These have narrow *cum* elongated about 1.5–6cm long bulb surrounded by the layers of clear skin with only a few fibres at the base (Danuta Sobolewska *et al.*, 2015). The plants reproduce primarily by seed but also by daughter bulbs that are sometimes formed around the main bulb. Contractile roots start developing when plant becomes nearly 3-years old (Eggert, 1992; Ernst, 1979; Oborny *et al.*, 2011) and due to contraction phenomenon the bulbs can reach down even up to 27cm deep in the soil after 10 years if the soil is soft enough to allow the roots going deeper (Ellenberg, 1988).

The inflorescence comprises of a semispherical umbel bearing 3-30 star-shaped snowy-white flowers which appear on the top of the solid floral stem (scape) that is triangular in cross-section (Danuta Sobolewska *et al.*, 2015). Flowering appearing usually from April to mid-May (Danuta Sobolewska *et al.*, 2015), surrounded by 2-3 spathal bracts until anthesis, comprising of 6 white tepals some 1.6–2cm in diameter and stamens shorter than perianth (Clapham *et al.*, 1981). Its seeds are black, subglobose, 2–3mm wide and in a trichotomic capsules (Hermy *et al.*, 1999; Sendl, 1995).

Plants that look similar to *A. ursinum* and may be mistaken for it include lily of the valley (*Colchicum autumnale*, *Arum maculatum* and *Veratrum viride* or *V. album*) though on crushing not emitting any garlic- or onion-like smell, albeit they are poisonous (Irene Gilotta and Miran, 2010).

References

Aiton, W. 1789. *Hortus Kewensis*. Vol. 1. Printed for George Nicol, Bookseller to his Majesty, UK, 428p.

Ankri, S. and D. Mirelman, 1999. Antimicrobial properties of allicin from garlic. *Microbes Infect.*, **1**(2):125–9. doi: 10.1016/S1286-4579(99)80003-3.

Błazewicz-Wozniak, M. and A. Michowska, 2011. The growth, flowering and chemical composition of leaves of three ecotypes of *Allium ursinum* L. *Acta Agrobot.*, **64**(4):171–80. doi: 10.5586/aa.2011.058.

Boscher, J., J. Auger, N. Mandon and S. Ferary, 1995. Qualitative and quantitative comparison of volatile sulphides and flavour precursors in different organs of some wild and cultivated garlics. *Biochem. Syst. Ecol.*, **23**(7/8):787–91. doi: 10.1016/0305-1978(95)00072-0.

Carotenuto, A., V. De Feo, E. Fattorusso, V. Lanzotti, S. Magno and C. Cicala, 1996. The flavonoids of *Allium ursinum. Phytochem..* **41**: 531-6.

Clapham, A.R., T.G. Tutin and E.F. Warburg, 1981. Excursion Flora of the British Isles (3rd ed.). Cambridge University Press. ISBN 0-521-23290-2. Retrieved from "https://en.wikipedia.org/wiki/Allium_ursinum" on 9 February, 2023.

Core, E.L. 1973. Cult of the Ramp Eaters. West Virginia Through the Seasons. Seneca Books, pp. 46-51. Grantsville, West Virginia, USA.

Dabeek, W.M., N. Kovinich, C. Walsh and M. Ventura Marra, 2019. Characterization and Quantification of Major Flavonol Glycosides in Ramps (*Allium tricoccum*). *Molecules*, **24**(18):3281. https://doi.org/10.3390/molecules2418328. Published: 9 September 2019. Retrived from https://www.mdpi.com/1420-3049/24/18/3281 on 10 February, 2023.

Danuta Sobolewska, Z. Janeczko, W. Kisiel, I. Podolak, A. Galanty and D. Trojanowska, 2006. Steroidal glycosides from the underground parts of *Allium ursinum* L. and their cytostatic and antimicrobial activity. *Acta Pol. Pharm. Drug Res.*. **63**: 219–23.

Danuta Sobolewska, Irma Podolak and Justyna Makowska-Wąs, 2015. *Allium ursinum*: botanical, phytochemical and pharmacological overview. *Phytochem. Rev.*, **14** (1): 81–97. doi:10.1007/s11101-013-9334-0. ISSN 1568-7767. PMC 4352197. PMID 25774103.https://www.ncbi.nlm.nih.gov/pmc/articles/PMC4352197.

Davis, J. and Jackie Greenfield. 2023. Cultivation of Ramps (*Allium tricoccum* and *A. burdickii*). Horticulture Information Leaflets, retrieved on Feb. 7, 2023. N.C. State Extension. https://content.ces.ncsu.edu/cultivation-of-ramps-allium-tricoccum-and-a-burdickii.

Densmore Frances, 1928. Uses of Plants by the Chippewa Indians. *SI-BAE Ann. Rep.*, **44**:273-379.

Djurdjevic, L., A. Dinic, P. Pavlovic, M. Mitrovic, B. Karadzic and V. Tesevic, 2004. Allelopathic potential of *Allium ursinum* L. *Biochem. Syst.* Ecol., **32**:533–44. https://www.ncbi.nlm.nih.gov/pmc/articles/PMC4352197. doi: 10.1016/j.bse.2003.10.001.

Eggert, A. 1992. Dry matter economy and reproduction of a temperate forest spring geophyte *Allium ursinum. Ecography*, **15**:45–52. www.ncbi.nlm.nih.gov/pmc/articles/PMC4352197. doi: 10.1111/j.1600-0587.1992.tb00007.x.

Ellenberg, H.H. 1988. Vegetation Ecology of Central Europe. Cambridge University Press, Cambridge and New York. https://www.ncbi.nlm.nih.gov/pmc/articles/PMC4352197.

en.wikipedia.org/wiki/Allium_tricoccum

en.wikipedia.org/wiki/Allium_ursinum

Ernst, W.H.O. 1979. Population biology of *Allium ursinum* in northern Germany. *J. Ecol.*, **67**:347–62. doi: 10.2307/2259355. www.ncbi.nlm.nih.gov/pmc/articles/PMC4352197.

FNA, 2021. *Allium tricoccum*. Flora of North America, vol. 26, pp. 224, 225, 226, 234. beta.floranorthamerica.org/Allium_tricoccum; www.efloras.org/florataxon.aspx?flora_id=1&taxon_id=242101410. Retrieved 3 January 2021.

Gleason Henry, A. (*Aka* Henry Allan), 1963 (1882-1975). Manual of Vascular Plants of Northeastern United States and Adjacent Canada. Arthur Cronquist, Princeton, N.J., 203p. ISBN 0-442-02722-2. OCLC 243396.

GRIN (Germplasm Resources Information Network), 2012. *Allium tricoccum*. Agricultural Research Service (ARS), United States Department of Agriculture (USDA). Retrieved 2 February 2012.

Hamel Paul, B. and U. Chiltoskey Mary, 1975. Cherokee Plants and Their Uses -- A 400 Year History, 52p. *Sylva, N.C. Herald Publishing Co.,* Sylva.

Hermy, M., O. Honnay, L. Firbank, C. Grashof-Bokdam and J.E. Lawesson, 1999. An ecological comparison between ancient and other forest plant species of Europe, and the implications for forest conservation. *Biol. Conserv.*, **91**(1):9–22. doi: 10.1016/S0006-3207(99)00045-2.

Irene Gilotta and Miran Brvar, 2010. Accidental poisoning with Veratrum album mistaken for wild garlic (*Allium ursinum*). *Clinical Toxicol.*, **48** (9): 949–52. doi:10.3109/15563650.2010.533675. ISSN 1556-3650. PMID 21171854. S2CID 207657813.

Ivanova, A., B. Mikhova, H. Najdenski, I. Tsvetkova and I. Kostova, 2009. Chemical composition and antimicrobial activity of wild garlic *Allium ursinum* of Bulgarian origin. *Nat. Prod. Commun.,* **4**: 1059-62.

Jandl, R., H. Kopeszki and G. Glatzel, 1997. Effect of dense *Allium ursinum* (L.) ground cover on nutrient dynamics and mesofauna of a *Fagus sylvatica* (L.) woodland. *Plant Soil*, **189**:245–55. doi: 10.1023/A:1004223011834. https://www.ncbi.nlm.nih.gov/pmc/articles/PMC4352197.

Karpaviciene, B. 2006. Distribution of *Allium ursinum* L. in Lithuania. *Acta Biol. Univ*. Daugavp., **6**(1–2):117–22.

Keith, J. 2023. How to Grow Ramps and Ramsons. https://blackgold.bz/how-to-grow-ramps-and-ramsons/. Retrieved 8 Februsry, 2023.

Landshuter, J., E.M. Lohmuller and K. Knobloch, 1994. Purification and characterization of C-S-lyase from ramson, the wild garlic, *Allium ursinum*. *Planta Med.*, **60**:343–7. doi: 10.1055/s-2006-959497.

McNeal, Jr., W. Dale and T.D. Jacobsen, 2002. *Allium tricoccum*. In: Flora of North America North of Mexico, edited by FNA Editorial Committee, vol. 26. New York and Oxford. Retrieved 21 August 2016 – via eFloras.org, Missouri Botanical Garden, St. Louis, MO & Harvard University Herbaria, Cambridge, MA. http://www.efloras.org/florataxon.aspx?flora_id=1&taxon_id=242101410.

Michael Hough, 2004. *Allium tricoccum*. Plants of the Northeastern United States. Retrieved 30 July 2018. http://www.thismia.com/A/Allium_tricoccum.html.

Niering, W.A. and C. Olmstead Nancy, 1985. The Audubon Society Field Guide to North American Wildflowers, Eastern Region, 595p. *Knopf.* ISBN 0-394-50432-1.

Oborny, B., Z. Botta-Dukat, K. Rudolf and T. Morschhauser, 2011. Population ecology of *Allium ursinum,* a space-monopolizing clonal plant. *Acta Bot. Hung.,* **53**(3–4):371–88. doi: 10.1556/ABot.53.2011.3-4.18. https://www.ncbi.nlm.nih.gov/pmc/articles/PMC4352197.

POWO, 2023a. *Allium tricoccum* Aiton. Plants of the World Online. Royal Botanic Gardens, Kew. Retrieved 7 February, 2023. https://powo.science.kew.org/taxon/urn:lsid:ipni.org:names:8906-2.

POWO, 2023b. *Allium ursinum* L. Plants of the World Online. Royal Botanic Gardens, Kew. Retrieved 7 February, 2023. https://powo.science.kew.org/taxon/urn:lsid:ipni.org:names:529088-1.

Rola, K. 2012. Taxonomy and distribution of *Allium ursinum* (Liliaceae) in Poland and adjacent countries. *Biologia,* **67**(6):1080–87. doi: 10.2478/s11756-012-0101-2.

Schmitt, B., H. Schulz, J. Strosberg and M. Keusgen, 2005. Chemical characterization of *Allium ursinum* L. depending on harvesting time. *J. agric. Food Chem.,* **53**:7288–94. doi: 10.1021/jf0504768.

Seidemann, J. 2005. World Spice Plants. *Springer,* p. 27. ISBN 978-3-540-22279-8. Retrieved 13 April 2011.

Sendl, A. 1995. *Allium sativum* and *Allium ursinum*: part 1. Chemistry, analysis, history and botany. *Phytomed.,* **1**(4):323–9. doi: 10.1016/S0944-7113(11)80011-5.

Sendl, A. and H. Wagner, 1991. Isolation and identification of homologues of ajoene and alliin from bulb-extracts of *Allium ursinum. Planta Med.,* **57**(4):361–2. doi: 10.1055/s-2006-960118.

Shmanova, I.W. and W.W. Krichfalushii, 1995. Biomorphological and ecologo-coenotic characteristics of *Allium ursinum* L. in Carpathians. *Rast. Res.,* **31**(3):1–18. https://www.ncbi.nlm.nih.gov/pmc/articles/PMC4352197.

Smeets, K., E.J. van Damme, F. van Leuven, Fred van Leuven and W.J. Peumans, 1997. Isolation, characterization and molecular cloning of a leaf-specific lectin from ramsons (*Allium ursinum* L.). *Plant Mol. Biol.,* **35**: 531–5.

Stajner, D. and V.I. Szollosi, .2003. An evaluation of the antioxidant abilities of *Allium* species. Acta Biol. Sze-ged.,. **47**: 103–6.

Szafer, W., S. Kulczynski, B. Pawłowski, Rośliny Polskie and P.W.N. Warszawa, 1988. https://www.ncbi.nlm.nih.gov/pmc/articles/PMC4352197.

Wagner, H. and A. Sendl, 1990. Barlauch und Knoblauch. Vergleichende chemische und pharmakologische Untersuchungen von Barlauch- und Knoblauchextrakten. *Dtsch. Apoth. Ztg.,* **33**:1809–15.

Wikispecies, 2023. *Allium tricoccum*. https://species.wikimedia.org/wiki/Allium_tricoccum; https://species.wikimedia.org/wiki/Allium_ursinum. Retrieved on 12,February, 2023.

William Herrick James , 1977. Iroquois Medical Botany. PhD Thesis, State University of New York, Albany, 281p.

Wu, H., S. Dushenkov, C.T. Ho and S. Sang, 2009. Novel acetylated flavonoid glycosides from the leaves of *Allium ursinum. Food Chem.,* **115**: 592–5.

Yoo, K.S. and L.M. Pike, 1998. Determination of flavor precursor compound S-alk(en)yl-L-cysteine sulfoxides by an HPLC method and their distribution in *Allium species. Sci. Hortic.,***75**(1–2):1–10

9

Wild Alliums With Aerial Bulbils

Sand leek (*Allium scorodoprasum* L.)

Subgenus: *Allium*

Section: *Allium* (Wikispecies, 2023)

Synonyms: *Allium arenarium* L., *A. contortum* Stokes,. *A. neglectum* Wender., ,*A. obscurum* M.Bieb. ex Schult. & Schult.f., *A. persicum* Fisch. ex Regel, *A. scorodoprasum* var. *multibulbillosum* Y.N.Lee, *A. supranisianum* Sailer, *Ascalonicum scorodoprasum* (L.) P.Renault, *Porrum arenarium* (L.) Rchb., *P. scorodoprasum* (L.) Rchb.

Common names: Korean pickled-peel garlic, Rocambole, Sand leek. (POWO, 2023a)

Introduction and origin

Sand leek is a wild Eurasian allium native to much of Europe ~ Norway to Britain and France, east to Russia, Ukraine and Greece; and West Asia ~ Turkey south to Israel, east to the Caucasus (POWO, 2023a; TPO: 2023). It has a long history of being used like garlic in the native places (NatureGate, 2023). The natural habitat of *A. scorodoprasum* is damp broad-leaved woodland, forest margins, shores, hillside meadows, hedgerows, and sometimes can be found near old habitations (NatureGate, 2023).

Botany

Allium scorodoprasum is a stiff perennial herb growing to a height of 30 to 90 cm, with cylindrical stem and leafless upper half (NatureGate, 2023) growing from an egg-shaped bulb (Stace, 2010) with 1-2 cm diameter and mild garlic-like flavour (Pfaf, 2023). It bears 2-5, flat, 7-20 mm wide (Stace, 2010), broadly linear, slightly keeled, mid-green leaves having sheath-like bases and are rough to touch on the edges and central veins (Shoot, 2023).

Inflorescence

The inflorescence is a globular cluster surrounded by membranous bracts in bud which wither when the flowers open (NatureGate, 2023). The inflorescence appears in summer bearing stalkless purple bulbils together with long-stalked pink-purple flowers (Shoot, 2023; NatureGate, 2023). Each individual flower has a 4 to 7 mm long purple perianth, six tepals, six stamens and a pistil formed from three fused carpels (NatureGate, 2023). The fruit is a capsule, but seldom set seeds, and a new plant is usually started from the bulbils when they break off and fall to the ground (NatureGate, 2023), or through lateral offsets.

Hegi (1909) recognized four varieties and one forma of *A. oleraceum* in Central Europe: 1) var. *angustifolium* Koch is widespread and rather tall, stout, with round, not flat leaves; 2) var. *alpicola* Aschers. & Graebner grows in mountains and is short, rarely taller than 30 cm, leaves round, with rough margins, flowers rose coloured; 3) var. *complanatum* Fries is tall, stout, leaves flat, up to 4 mm broad, groved, margins only weakly rough; 4) var. *pauciflorum* Aschers. & Graebner has only 2–6 flowers; and 5) –f. sterile Beck is without flowers.

Component profiles

Component profiles of ethanolic extracts of *Allium scorodoprasum* L. showed sharp peaks for vanillic acid, protocatechuic acid, p-hydroxybenzoic acid, gallic acid, protocatechuic aldehyde, rutin, m-coumaric acid, p-coumaric acid, caffeic acid, quercetin, catechin, ferulic acid, naringenin, kaempferol, chlorogenic acid and rosmarinic acid. The estimated number of detected constituents in the extract are 5.56 mg/g caffeic acid, 2.09 mg/g catechin, 8.26 mg/g chlorogenic acid, 2.65 mg/g ferulic acid, 4.73 mg/g gallic acid, 8.11 mg/g kaempferol, 5.84 mg/g m-coumaric acid, 6.65 mg/g naringenin, 5.31 mg/g p-coumaric acid, 9.17 mg/g p-hydroxybenzoic acid, 5.09 mg/g protocatechuic acid, 6.36 mg/g protocatechuic aldehyde, 10.93 mg/g quercetin, 4.76 mg/g rosmarinic acid, 3.09 mg/g rutin, 6.73 mg/g vanillic acid, 8.1 mg/g Σ HBA (sum of benzoic acid derivatives), 6.9 mg/g Σ HCA (sum of cinnamic acid derivatives) and 10.7 mg/g Σ PHA (sum of phenolic acids) respectively (Ekici *et al.*, 2022).

Uses

Bulbs and leaves can be consumed raw or cooked, but flowers are usually used raw as a garnish on salads (Pfaf, 2023). The sweet and light garlicky taste of sand leek is used to flavour salads, soups and stir-fries, and can be used as a substitute to not only garlic but can also replace onions in recipes. Bulbils can be eaten raw or cooked.

Although no specific mention of medicinal uses has been seen for this species, members of this genus are in general very healthy additions to the diet due to presence of sulphur compounds along with several non-sulphur compounds. When added to the diet on a regular basis they help reduce blood cholesterol levels, act as a tonic to the digestive system and also tonify the circulatory system.

Cultivation

Sand leek rarely produces seeds. It is usually propagated vegetatively by division of off-sets and by bulbils, similar to tree onions, shallots and hardneck garlic varieties (NatureGate, 2023). Like many other alliums, it too grows in fertile, well-drained soil in full sun. Bulbs should be planted 5-10 cm deep in autumn. Cultivation of sand leek is almost same as for shallots.

Field garlic (*Allium oleraceum* L. 1753 not Des Moul. 1840)

Subgenus: *Allium*

Section: *Codonoprasum* (Wikispecies, 2023)

Synonyms: *Allium complanatum* Boreau, *A. intermedium* G.Don, *A. oleraceum* var. *carinatum* Nyman, *Allium oleraceum* var. *latifolium* W.D.J.Koch, *A. oxypetalum* G.Don, *A. pallens* var. *pseudooleraceum* Seregin, *A. parviflorum* Thuill., *A. scabrum* Gilib., *A. virens* Lam., *A. virescens* Lam., *Cepa oleracea* (L.)

Bernh., *Codonoprasum alpicola* Jord. & Fourr., *C. complanatum* (Boreau) Fourr., *C. intermedium* Rchb., *C. oleraceum* (L.) Rchb., *C. viridiflorum* Schur, *Porrum oleraceum* (L.) Moench, *Raphione oleracea* (L.) Salisb. (POWO, 2022a)

Introduction and origin

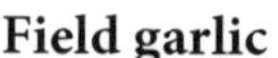

Field garlic

By Ivar Leidus - Own work, CC BY-SA 3.0, https://commons.wikimedia.org/w/index.php?curid=27580062

The specific epithet oleraceum means "vegetable/herbal" in Latin and is a form of holeraceus (oleraceus) (Parker, 2018; Whitney, 1899). Field garlic is a wild Eurasian species native from Europe to Caucasus (POWO, 2022a) and sparingly naturalised in scattered locations in North America (Gleason and Cronquist, 1991). In the United Kingdom, it is found in dry, grassy places, usually steeply sloping and calcareous soils, and on open sunny banks in river floodplains (Stace, 2010). It is scattered throughout England and very scattered in Wales, Scotland and Ireland (Stace, 2010). It gets spread easily by the bulbils (efloras.org, 2023a). *Allium oleraceum* subsp. *girerdii* was formerly included, but is now classified as *Allium oporinanthum* (POWO, 2022b).

Botany

Allium oleraceum is a perennial herb having one or more ovoid and up to 2 cm in diameter bulbs (McClintock, 1961) enclosed in fibrous, brown to greyish brown outer coats; inner coats are white to light-brown (efloras.org, 2023a). The rounded, but occasionally flattened scape (McClintock, 1961) is persistent, solitary, erect, terete, and 25–100 cm long by 4–8 mm wide (efloras.org, 2023a), and bears 2-4, alternate, basal (nativeplanttrust.org, 2023) leaves growing to a height of more than half the height of scape and wither from tip by anthesis (efloras.org, 2023a). The linear to filliform and 1.5–2.5 cm × 0.5–5 mm leaf blades are fistulose proximally, solid distally, terete, prominently ribbed proximally, channeled distally; margins and veins usually scabrid with minute teeth, apex acute (nativeplanttrust.org, 2023; efloras.org, 2023a). Inflorescence is a persistent, erect, subglobose umbel (efloras.org, 2023a) comprising a number of small, stalked, pinkish-brown flowers (McClintock, 1961) and few-to-many bulbils; sometimes only flowers or only bulbils occur (efloras.org, 2023a). The papery bracts have long points which often overtop the flowers, the stamens of which do not protrude (McClintock, 1961). Flowers usually abort before capsules mature, if present, campanulate, 6–8 mm; tepals erect, whitish or pinkish to purple, outer narrowly obovate, inner ± elliptic, unequal, margins entire, apex obtuse; stamens included; anthers yellow to reddish; pollen yellow; ovary crestless; style linear, equaling stamens; stigma capitate, unlobed; pedicel 15–60 mm (efloras.org, 2023a). Seed-coat unknown; capsules only rarely produced (efloras.org, 2023a).

Cultivation and uses

Field garlic reproduces by seed, bulbs and by the production of small bulbils in the flower head (similarly to *Allium vineale*). Although not cultivated, it can be cultivated like sand leek and similarly used. It has been used as a food, spice and medicinal plant during the first millennium A.D. and during the medieval time (Pettersson, 1943; Seppanen, 1988). According to Nyman (1868), tender leaves of *A. oleraceum* has been used together with green cabbage. The young leaves have been used as a spice in wintertime (Hulten *et al.*, 1958). Hoeg (1976) mentioned that it has been cultivated next to the houses in Kvikne in Norway, and that it has been used as a spice in fish soup in the same way as *Allium schoenoprasum*.

Crow garlic (*Allium vineale* L.)

Subgenus: *Allium*

Section: *Allium* (Wikispecies, 2023)

Common names: Crow garlic, Onion grass, Stag's garlic or wild garlic.

Synonyms: *Allium affine* Boiss. & Heldr.; *A. arenarium* Wahlenb. 1828, illegitimate homonym not L. 1753; *A. assimile* Halacsy; *A. campestre* Schleich. ex Steud.; *A. compactum* Thuill.; *A. descendens* W.D.J.Koch 1837, illegitimate homonym not L. 1753; *A. laxiflorum* Tausch; *A. littoreum* Bertol. 1827, illegitimate homonym not Bertol. 1819; *A. margaritaceum* var. *bulbiferum* Batt. & Trab.; *A. nitens* Sauze & Maill.; *A. purshii* G.Don; *A. rilaense* Panov; *A. rotundum* Wimm. & Grab. 1824, illegitimate homonym not L. 1762; *A. sphaerocephalum* Crome ex Schltdl. 1824, illegitimate homonym not L. 1753; *A. subvineale* Wendelbo;

A. vineale subsp. affine (Regel) K.Richt.; *A. vineale* subsp. *asperiflorum* (Regel) K.Richt.; *A. vineale* subsp. *capsuliferum* (Syme) K.Richt.; *A. vineale* subsp. *compactum* (Thuill.) K.Richt.; *A. vineale* subsp. *kochii* (Lange) Nymaṇ; *A. vineale* var. *affine* Regel; *A. vineale* var. *asperiflorum* Regel; *A. vineale* var. *bulbiferum* Syme; *A. vineale* var. *capsuliferum* Syme; *A. vineale* var. *compactum* (Thuill.) Lej. & Courtois; *A. vineale* var. *descendens* Nyman; *A. vineale* var. *kochii* Lange; *A. vineale* var. *multiflorum* Baguet; *A. vineale* var. *nitens* (Sauze & Maill.) Nyman; *A. vineale* var. *purshii* (G.Don) Regel; *Getuonis vinealis* (L.) Raf.; *Porrum capitatum* P.Renault; *P. vineale* (L.) Schur. (POWO, 2022c)

Introduction and origin

Crow garlic is a wild allium native to Europe, Mediterranean to North Iran and introduced in Australia and North America, where it has become a noxious weed (POWO, 2022c). Its small, wheat-sized bulbils frequently contaminated wheat grown in infested areas (POWO, 2022c). Since all parts of the plant have a strong garlic odour, the bread made from such contaminated wheat and the milk of cows grazing in infested pastures has garlic-flavour (efloras.org, 2023b).

Botany

This possesses 2n=32 and 40 chromosome numbers. Crow garlic produces a cluster of 5-20, ovoid and hard-shelled bulbs that are 1-2 cm in diameter and have fibrous, brownish to yellowish outer coats and white to light brown inner coats. The 2–4 green leaves with waxy texture are approximately ½ the length of scape and remain green at anthesis; leafblade is 20–60 cm × 2–4 mm, hollow below middle, terete, cylindric or filiform, not carinate, margins entire. The solitary, erect, terete scape is 30–120 cm × 1.5–4 mm bearing an epical inflorescence having 2-5 cm diameter.

Crow garlic	Crow garlic
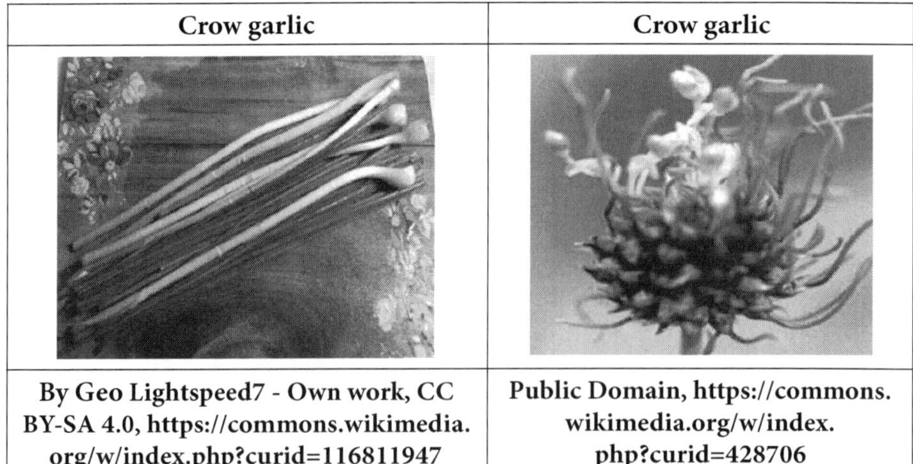	
By Geo Lightspeed7 - Own work, CC BY-SA 4.0, https://commons.wikimedia.org/w/index.php?curid=116811947	**Public Domain, https://commons.wikimedia.org/w/index.php?curid=428706**

Inflorescence is a persistent, erect, compact, and subglobose to ovoid or hemispheric umbel comprising none to a few flowers (efloras.org, 2023b) and several small wheat-sized bulbils that are elliptic, yellowish brown (NatureGate, 2022), basally narrowed and 4–6 × 2–3 mm (efloras.org, 2023b). Umbel is enclosed in bud within one membranous bract (NatureGate, 2022) that is caducous, 1, 2–several-veined, ovate, apex caudate, beaked, beak ± equaling or longer than base (efloras.org, 2023b). If present, the stalked flowers have pinkish-green perianth and are campanulate, 3–4 mm long (efloras.org, 2023b); tepals are 6 in number (NatureGate, 2022), erect, greenish to purple, elliptic-lanceolate, ± equal, withering in fruit, margins entire, apex obtuse (efloras.org, 2023b); stamens are 6 in number (NatureGate, 2022) and have same length as perianth or protruding, exserted (NatureGate, 2022), outer 3 filaments without appendages, inner 3 filaments with 2 prominent lateral appendages; anthers purple; pollen white; ovary crestless; style exserted, linear, ± equaling stamen; stigma capitate, scarcely thickened, unlobed; flowering pedicel 10–20 mm (efloras.org, 2023b). The fruit is a capsule but seldom set seeds; seed coat shining; cells smooth (efloras.org, 2023b).

Cultivation and uses

Crow garlic doesn't usually produce seeds and sometimes the flowers are completely lacking. It propagates itself mainly through bulbils and lateral segments of the bulb (NatureGate, 2022). Its bulbs, leaves as well as bulbils can be consumed raw and can be used as a substitute to garlic.

Canada onion (*Allium canadense* L.)

Subgenus: *Amerallium*

Section: *Amerallium* (Wikispecies, 2023)

Other common names: Canadian garlic, Meadow garlic, Meadow leek, Rose leek, Wild garlic and Wild onion (GRIN, 2023; Tropicos, 2023).

Synonyms: *Kalabotis canadensis* (L.) Raf. (POWO, 2023b).

Introduction and origin

Canada onion is native to East Canada to Central & East USA and also reportedly introduced in Cuba (POWO, 2023b). Although formerly collected from the wild to be eaten by Native Americans and by European settlers (Hanelt, 2001), it is now cultivated in many regions, particularly in Cuba, as an ornamental and as a garden culinary herb (Izel, 2014). Its bulbs and leaves have a typical oniony flavour and fragrance and can be similarly used raw or cooked, but unlike commercial varieties it has a thick, fibrous skin (Izel, 2014). It is considered a noxious weed in several states.

Canada onion

By Eric Hunt - Own work, CC BY-SA 4.0,

Botany

Canada onion produces 1-4 or more bulbs with or without basal bulbils, often clustered, ovoid, 1–2.5 × 0.6–3 cm (efloras.org, 2023c). The bulb's outer coat enclosing 1 or more bulbs is brownish or grayish, reticulate and fibrous, whereas inner coat is white (efloras.org, 2023c). The grass-like foliage appears from the base (Izel, 2014); blade solid, flat, channeled, 20–50 cm × 1–7 mm, margins entire or denticulate, apex acute to obtuse (efloras.org, 2023c). Scape is usually solitary, erect, terete, 10–60 cm × 1–5 mm, and bears hemispheric to globose umbel comprising 0-60 flowers with white to pink or lavender tepals that are lanceolate to elliptic (efloras.org, 2023c). These flowers may be partially or entirely replaced by bulbils (efloras.org, 2023c). When present, the flowers are hermaphroditic (both male and female organs) and are pollinated by American bees (not honeybees) and other insects. It typically flowers in the spring and early summer, from May to June (Schwegman, 1991), though can bloom as early as April and as late as July (Izel, 2014)

Varieties

Five to six varieties of the species are recognized by POWO (2023b) and/or by efloras.org (2023c).

- *Allium canadense* var. *canadense* (POWO, 2023b; POWO, 2023c; efloras.org, 2023c) – native to E. Canada to Central & E. U.S.A.; most pedicels replaced by bulbils, rarely producing fruits or seeds.
- *Allium canadense* var. *ecristatum* Ownbey (POWO, 2023b; POWO, 2023d; efloras.org, 2023c) – native to SE. Texas where it is found in coastal plain; tepals deep pink and rather thick.

- *Allium canadense* var. *fraseri* Ownbey (efloras.org, 2023c) – from Texas to Kansas; flowers white; Great Plains.
- *Allium canadense* var. *hyacinthoides* (Bush) Ownbey (POWO, 2023b; POWO, 2023e; efloras.org, 2023c) – native to Oklahoma to N. Central Texas; tepals pink, thin, flowers fragrant.
- *Allium canadense* var. *lavandulare* (Bates) Ownbey & Aase (POWO, 2023b; POWO, 2023f; efloras.org, 2023c) – native to N. Central U.S.A. to NW. Arkansas; flowers lavender, not fragrant.
- *Allium canadense* var. *mobilense* (Regel) Ownbey (POWO, 2023b; POWO, 2023g; efloras.org, 2023c) – native to Central & SE. U.S.A; flowers lilac, pedicels thread-like.

Cultivation and uses

Canada onion propagates by division when dormant, by seed-like bulbits or by seeds. Bulbils cannot be stored and must be planted as soon as they ripen. This hardy, easy to grow and drought tolerant plant grows best in moist, but very well-drained soils. Persistently wet conditions will lead to root rot (Izel, 2014)

The whole plant can be eaten raw, with the tougher outer layers removed, and can also be cooked and included in any recipe as a sunstitute for onions (Elias and Dykeman, 2009). Various Native American tribes also used the plant for other purposes: for example, rubbing the plant on the body for protection from insect, lizard, scorpion, and tarantula bites (Moerman, 1998). It has also been used medicinally to treat insect stings, control coughs and vomiting, sooth eye and ear infections (Izel, 2014).

References

Efloras.org, 2023a. *Allium oleraceum* Linnaeus. Flora of North America. *FNA Volume* 26. Page: 225, 226, 238. www.efloras.org/florataxon.aspx?flora_id=1&taxon_id=242101382. Retrieved on 12 February, 2023.

Efloras.org, 2023b. *Allium vineale* Linnaeus. Flora of North America. *FNA Volume* 26. Page: 226, 237, 238. http://www.efloras.org/florataxon.aspx?flora_id=1&taxon_id=242101415. Retrieved on 12 February, 2023.

Efloras.org, 2023c. *Allium canadense* Linnaeus. Flora of North America. *FNA Volume* 26. Page: 226, 227, 235. http://www.efloras.org/florataxon.aspx?flora_id=1&taxon_id=242101341. Retrieved on 12 February, 2023.

Ekici M., & Gungor H., Demir T., Ateş M.B. and & Kumru A. (2022). ANTI-INFLAMMATORY AND ANTIOXIDANT EFFECT OF ALLIUM SCORODOPRASUM L. ETHANOLIC EXTRACT (ASE) IN MICE. *J. Animal & Pl. Sci.,* **32**. 10.36899/

JAPS.2022.5.0529. https://www.researchgate.net/publication/360218314_ANTI-INFLAMMATORY_AND_ANTIOXIDANT_EFFECT_OF_ALLIUM_SCORODOPRASUM_L_ETHANOLIC_EXTRACT_ASE_IN_MICE

Elias Thomas S. and Peter A. Dykeman, 2009 [1982]. Edible Wild Plants: A North American Field Guide to Over 200 Natural Foods. *New York: Sterling*. pp. 58, 61. ISBN 978-1-4027-6715-9. OCLC 244766414.

Gleason H. A. & A.J. Cronquist. 1991. Manual of the Vascular Plants of Northeastern United States and Adjacent Canada (ed. 2) i–910. New York Botanical Garden, Bronx. https://en.wikipedia.org/wiki/Allium_oleraceum.

GRIN, 2023. USDA, Agricultural Research Service, National Plant Germplasm System. 2023. Germplasm Resources Information Network (GRIN Taxonomy). National Germplasm Resources Laboratory, Beltsville, Maryland. URL: http://npgsweb.ars-grin.gov/gringlobal/taxon/taxonomydetail?id=316541. Accessed 13 February 2023.

Hanelt Peter (2001). "Alliaceae". In P. Hanelt (ed.). Mansfeld's Encyclopedia of Agricultural and Horticultural Crops (except ornamentals). Berlin: Spring-Verlag. p. 2250. ISBN 3-540-41017-1.

Hegi G. 1909: Illustrierte Flora von Mittel-Europa. Mitbesonderer Berucksichtigung von Deutschland, Oesterreich und der Schweiz. II. — 405 pp., plates 42-72. J. F. Lehmann's Verlag. Munchen. https://www.researchgate.net/publication/279605034_Allium_oleraceum_Alliaceae_in_Finland_Distribution_habitats_and_accompanying_vascular_plant_species.

Hoeg O. A. 1976: Planter og tradisjon. Floraen i levendetale og tradisjon i Norge 1925–1973. — Universi-tetsforlaget, Oslo, Bergen & Tromso. VIII + 751 pp. https://www.researchgate.net/publication/279605034_Allium_oleraceum_Alliaceae_in_Finland_Distribution_habitats_and_accompanying_vascular_plant_species.

Hulten E., Faegri K. and M. Skytte Christiansen, 1958: Varsvenska flora i farg jamte ett urval vaxter fran de nordiska grannlanderna. [Band I.] — AB Svensk Litteratur. 380 + 176 pp. https://www.researchgate.net/publication/279605034_Allium_oleraceum_Alliaceae_in_Finland_Distribution_habitats_and_accompanying_vascular_plant_species.

Izel, 2014, Native Plants For Your Garden, Washington DC, *Allium canadense.* Archived from the original on March 16, 2014. Retrieved March 16, 2014. http://www.izelplants.com/plants/mapsearch/perennials/item/allium-canadense.

McClintock David and R.S.R. Fitter 1961. *The Pocket Guide to Wild Flowers.* London: Collins. p. 205. https://en.wikipedia.org/wiki/Allium_oleraceum.

Moerman David, E. 1998. Native American Ethnobotany. Portland, OR: Timber Press. p. 57. ISBN 0-88192-453-9. https://en.wikipedia.org/wiki/Allium_canadens.

Nativeplanttrust.org, 2023. *Allium oleraceum* — field garlic. Native Plant Trust. Go Botany. https://gobotany.nativeplanttrust.org/species/allium/oleraceum/. Retrieved on 23 February, 2023.

NatureGate, 2022. Crow Leek (*Allium vineale*). https://luontoportti.com/en/t/2275/wild-garlic. Retrieved 30 December, 2022.

NatureGate, 2023. Sand Leek (*Allium scorodoprasum*). https://luontoportti.com/en/t/1288/sand-leek. Retrieved 10 February, 2023.

Nyman C. F. 1868: Utkast till svenska vaxternas natur-historia eller Sveriges fanerogamer skildrade i korthetmed deras vaxtstallen och utbredning m.m., derasegenskaper, anvandning och historia i allmanhet.Senare delen. — Abr. Bohlin, Orebro. 566 pp. https://www.researchgate.net/publication/279605034_Allium_oleraceum_Alliaceae_in_Finland_Distribution_habitats_and_accompanying_vascular_plant_species.

Parker Peter, 2018. A Little Book of Latin for Gardeners. *Little Brown Book Group*. p. 328. ISBN 978-1-4087-0615-2. *oleraceus, holeraceus* = relating to vegetables or kitchen garden. https://en.wikipedia.org/wiki/Allium_oleraceum.

Pettersson B. 1943: Av manniskan inforda vaxter i florankring vara fornborgar och medeltidsfasten. — Nordenskiold-samfundets tidskrift 3: 33–84. https://www.researchgate.net/publication/279605034_Allium_oleraceum_Alliaceae_in_Finland_Distribution_habitats_and_accompanying_vascular_plant_species.

Pfaf, 2023. Plants For A Future, 2023. *Allium scorodoprasum* L. https://pfaf.org/user/Plant.aspx?LatinName=Allium+scorodoprasum. Retrieved 10 February, 2023.

POWO, 2022a: *Allium oleraceum* L. Plants of the World Online. Royal Botanic Gardens, Kew. Retrieved 30 February, 2022. https://powo.science.kew.org/taxon/urn:lsid:ipni.org:names:528834-1.

POWO, 2022b: *Allium oleraceum* subsp. *girerdii*. Plants of the World Online. Royal Botanic Gardens, Kew. Retrieved 30 February, 2022. https://powo.science.kew.org/taxon/urn:lsid:ipni.org:names:974397-1.

POWO, 2022c: *Allium vineale* L. Plants of the World Online. Royal Botanic Gardens, Kew. Retrieved 30 February, 2022. https://powo.science.kew.org/taxon/urn:lsid:ipni.org:names:529111-1.

POWO, 2023a: *Allium scorodoprasum* L. Plants of the World Online. Royal Botanic Gardens, Kew. Retrieved 7 February, 2023. https://powo.science.kew.org/taxon/urn:lsid:ipni.org:names:528834-1.

POWO, 2023b. *Allium canadense* L. Plants of the World Online. Royal Botanic Gardens, Kew. Retrieved 7 February, 2023. https://powo.science.kew.org/taxon/urn:lsid:ipni.org:names:527759-1.

POWO, 2023c. *Allium canadense* var. canadense. Plants of the World Online. Royal Botanic Gardens, Kew. Retrieved 7 February, 2023. https://powo.science.kew.org/taxon/urn:lsid:ipni.org:names:77171983-1.

POWO, 2023d. *Allium canadense* var. *ecristatum* Ownbey. Plants of the World Online. Royal Botanic Gardens, Kew. Retrieved 7 February, 2023. https://powo.science.kew.org/taxon/urn:lsid:ipni.org:names:8709-2.

POWO, 2023e. *Allium canadense* var. *hyacinthoides* (Bush) Ownbey. Plants of the World Online. Royal Botanic Gardens, Kew. Retrieved 7 February, 2023. https://powo.science.kew.org/taxon/urn:lsid:ipni.org:names:8709-2.

POWO, 2023f. *Allium canadense* var. *lavandulare* (Bates) Ownbey & Aase. Plants of the World Online. Royal Botanic Gardens, Kew. Retrieved 7 February, 2023. powo.science.kew.org/taxon/urn:lsid:ipni.org:names:8715-2.

POWO, 2023g. *Allium canadense* var. *mobilense* (Regel) Ownbey. Plants of the World Online. Royal Botanic Gardens, Kew. Retrieved 7 February, 2023. https://powo.science.kew.org/taxon/urn:lsid:ipni.org:names:8717-2.

Schwegman, J. E 1991. "The Vascular Flora of Langham Island, Kankakee County, Illinois". *Erigenia*, **11**: 1–8. https://en.wikipedia.org/wiki/Allium_canadense.

Seppanen K. 1988: Backloken (*Allium oleraceum*) iPaijat-Hame, en nyttovaxt fran jarnaldern. — Museo-uutiset. Paijat-Hameen museoiden aanenkannattaja 1988(1): 19–20. Lahti. https://www.researchgate.net/publication/279605034_Allium_oleraceum_Alliaceae_in_Finland_Distribution_habitats_and_accompanying_vascular_plant_species.

Shoot, 2023. *Allium scorodoprasum* (Sand Leek). https://www.shootgardening.co.uk/plant/allium-scorodoprasum. Retrieved 10 February, 2023.

Stace C.A. 2010. New Flora of the British Isles (Third ed.). Cambridge, U.K.: Cambridge University Press, p. 902-903. ISBN 9780521707725.

TPO: 2023. Temperate Plants Database, Ken Fern. temperate.theferns.info. 2023-02-11. <temperate.theferns.info/plant/Allium+scorodoprasum>

Tropicos, 2023. *Allium canadense* L. Missouri Botanical Garden. http://legacy.tropicos.org/Name/18401640?projectid=0. Retrieved 14 February, 2023.

Whitney, William Dwight (1899). The Century Dictionary and Cyclopedia. Century Co. p. 2856. L. holeraceus, prop. oleraceus, herb-like, holus, prop. olus (oler-), herbs, vegetables.

Wikispecies, 2023. https://species.wikimedia.org/wiki/Allium_scorodoprasum; https://species.wikimedia.org/wiki/Allium_oleraceum; https://species.wikimedia.org/wiki/Allium_vineale; https://species.wikimedia.org/wiki/Allium_canadense. Retrieved on 12,February, 2023.

10

Jimbu/Jambu/Faran and Other Similar Alliums

Common names: Faran, Jamboo or Jambu in India and Jimbu in Nepal.

Introduction

Jambu, Jimbu or Faran is the dried product obtained from one or more of the several species of *Allium* widely distributed in mountain areas in the Himalayas of India, Tibet, Nepal, Pakistan, Mongolia and China (FOC, 2023a; FOI, 2015) at altitudes of 2000-4800m (FOI, 2015). While most literatures refer Jimbu, Jambu or pharan/Faran/Fran/Phran to be a spice obtained from two allium species, *Allium hypsystum* Stearn and *A. przewaliskianum* Regel, but it is actually one of several vernacular names of many Himalayan species growing wild (only rarely cultivated) at altitudes of 2000-5500m and whose leaves (occasionally flowers and bulbs) are dried and sold as spices in local markets for seasoning and garnishing. Besides widely cultivated plants like garlic, Indian leek, leek, nodding onion, onion, shallot, etc., these Himalayan *Allium* species are partially cultivated by the natives in their kitchen garden or backyards for green vegetables, spices and condiments.

Some of these jambu species include *Allium stracheyi* Baker, *A.wallichii* Kunth, *A. hypsystum* Stearn, *A. przewaliskianum* Regel and a new type *A. negianum* (quite similar to *A. przewaliskianum*) (Anjula Pandey *et al.*, 2021b;

IUCN, 2004; Nepal, 2006). These perennial herbs grow in clumps in high-arid regions with well drained sandy soils (Nepal, 2006). They are recorded wild from Nepal as well as from temperate Indian hills where these are used in Nepalese and Indian cuisines (Nepal, 2006). Many *Allium* species have been identified for similar use in the Indian Himalaya and have got regional recognition as 'jambu' or 'faran'. In trade many other Allium species are used for similar purpose, more often as an adulterant to 'jambu' obtained from *A. stracheyi.*

Although these plants are usually harvested from the wild but there are reports for large scale cultivation of a newly found *A. negianum* in Niti valley (Uttarakhand) as 'seasoning allium spice' called 'jambu' and 'phran' (Anjula Pandey *et al.,* 2021a). These species can be distinguished based on leaf size (Anjula Pandey *et al.,* 2021b).

Vernacular names

Allium stracheyi, A.wallichii and *A. humile* are known by vernacular names 'jambu', 'jamboo', 'dhungar' (by 'Bhotias' of Johar and Dharma valleys, Pithoragarh, Kumaon region) and 'pharan/faran/fran/phran', 'koch', 'ladum', 'laiku', 'dodoo' (by 'Marchha', 'Tolchha', 'Bhotias' of Niti valley in Garhwal region) and 'jamboo', 'jimbu', 'doodli' by local inhabitants of Pandukeshwar, Hanuman Chatti, Bhyundar and Mana areas of Chamoli, Garhwal region of Uttarakhand (Anjula Pandey *et al.,* 2021b). Similarly, *A. przewaliskianum, A. stracheyi, A.roylei* in Kashmir (J&K) and *A. przewaliskianum* and *A. hypsystum* (jimbu) in the Nepal Himalaya are known for their regional values as seasoning spices (Anjula Pandey *et al.,* 2021b).

Uses

Jambu is used as a green leafy vegetable and its hammered, dried, processed leaves are used as a spice and condiment in the Himalayan regions. The commercial product is known as Jambu, Faran or *Jambu-Faran.* It is used in some regions of Nepal, where it is called Jimbu, and central Himalayan states of India such as Uttarakhand, H.P. and J&K (en.wikipedia.org/wiki/Jimbu) where it is called by several different names like Pharan, Jambu, *etc.* Depending on the species, area of cultivation and cultural practices, the flavour of the herb is in between onion and chives or somewhat similar to garlic and shallots (en.wikipedia.org/wiki/Jimbu; Anjula Pandey *et al.,* 2021b, 2022; Nepal, 2006). In India, Jambu is a substitute to onion and garlic for those who religiously do not use these.

The above ground parts of these plants are used for spices and traditional medicines. Use of spice is primarily for flavouring and seasoning. The dried leaves are fried in butter oil (*ghee*) to develop their pungent flavour much like garlic and shallot, to be used in black gram pulse (*urad daal*) or lentils, vegetables, curries, soups, pickles and meat (Nepal, 2006; en.wikipedia.org/wiki/Jimbu). It is considered nutritious by people in the Upper Mustang region of Nepal (Anjula Pandey *et al.*, 2021b). In Nepal and parts of Northern India, it is extensively used in cuisine and for medicinal purposes. Jimbu is also considered to be of high value medicinal plant. In Upper Mustang and other North-Central parts of Nepal, households use Jimbu to cure various illnesses such as flues, coughs and stomach pains (IUCN Nepal, 2000). The pungent sulphur smell of the leaves repel insects and is used as an insect repellent in the region.

Cultivation

Among the above mentioned species, only *Allium negianum* was found in large scale cultivation. Therefore, cultivation practices would be more in the context of this species.

Propagation

These Jambu onions are propagated in April-May through bulbs and seeds. Growers in higher elevation areas of 2500-3500m (Malari, Kailashpur, Gamshali and Niti) prefer growing these alliums from seeds due to rainfed and high intensity of sunlight with low humidity. Mature umbels are left on plant to dry (August–September) and then buried into the soil and seedling emergence happens later. Whereas, growers in lower elevation areas of 1800-2500m (Tolma, Fagti, Sukhi, Tamak) prefer to grow the crop using bulbs due to the microclimate, *i.e.* low intensity of sunlight and high humidity. Planting using clustered intact bulbs leads to slow plant growth, therefore the outer reticulate tunic should be separated from bulb clusters and individual bulb should be planted (Anjula Pandey *et al.*, 2021b).

Cultural practices

Before planting or sowing, if need be, organic manure @5-10t/ha may be mixed in the soil. However, weeding should regularly be carried out at every 20-30 days, especially during rainy season. The crop is usually rainfed and can tolerate drought condition well during the course of vegetative growth (Anjula Pandey *et al.*, 2021b). The plants remain economic up to 5–6 years after plantation and then these are replanted. Replantation is done at periodic intervals by rearing the crop on fertile sandy loam under good sunlight.

Harvesting

In Nepal, people collect the aboveground vegetative part of jimbu herb (*A. przewaliskianum* and *A. hypsystum*) from the wild. While harvesting wild jambus, the villagers carefully handpick the green and tender foliage from the plants between June and September. Though they collect both the parts ~ flowering and non-flowering, as best flavours come from buds that are not fully open. The leaves and tender stems are carefully picked and dried though this way flavour is weakened but when these are browned in hot water then flavour is found increased. After drying, jambu loses its green colouration and looks like a brownish-green dried grass (Nepal, 2006).

In case of *Allium negianum*, the finest quality 'jambu' includes the aerial parts consisting of young leaves and stem. However, in case of *Allium stracheyi*, dried leaves with flowers is the major part but sometimes bulbs are also adulterated especially for self consumption (Anjula Pandey *et al.*, 2021b).

While most Jambo species such as *Allium stracheyi* and *A. wallichii* are harvested once a year in September (Anjula Pandey *et al.*, 2021b), *A. przewaliskianum* and *A. hypsystum* during June-September (Nepal, 2006) though main harvest of *Allium negianum* is generally performed between July and October. Biomasses usually above 5–8cm from ground are harvested through a sickle to leave the underground parts intact for producing subsequent crops. Some people prefer jambu with leaves and flowers to be best for seasoning while others believe that leaves harvested before the flower initiation is of better quality and aroma (Anjula Pandey *et al.*, 2021b).

At higher altitudes (2500-3500m), *A. negianum* is harvested twice a year, first in the month of May-July and second in September-October. First harvest (May) is considered to be best and mostly preferred for domestic uses. At lower altitudes (1800-2500m) usually it is harvested three times in a year ~ first in May, second in July and finally in October, the latest one being probably of best quality probably due to processing of product in clean weather, higher variation in day night temperature followed by low humidity. Certain areas such as in Tolma the crop is harvested four times in a year because of the existence of most favourable microclimatic conditions that favour fast vegetative growth (Anjula Pandey *et al.*, 2021b).

Post-harvest, processing and storage

Being a seasonal herb, people dry and store dried jambus for later use. The processing of the harvest from all the species is practiced in a same way and these after cleaning may or may not be cut into pieces for closed area complete air drying in a shed for several days (Nepal, 2006; Anjula Pandey *et al.*, 2021b;

en.wikipedia.org/wiki/ Allium_przewalskianum). Dried leaves are stored in the cotton or plastic bags in moisture-free conditions. Dried leaves can be stored for 4–6 months at 20°C temperatures with 60% relative humidity (Anjula Pandey *et al.*, 2021b).

Allium hypsistum Stearn

Vernacular names: In Gurung: Dzim; Sherpa: Rukpa; Nepali: Zimbu; Tibetan: Jimnak (Bhetuwal, 2014).

Introduction

Allium hypsistum, endemic to Nepal (IUCN, 2004), is one of the two species that are used to prepare jimbu spices in Nepal, the other is *A. przewalskianum*. Usually these are collected from the wild from central Nepal, *i.e.* north of the main Himalayan range and adjoining the Tibetan plateau. It has been recorded at about 6-7 kilometers southwest of Saldanggaon (NC Nepal) at about 5500m a.s.l (Stearn, 1960). Its dried leaves are used as spices for seasoning various cuisines (Nepal, 2006). It is used as tonic and in treatment of cough, cold and gastritis. The leaves are used as diuretic and carminative (Shrestha *et al.*, 1995; Chhetri, 1999).

Botany

Its reticulately fibrous bulb-coats, 4-6 narrow linear leaves, very short pedicels, slightly dentate tepals and simple included filaments together distinguish it from other Himalayan species (HMG, 1982). Its atems are cylindrical and its pinkish to white flowers occur from July to August (HMG, 1982) though stalks are found clustered at the terminal of the scape (Bhetuwal, 2014).

Allium negianum A.Pandey, K.M. Rai, P.K. Malav & S. Rajkumar

Subgenus: *Rhizirideum.*

Section: *Eduardia.*

Vernacular names: Phran, Jambu, Sakua, Sungdung and Kacho (Anjula Pandey *et al.*, 2022).

Introduction

Specific epithet *negianum* is given in honour of Late Dr. Kuldeep Singh Negi, an eminent Indian explorer who had dedicated his life in collection of germplasm of indigenous *Allium* species along with associated indigenous knowledge across the country. *Allium negianum*, a new species described by the ICAR-NBPGR, is native to India and grows on higher altitudes of Sumna and Niti valley near Malari glacier and other adjoining regions in Uttarakhand where it is found wild as well as cultivated. *A. negianum* has never been collected from any other area in the country or abroad. Therefore it is said to be localized in distribution. Its habitats include open grassy meadows, sandy soils along rivers and streams occurring in the snow pasture lands along the alpine meadows (locally known as '*bugyal/bugial*') between 3000–4800m a.s.l. in synanthropic habitats. It was reported growing wild in Darma valley of Pithoragarh along Gori Ganga (also Gori Gad) river in the Munsiyari (Pithoragarh district), in Milam Glacier, in northeast of Nanda Devi (Uttarakhand), India (Anjula Pandey *et al.*, 2021a, b).

The leaves of *Allium negianum* have strong onion-garlic aroma which are used as a seasoning spices and condiments, fresh or dried. The species has high commercial market in Uttarakhand. It can be propagated both by seeds and bulbs. *A. negianum* has robust plant habit and stronger plant aroma in wild habitat as compared to plants growing under cultivation (Anjula Pandey *et al.*, 2021a).

Allium negianum

Bulbs and root | **Semi-opened inflorescence** | **Opened inflorescence**

Photo courtesy: Anjula Pandey *et al.*, 2022

Botany

It has 2n=2x=16 chromosome numbers. *Allium negianum* is a 27-50cm tall semi-erect herb with ovoid-cylindric bulbs clustered on oblique, condensed and 6.5-8.5mm long rhizomes. The outer tunic of bulbs is finely reticulate and reddish to dark-brown while inner one membranous, light-brick red or orange-red. The plant has up to 8 dark green leaves that are shorter than scape, 12-40cm long and 1.0-3.2mm wide. Its stout but terete scape is solid in cross-section (hollow when mature) (Anjula Pandey *et al.*, 2021a, 2022).

Flowering occurs from June to September, at lower elevations (1800–2500m) in June-July and at higher elevations (2500–3500m) in August–September. The hemispherical umbel bears very long and narrow-beaked spathe, more than 40 lax flowers, deep purple tepals with distinct green mid-line, asynchronous flowering and inner stamen filaments have longer and sharp teeth. Yellowish-purple and oblong-lanceolate inner tepals are little longer than outer ones with acute apices (Anjula Pandey *et al.*, 2021a, 2022). Filaments are purple, slightly exserted, anthers oblong, yellowish-purple (on maturity), 2.3-2.6mm long, connate at base and adnate to perianth segments, outer ones subulate, ovary sub-globose, tinged with purple, the styles are much exserted, ovules 2 per locule, the seeds are obovate with a prominent notch on one side, 3.21-4.06 x 1.92-1.97mm and deep black (Anjula Pandey *et al.*, 2021a, 2022).

Comparision of *A. przewalskianum* and *A. negianum*

Traits	*A. przewalskianum* Regel	*A. negianum*
Subgenus	Rhizirideum	Rhizirideum
Section	Eduardia	Eduardia
Chromosome number	2n = 16, 32	2n = 16
Habitat; altitudes	Scrub, dry slopes, plains, rock crevices, carbonaceous slates or gravel; 3300–5200m	Grassy meadows, open sandy slopes, along rivers/streams; 3000–4800m
Distribution world	India, Nepal, Tibet, China	India
Distribution India	Himachal Pradesh, Jammu & Kashmir, Ladak and Western Himalayas	Higher altitudes of Sumna and Niti valley, near Malari glacier and other adjoining regions in Uttarakhand
Status occurrence	Common; threatened in Nepal	Common
Cultivation/wild	Cultivated occasionally in kitchen garden; and wild	Frequently cultivated; large scale cultivation; and wild (above 3000m)
Commercial value; area of local use	Commercialized; Ladakh, Nepal	Commercialized; Uttarakhand
Propagation	Seed and bulb	Seed and bulb
Closest identical relative	*Allium negianum*	Outer tunic is morphologically similar to *A. stracheyi*, *A. auriculatum* and *A. przewaliskianum*
Plant growth (under experimental condition)	Robust, shorter	Taller, plants and leaves
Plant height (cm)	20–45	27–50
Pseudostem-length	Developed	Developed
Bulbs no. in cluster	2–4	2–7

Traits	*A. przewalskianum* Regel	*A. negianum*
Bulb no., shape	Clusters 3–4; cylindrical-narrowly ovoid	Clusters 4–8; cylindrical-narrowly ovoid
Bulb length (cm)	10.2–12.5	6.8–12
Bulb diameter (cm)	0.6–0.7	0.8–1.2
Tunic outer**	Finely reticulate; reddish-orange (rarely light brown)	Finely reticulate; reddish-dark brown (rarely lighter-brown)
Tunic inner	Membranous, brown-red	Membranous, orange-red
Rhizome type; size (mm)/root-type	Vertical, short; 3–5/non fleshy	Oblique; 7–12/non fleshy
Leaf no., colour	3–5, lighter brown-green	4–6, dark green
Leaf *vs.* scape	Shorter to slightly longer	Slightly shorter than scape
Leaf-shape/type	Fistulose, semiterete-terete (4–5 angled)	Fistulose, semiterete-terete
Leaf length (cm)	15–30	12–40
Leaf width (mm)	2.0–2.5	1–3.2
Leaf erectness	Erect	Erect to semi-erect
Leaf waxiness	Non-waxy	Waxy
Leaf cross section	Circular	Circular
Scape	Stout, terete, solid, erect, central; 30–40 × 0.2–0.35	Stout, terete, solid, erect to semi-erect, lateral-central; 20–50 × 0.36–0.48; 1/3 of the base
Spathe valve if persistent	1(2)-valved, persistent	1-valved, persistent
Spathe valve shape, size	Ovate	Ovate-oblong
Spathe size (cm)	2–3 (two times the base; short, blunt)	4–6 (long narrow beak; 3 times the base)
Pedicel *vs.* perigone	Subequal	2–3 times longer
Umbel flower opening pattern	Synchronous (80 per cent)	Asynchronous (30–40 per cent)
Umbel shape	Spherical-hemispherical, densely flowered, compact	Hemispherical, lax, loosely flowered
Flowering time	June-September	June-September
Umbel diameter (mm)	28.5–30.2	25.1–42.0
Umbel flower (no.)	25–40	30–40
Pedicel vs. perianth	Subequal	2–2.5 times longer
Peduncle size (cm)	0.5–1.0	0.8–2.5
Flower size (cm)	0.4–0.5×0.3	0.5–0.5× 0.4
Flower colour	Pale red-purple pink (variable)	Dark purple (as recorded now)
Perigonium shape and colour	Campanulate, pink-dark purple, tepal wide open	Campanulate, lilac, light to dark purple, tepal partly opened

Traits	*A. przewalskianum* Regel	*A. negianum*
Tepal shape	Ovate-lanceolate, apex obtuse	Elliptic, ovate-lanceolate; apex-acuminate-mucronate
Tepal inner size length × width (cm)	0.3–0.4 ×0.2–0.3	0.5–0.6 × 0.3–0.4
Tepal outer size length × width (cm)	0.5–0.7 × 0.2–0.3	0.6–0.7 × 0.3–0.5
Tepal apex shape	Acute-acuminate	Acute, mucronate
Tepal maturity	Curved outwards	Slightly inwardly curved/rolled
Tepal mid-vein	Non-conspicuous; purple green-dark purple	Very conspicuous; green-light green
Anther-lobe	Exserted; outer linear, broad at base, inner oblong	Lightly exserted; outer linear, broad at base; inner oblong
Anther length (mm)	6.1–9.3	6.8–8.5
Anther lobe length (mm)	Oblong-ovate, 1–2	Oblong, 1–2
Anther lobe colour	Yellow-purple	Yellow-purple
Filament colour	Yellowish-purple	Greenish yellow-purplish green
Stigma vs. stamen	Subequal stigma	Slightly longer than stigma
Filament length, position	Double the size of tepal; exserted	Half the size of tepal; slightly exserted
Filament inner and outer anther	Inner – two sharp teeth up to 1/2 to 1/4 length of filament with broader base; outer narrower base	Inner – two shallower-sharper teeth up to 1/2 to 2/3 length of filament with base as wide as tepal; outer narrower base
Ovary shape	Ovoid – globose, wrinkled; without concave nectaries at base	Obovoid – subglobose
Ovary colour	Purple green, tinged with purple	Dark-pale purple
Ovary style *vs.* anther (after pollination)	Much exserted, longer than ovary	Slightly exserted or equal
Ovary stigma tip	Acuminate-acute	Acuminate
Stigma *vs.* stamen	Sub-equal	Slightly longer
Capsule	Obovoid, 3-4 mm, umbelicate, light brownish yellow, conspicuously trilobed with persistent tepals, splitting from the top	Sub-globose
Seed length (mm)	2.75–2.96	3.21–4.05
Seed width (mm)	1.55–1.59	1.92–1.97
Seed colour	Dull black	Shiny black
Seed no./locule	2	2
1000 seed wt (g)	2.12	2.73
Odour when crushed	Strong onion-light garlic	Strong onion-garlic
Source: *Anjula Pandey et al., 2021a, 2022.*		

Allium przewalskianum Regel

Subgenus: *Rhizirideum.*

Section: *Rhizirideum.*

Other common names: Chives, Ladakh Onion, Flowering Onion, Przewalski's Onion, Jangli pyaz, Jambu, Jimbu, Pharan (FOI, 2015; Anjula Pandey *et al.*, 2022).

Synonyms: *A. jacquemontii* Regel, *A. jacquemontii* var. *parviflorum* (Ledeb.) Aswal, *A. junceum* Jacquem. ex Baker, *A. przewalskianum* var. *planifolium* Regel, *A. rubellum* var. *parviflorum* Ledeb., *A. stenophyllum* Wall., *A. stoliczkii* Regel.

Introduction

Allium przewalskianum, named in honour of 19th century Russian explorer and naturalist Nicolai Przewalski (FOI, 2015), is widely distributed in mountain areas in the Himalayas of India, Tibet, Nepal, Pakistan, Mongolia and China (FOC, 2023a; FOI, 2015) at altitudes of 2000-4800m (FOI, 2015). It is one of the two species referred to as Jimbu in Nepal; the other is *A. hypsistum* (Nepal, 2006). It is also one of the several species that are called Jambu or Faran in India. Species is considered as vulnerable in regard to its conservation status (Shrestha and Joshi, 1996; Nepal, 2006).

It is largely used as seasoning spice in Nepal and India (Uttarakhand and Ladakh). Whole plant is smashed into a semi-pulp casted as big fist shape, strung through the middle in a string and sold in the market (Anjula Pandey *et al.*, 2022).

Allium prezwalskianum

Uprooted bulb and root | **Inflorescences** | **Dried processed leaves**

Source: *Anjula Pandey et al. (2022)*

Botany

This is a diploid as well as tetraploid species having 2n=4x=32 chromosome numbers. Ladakh onion is an erect herb with slender and tightly clustered bulbs having bright orange bases some 1cm across (FOI, 2015) *cum* finely reticulate bulb-coat (Stearn, 1960). Outer bulb tunic is tinged red while inner-tunic red-orange (very rarely light-brown) (Anjula Pandey *et al.*, 2022). The terete stem is upto 3mm in diameter. The fine, narrow and cylindrical leaves are 0.5-1.5mm wide and almost of the size of scape that are up to 40cm long (FOI, 2015). Spathe is 1-valved, persistent and with shorter beak (Anjula Pandey *et al.*, 2022). Each scape carries a round head consisting of numerous bright purplish flowers (FOI, 2015; en.wikipedia.org/wiki/Allium_przewalskianum; Regel, 1875) with protruding stamens, the inner three filaments having a broad oblong base with a tooth on each side at the top, and a slender upper part while three outer filaments are subulate (Stearn, 1960).

Umbel is compactly hemispheric-globose, tepal segment pale-red to dark purple; filaments longer than perianth segments and the style is much exserted (Anjula Pandey *et al.*, 2022) and quite longer than ovary (FOI, 2015). Flowering occurs from June to September (FOI, 2015).

Allium stracheyi Baker

Subgenus: Polyprason.

Section: Orioprason.

Synonym: *Allium longistaminum* Royle.

Common names: Jambu/Jamboo, Jimbu, Dhungar, Pharan, Koch, Ladum, Laiku, Dodoo and Doodle.

Introduction

Allium stracheyi and some other alliums are found growing wildly in Tibet and its adjoining borders in Nepal (Anjula Pandey *et al.*, 2022) and India (Uttarakhand) (Tiwari *et al.*, 2014). *A. stracheyi*, locally known as Jamboo/Dhungar (in Kumaun region) and Pharan (in Garhwal region), occurs at high altitude areas in Jammu & Kashmir, Himachal Pradesh and Uttarakhand *cum* Nepal (Dasgupta, 2006; Rawat *et al.*, 2016; Tiwari *et al.*, 2014). Its habitat includes rocky soil near water streams or dry open slopes (Hanelt, 2001) at altitudes of 2000-3800m (Anjula Pandey *et al.*, 2022). It is usually harvested from the wild though occasionally cultivated in kitchen garden by Bhotias in Uttarakhand.

Allium stracheyi is one among the 81 medicinal plants used for industrial trading (Chandrasekaran *et al.*, 2020). Its flowers, leaves, roots, bulbs and whole plant are traditionally used as spices, condiments and medicine by tribes (Anjula Pandey *et al.*, 2008). In North Western Himalaya, the fleshy fibrous roots of *A. stracheyi* are consumed as vegetable (Hanelt, 2001), soup and for prepareing pickles (Anjula Pandey *et al.*, 2008). It is processed by hammering and drying to prepare spice/condiment called *Jambo Faran* which is of economic value, especially in Uttarakhand, where it is extensively used for seasoning, garnishing and also in traditional medicines to treat various ailments.

Botany

Anjula Pandey *et al* (2022) reported this species having diploid, 2n=14 chromosome numbers. *Allium stracheyi* is a small perennial herb growing up to 35cm tall (Gautam *et al.*, 2019) with stem that is leafy but glabrous at the base. The leaves are linear, flattened and narrow (Shah, 2014). Heads are globose, pedicel is shorter than the flower, and the tepals are dark pink-reddish (Shah, 2014).

Allium stracheyi

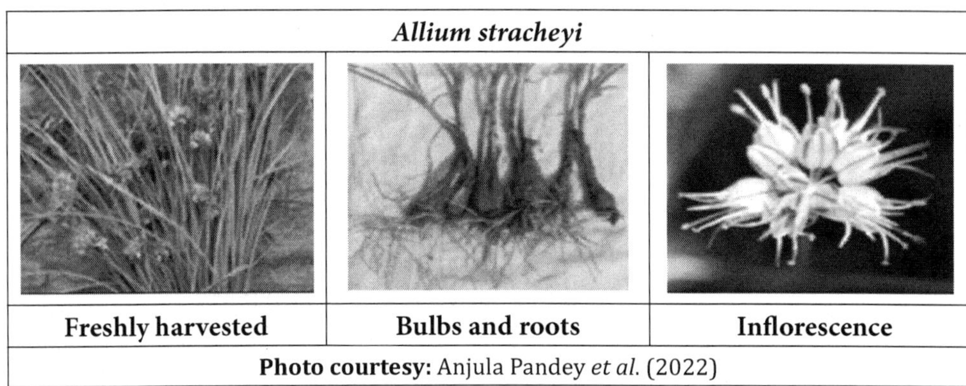

Freshly harvested | **Bulbs and roots** | **Inflorescence**

Photo courtesy: Anjula Pandey *et al.* (2022)

Major characteristics of this species are given in the following table:

Traits	***Description***
Propagation	Seed and bulb
Closest identical relative	*A. consanguineum* Kunth; except for plant height and colour of flower
Plant height (cm)	20–35
Pseudostem-length	Developed
Bulbs	Small; clustered; cylindric-narrowly ovoid; solitary with outer tunic fibrous, coriaceous, brown striate and inner red membranous tunic, produced into a neck
Roots	Non-fleshy roots
Leaf	3–4, green, erect-semierect, flattened, 20-30cm long, fistulose, linear, length almost equal to scape
Scape	Slender, round in cross section and nearly equal to leaves
Spathe	2-valved, persistent, bifid, ovate, acuminate
Pedicel vs. perigone	pedicels 3-5mm long; smaller than flower (5-7mm)
Umbel	Hemispheric, 20-30cm diameter, with 8–10 laxy, 0.5-0.7mm long flowers
Flowering time	July-September (rarely in October)

Traits	*Description*
Pedicel vs. perianth	Shorter
Peduncle size (cm)	0.3-0.5
Flower colour	Purplish tinge-dark pink-rosy-purple-yellow (white in Kashmir samples)
Perigonium shape and colour	Oblong-campanulate; pouched at the base
Tepal	Broadly oblong; outer convex, apex obtuse; 0.5-0.7mm long; inner slightly bigger; curved inwards when mature
Tepal mid-vein	Prominent pink mid-veins
Anther	Very much exserted, oblong; anther lobe yellow
Filament	Linear, much exserted, purple, free, adherent to perianth at base up to 1mm
Stigma vs. stamen	Subequaling stigma
Ovary	Oblong, deeply trilobed; Style: purple, lenear, stigma obscure
Capsule	Globose-oblong, thin pericarp
Seed no./locule	2
Odour when crushed	Sweet onion-mild garlic
Source: Anjula Pandey *et al.*, 2021a, 2022.	

Nutritional value

Allium stracheyi has higher amount of fibre and protein contents with lower fat and carbohydrates (Gautam *et al.*, 2019). It possesses protein (4.26%), fat (0.1%), fibre (79.02%), carbohydrate (3.18%), calcium (0.8 mg), phosphorus (0.05 mg), iron (0.50 mg), magnesium (0.82 mg) and potassium (0.95 mg) per 100 mg (Dobhal, 2015).

Phytochemicals

Allium stracheyi leaves are reported to contain alkaloids, saponins, fixed oils, phytosterols, phenolics & flavonoids, tannins and steroids (Gautam *et al.*, 2019). It contains different types of compounds like hydrocarbons, terpenes, terpenoid, *etc.* in which most of the compounds have sulphur as major content like 1,2-bis(methylthio) ethene (1), 2,4-dimethylthiophene (2), dimethyl disulphide (3) and dimethyl trisulphide (4) as the major volatile components (Shah, 2014; Gautam *et al.*, 2019). These sulphur rich compounds have antioxidant, anti-inflammatory and antimicrobial properties, and are reported in reducing blood cholesterol (Abuajah *et al.*, 2015; Gautam *et al.*, 2019).

Through another study on the volatile composition of *A. stracheyi*, it reported 16-hentriacontanone (50.57%), 4-nitrophenyl ester hexadecanoic acid (31.59%), heneicosane (28.84%), (Z,Z,Z)-9,12,15-octadecatrienoic acid

(24.44%), isobutyl octadecyl ester oxalic acid (47.51%), 1-heneicosyl formate (13.79%), vitamin E (12.85%), phytol (9.42%), 12-hydroxydodecanoic acid (8.74%), (2-bromocyclopropyl) benzene (8.25%), hexacosane (7.76%), tricosane (5.37%), 1,5,9,13-tetradecatetraene (5%), (Z,Z,Z)-9,12,15-octadecatrien-1-ol (4.14%), 11-hexacosyne (3.09%) and 3-phenylthiolane 1,1-dioxide (5) as major constituents (Joshi, 2016; Gautam *et al.*, 2019). The identified volatile compounds, i.e. (Z,Z,Z)-9,12,15-octadecatrienoic acid also has been reported in curing cancer (Kim *et al.*, 2011; Gautam *et al.*, 2019) as have the anti-inflammatory and antioxidant properties (Kim *et al.*, 2011; Agoramoorthy *et al.*, 2007), palmitone the anxiolytic-like effect (Gonzalez-Trujano, 2006) and provide relief in spasmodic and arthritic pain (Carballo *et al.*, 2010).

Two new bioflavonoids allistatin I and allistatin II are reported to be present in *Allium stracheyi* (Etewa and Abaza, 2011; Rajan *et el.*, 2014). These were first reported to be isolated from garlic and are reported to be present in *A. cepa* (Arshad *et al.*, 2017; Alamgir, 2017) as well as other plants like *Zingiber officinale* (Helen *et al.*, 2020). Many scholars stated allistatin as a non-sulphur-containing bioflavonoid (Arshad *et al.*, 2017; Farooqui and Farooqui, 2017; Gregrova *et al.*, 2013; Ebadi *et al.*, 2014) while a few researchers relate it to well-studied allicin (a sulphur-containing allyl compound) (Cundell, 2014).

Allium wallichii Kunth

Subgenus: Amerallium.

Section: Bromatorrhiza.

Synonyms: *Allium bulleyanum* Diels; *A. caeruleum* Wall.; and *A. liangshanense* Z.Y.Zhu

Common names: Himalayan onion (Anjula Pandey *et al.*, 2022) and Wild garlic in English; in Greek: Palengu; Kham and Sherpa: Gokpa, Kojerogokpa; Nepali: Ban lasun, Ban pyaj, Jimbu, Jimbujhar; Rai: Bathatuva; Tibetan: Zinmunagpo, Nimnak (Bhetuwal, 2014), Khasi: Ujnad, Uttarakhand: Paniol, Jimbur (Anjula Pandey *et al.*, 2022).

Other names: Jambu/Jamboo, Dhungar, Faran, Koch, Ladum, Laiku, Dodoo and Doodli.

Introduction

Allium wallichii, the only species found fully open to monsoon rains, is commonly found growing wild in the Himalayan regions of Bhutan, China, Myanmar, Nepal, Tibet and India (Assam, Kashmir, Meghalaya, Sikkim and

Uttarakhand). Its growing habitat is running water, forest margins, hilly slopes or rocky areas in open conifer forests, scrub, moist meadows, and stream banks at altitudes of 2800-3700m. It is usually harvested from the wild, though some people grow it in their kitchen garden. The plant is of economic value in Uttarakhand (Anjula Pandey *et al.*, 2022). Its leaves have strong garlic/leek smell and are cooked with fried fish or roasted meat to impart flavour. It has been used as spice and traditional medicine in the native regions. Its whole plant or part thereof is taken in abdominal pain and the roots are used as tonic and aphrodisiac (Ghimire *et al.*, 2001). Also it is effective in treating headache (Joshi and Edington, 1990).

Botany

Its diploid to tetraploid chromosome number is 2n=14 and 28. *Allium wallichii* grows to a height of 40-65cm having solitary to clustered cylindrical cum elongated bulbs which have profuse *cum* thick basal rooting, tunic being yellowish-brown and laciniate or fibrous to sub-reticulate, and scapes lateral. Pseudostem base is covered by old fibrous leaf bases. It has strong smelling similar to leek. Its leaves are linear to oblong-lanceolate or lanceolate, midvein distinct and base narrowing mostly into a petiole. Pedicels are subequal, ebracteolate, and laxly or densely flowered (Anjula Pandey *et al.*, 2022).

Major characteristics of this species are given in the following table:

Characteristics	***Description***
Propagation	Seed and bulb
Plant height(cm)	40–65
Bulb-no.	Solitary/clustered
Bulb-shape	Cylindric
Bulb-tunic colour (at maturity)	Yellowish-brown
Tunic-type	Fibrous-subreticulate
Root-type	Feshy-nonfleshy
Leaf-no.,colour	2–4; green
Leaf-shape/type	Flat, linear
Leaf vs. scape	Shorter-longer
Spathe	1–2 valved, lanceolate, deciduous
Umbel-shape, no.of flowers	Hemispheric, 20–30 densely flowered or laxy
Pedicel vs. perianth	Subequal
Flowering	July-October
Flower-shape	Broadly campanulate, stellate
Flower colour	Brilliant magenta to deep purple or pale red (rarely white)

Characteristics	*Description*
Tepal- size	Subequal
Tepal-shape	Ovate-lanceolate, apex-acute
Tepal-mid-line colour	Dark purple
Mid-line prominence	Not very prominent
Anther-filament	Shorter or subequal to perianth segments
Anther lobe	Not exserted; erect, oblong and purple
Tepals after maturity	Curved outwards
Ovary	Obovoid-globose, smooth
Style	Style linear, 2.5-4.5mm long and longer than ovary
Capsules	Broiwnish yellow, globose to obovoid and trilobed
Locules	Per locule 2 ovules
Seed	Brownish-black, oblong and 3.5-4 x 2mm
Leaf-aroma	Strong garlic
Uses	Leaves cooked with fried fish or roasted meat to impart flavour. Whole plant part is taken in abdominal pain (Bhetuwal, 2014). Used in the treatment of headache (Joshi and Edington, 1990). Roots are taken as tonic and aphrodisiac.
Source: Anjula Pandey *et al.,* 2021a, 2022.	

Allium wallichii

Field view | **Half-opened bud** | **Inflorescence** | **Root**

Photo courtesy: Anjula Pandey *et al.* (2022).

Nutritional value

Nutritional composition analysis of whole plant of *Allium wallichii.*

Parameters Value in per 100g of sample			
Nutritional value		**Chemicals content**	
Energy	262 Kcal	Steroids	Absent
Total	fat 0 g	Phenolic content	6.48g
Total carbohydrate	64.28 g	Carotenoids	3.72g

Parameters Value in per 100g of sample			
Protein	1.22 g	**Essential Amino Acid Analysis**	
Cholesterol	0 mg	Valine	1.47
Vitamins		Leucine	4.89
Vitamin A	0 IU	Iso-Leucine	11.25
Vitamin C	6.39 mg	Threonine	2.70
Vitamin B Complex	7.28 µg	Methionine	0.48
Vitamin D	1.17 µg	Lysine	0.78
Vitamin E	0.48 µg	Phenyl Alanine	1.89
Vitamin K	1.29 µg	Histidine	16.70
Minerals		Tyrosine	1.78
Calcium	137 mg	Arginine	2.89
Copper	0.24 mg	Cysteine	2.70
Iron	9.65 mg	Tryptophan	3.67
Magnesium	4.69 mg	**Non-Essential Amino Acid Analysis**	
Manganese	1.27 mg	Alanine	1.65
Phosphorus	6.38 mg	Glycine	10.13
Potassium	175 mg	Serine	7.27
Selenium	0.03 mg	Proline	1.89
Sodium	282 mg	Aspartic acid	10.65
Zinc	1.65 mg	Glutamic acid	16.90

Phytochemicals

The leaves are reported to contain phytochemicals such as flavonoids, reducing sugars, glycosides, steroids and terpenoids that are reported to have moderate anti-oxidant, cytotoxic and anti-cancer activity (Bhandari *et al.*, 2017). The most important volatile constituents of *Allium wallichii* were found to be 1, 2 bis (methylthio) ethane; 2, 4 di-methiophene; di-methyl disulphide and di-methyl trisulphide (Kattel and Maga, 1995). Diosgenin and tigogenin present in the bulbs of *A. wallichii* are reported as steroidal sapogenins that shows encouraging pharmacological potentials (Kamal and Sharma, 1984).

Comparision of *A. stracheyi*, *A. wallichii*, *A. przewaliskianum* and *A. negianum*

Comparitive morphology for selected characters of *A. stracheyi*, *A. wallichii*, *A. przewaliskianum* and *A. negianum* is shown below

Traits	*Allium stracheyi*	*Allium wallichii*	*Allium przewaliskianum*	*Allium negianum*
Plant height (cm)	20–30	40–65	20–45	30–50
Pseudostem-Length	Developed	Not developed	Developed	Developed
Bulb if developed	Developed	Not developed	Developed	Developed
Bulb-no.	Clustered	Solitary/clustered	Clustered	Clustered
Bulb-shape	Cylindric-narrowly ovoid	Cylindric	Cylindric-narrowly ovoid	Cylindric-narrowly ovoid
Bulb-tunic colour (at maturity)	Brown-darkest brown	Yellowish brown	Red (rarely light brown)	Reddish-dark brown (rarely lighter-brown)
Tunic-type	Coriaceous	Fibrous-subreticulate	Finely reticulate	Finely reticulate
Root-type	Nonfleshy	Fleshy- nonfleshy	Nonfleshy	Nonfleshy
Leaf-no., colour	3–4; green	2–4; green	3–5; light brown-green	4–6; dark green
Leaf- shape/ type	Fistulose; linear	Flat, linear	Fistulose, semiterete - terete (4–5-angled)	Fistulose, semiterete-terete
Leaf vs. scape	Shorter	Shorter-longer	Shorter or slightly longer	Shorter
Scape	Laxy	Stout (3-angled)	Stout, terete	Stout, terete, solid
Spathe	2-valved, persistent	1–2 valved, deciduous	1–2 valved, persistent	1-valved, persistent
Umbel-shape, no.of flowers	Hemispheric, 8–10 flowered, laxy	Hemispheric, 20–30 densely flowered, laxy	Hemispheric-globose, 25–40 densely flowered	Hemispheric, 30–40 flowered, laxy
Pedicel vs. perianth	Shorter	Subequal	Subequal	2–2.5 times longer
Flower-shape	Oblong-campanulate; pouched at the base	Broadly campanulate, stellate	Campanulate, tepals wide open	Campanulate, tepal partly open
Flower- colour	Purplish tinge-dark pink-rosy-purple-yellow (white- in Kashmir samples)	Brilliant magenta-deep purple-pale red (rarely white)	Pale red-dark purple-pink	Dark purple
Tepal- size	Inner slightly bigger	Subequal	Inner slightly Bigger	Inner slightly Bigger
Tepal- shape	Broadly ob-long -ob-oblong; outer convex, apex- obtuse	Ovate-lanceolate, apex-acute	Ovate-lanceolate, apex-obtuse	Ovate-lanceolate, apex- acuminate-mucronate
Tepal-midline colour	Dark pink	Dark purple	Dark purple-light green	Light green

Traits	*Allium stracheyi*	*Allium wallichii*	*Allium przewaliskianum*	*Allium negianum*
Mid-line prominence	Very prominent	Not very prominent	Not very prominent	Not very prominent
Anther- filament	Purple, linear, free, adnate at the base, long exerted after anthesis	Shorter or subequal to perianth segments	Equal, as long as perianth segments, connate at base,adnate to perianth segments; outer ones subulate; inner ones broadened for 1/3–1/2 conate at base forming a ring	As long as perianth segments, adnate to perianth segments; outer ones subulate; inner ones broadened for 1/3–1/2 conate at base
Anther-lobe	Very much exserted; yellow, oblong	Not exserted; erect, purple, oblong	Exserted; yellow-purple, outer linear, broad at base, inner oblong	Slightly exserted; greenish yellow-purple; outer linear, broad at base; inner oblong
Stigma vs. stamen	Sub-equaling stigma	Equal to stigma	Subequal stigma	Slightly longer than stigma
Tepals after maturity	Curved inwards	Curved outwards	Curved outwards	Curved inwards; margin rolled
Leaf-aroma	Sweet onion-mild garlic	Strong garlic	Strong onion-light garlic	Strong onion-garlic
Status occurrence	Rare; threatened	Common	Common	Common
Cultivation	Occasional; in kitchen garden	Rare; in kitchen garden	Occasional; in kitchen garden	Frequently cultivated; large scale cultivation
Commercial value; area of local use	Commercialized; Uttarakhand	Commercialized; Uttarakhand	Commercialized; Ladakh, Nepal	Commercialized; Uttarakhand

Source: Anjula Pandey *et al.*, 2021a, 2022.

Other similar Himalayan species

Allium auriculatum Kunth

Subgenus: *Cyathopora.*

Section: *Coleoblastus.*

Common names: Wild lahsun, Jangli lahsun.

Propagation: Seed and bulb.

Introduction and origin

Allium auriculatum is native to the Himalayan regions of India and Nepal where it is found growing wild on rocky slopes at altitudes of 3350-5480m. In India, it is distributed in Uttarakhand (Dharchula and Munsiyari), Himachal Pradesh, and Jammu & Kashmir (Kashmir). Though the leaf of *A. auriculatum* is very distinctly flat with obtuse tip but clearly differs from *A. przewalskianum* whose leaves are needle-shaped. It is often mistaken for *A. przewalskianum* due to bulb and tunic fibres (Anjula Pandey *et al.*, 2022).

Botany

Allium auriculatum is a perennial berb growing to a height of 12-35cm, forming cylindrical, 5-8 x 1-1.5cm bulbs with fibrous *cum* reticulate tunic. Its 3-4 linear, flat and obtuse leaves are 12-20cm long and 3-5mm across arising from basal half of stem. Internodes are 1-5cm long. Flowering and fruiting occur between July and September. Several campanulate flowers emerge out in a hemispherical umbel that is around 1.5-2.5cm in diameter. Spathe is lanceolate, bifid and 1-2 x 0.5cm. Pedicels are 2-8mm long. Red or purple tepals are broadly lanceolate, 4.5-10 x 1.5-3mm, acute, with mid-vein, scarious and basally pouched. Subulate filaments are 3.5-7mm x 1mm, connate up to 1mm at base and adherent to the perianth. Ovoid or oblong anthers are 1-1.7mm long. Ovary is obovoid, 1.5-2 x 1-1.7mm, triquetrous and stipitate. The style is 2-3.5mm long. The obovoid capsules are 5 x 4-5mm with papery pericarp. Black and non-lustrous seeds are 6 in number, oblong, concavo-convex and 2.5 4 x 1-2mm. The leaves are used as seasoning spices in traditional dishes of hilly areas of Uttarakhand (Anjula Pandey *et al.*, 2022).

Chemical constituents

The whole aerial parts of *Allium auriculatum* yield 0.1% pale-yellow oil with a pleasant aroma. The GC and GC-MS (gas chromatography-mass spectrometry) analysis of the essential oil revealed the presence of 9 compounds and all of them were identified by comparing their mass spectra with NIST05 MS library. The oil is rich in phthalates contributing 79.80% while the monoterpenoids contributed 4.07%. Major constituents are bis (2-ethylhexyl) phthalate (79.8%); palmidrol (7.45%); octanamide (6.83%) and d-limonine (2.65%) whereas minor constituents isolated are (s)-(+)-6,8-p-menthadien-2-one (1.34%); o-cymene (0.96%); γ-terpinene (0.46%); 9-hexacosene (0.27%) and 2-(4-methoxy-2,6-dimethylphenyl)-3-methyl-2h-benzo[g]indazole (0.24%) (Anjula Pande *et al.*, 2015). The main constituent's bis (2-ethylhexyl) phthalate was also investigated for the first time in *Allium* species.

As per Pande *et al.* (2015), the GC-MS analysis of powder of whole dried aerial parts of *Allium auriculatum* revealed the presence of 21 volatile organic components and all of them were identified by comparing their mass spectra with ms library except one. The dominant component of the oil was found to be a monoterpene hydrocarbon. The monoterpene hydrocarbon, aldehyde, sulphur containing compounds, oxygenated monoterpenes, hytrocyclic compounds, organic acids, fatty acids, sulphur, nitrogen containing compounds and aromatic compounds amounted to 61.39%, 17.33%, 5.54%, 4.02%, 3.44%, 2.36%, 1.68%, 1.14% and 0.46% respectively (Pande *et al.*, 2015).

The major constituents are limonene (60.05%); isovaleraldehyde (17.33%); (s)-carvone (4.02%); furfural (2.67%); acetic acid (2.36%); thiophene, 2,5-dimethyl (1.88%); l-proline, n-allyloxycarbonyl-, dodecyl ester (1.68%); disulphide dipropyl (1.12%) and β-pinene (1.05%), though minor constituents found were rhodanine, 3-amino (0.95%); dimethyl trisulphide (0.94%); furfural, 5-methyl (0.77%); cis-3,5-diethyl-1,2,4-trithiolane (0.71%); hyacinthin (0.46%); sulphide, allyl methyl (0.33%); 1,3-dithiolane (0.33%); α-pinene (0.29%); 1-propene-1-methylthio (0.23%); 1-butanamine, 4-methoxy (0.15%); and an unidentified constituent (0.99%) (Pande *et al.*, 2015).

The presence of seven sulphur containing and three monotepnoid hydrocarbon compounds make it a remarkable phytofood. Isovaleraldehyde, one of the most abundant headspace components, is used in perfumes, pharmaceuticals and for flavouring (Pande *et al.*, 2015). Isovaleraldehyde reacts with alcohols, aldehydes and thiols in food so consequently forming various kinds of flavouring substances in food (Yamanishi *et al.*, 1956).

Allium carolinianum DC. in Red.

Subgenus: *Polyprason.*

Section: Falcatifolia (Anjula Pandey *et al.*, 2022).

Synonyms: *Allium aitchisonii* Boiss.; *A. obtusifolium* Klotzsch; *A. platyspathum* var. *falcatum* Regel; *A. platystylum* Regel; *A. polyphyllum* Kar. & Kir.; *A. polyphyllum* var. *nudicaule* Regel; *A. thomsonii* Baker (WFOD, 2022).

Vernacular name: In Uttarakhand: Paniol (Anjula Pandey *et al.*, 2022); in Gurung (Manang): Rotangtea; in Kham: Dangsong, Gokpa, Zimril; in Nepali: Jangali Lasun, Kage Lasun; in Tibetan: Gok Pa, Lhun Gok, Rung Mar, Ruk Nak, Ruk Pa (Bhetuwal, 2014).

Introduction and botany

FOC (2002) cites that *Allium carolinianum* ~ a Himalayan species is not from Carolina. Foliage is quite remarkable being very thick and waxy particularly when first emerging (www.pacificbulbsociety.org/pbswiki/index.php/Allium_carolinianum). Garlic-flavoured young leaves and bulbs are used to flavour food (Anjula Pandey *et al.*, 2022). It is used tradiationally by the local Amchis and Vaidyas of Dolpa and Manang districts in the tratment of stomachache, headache, diarrhoea and dysentery. The leaf and bulb are locally used for preparation of tonic that imparts vitality to the body (Ghimire *et al.*, 2000, 2001; Lama *et al.*, 2001; Bhattarai *et al.*, 2006).

Allium carolinianum

Photo courtesy from http://www.fotomontaro.com/flora/allia/allium_carolinianum04.shtml. via Bhetuwal, 2014

Traits	***Description (as per Anjula Pandey** et al., **2022, and FOC, 2002)***
Chromosome number	2n = 32
Habitat; altitudes	Commonly grows in inner high ranges on rocky gravelly habitat, crevices, near streams (3000-5000m)
Distribution	Native to central and southern Asia [Xinjiang, Xizang (Tibet), Afghanistan, Bhutan, India, Kazakhstan, Kyrgyzstan, Nepal, Pakistan, Tajikistan, Uzbekistan] (en.wikipedia.org/wiki/Allium_carolinianum)
Cultivation/wild	Wild; cultivated (western Himalaya)
Propagation	Seed and bulb
Plant height (cm)	14-52
Bulbs	Usually paired, cylindrical to ovoid, 1-2.5cm in diameter
Tunic	Outer tunic coriaceous, dull brown, brown to yellowish brown, inner tunic membranous

Traits	*Description (as per Anjula Pandey et al., 2022, and FOC, 2002)*
Leaves	5-6, shorter than scape, broadly linear to lanceolate, usually falcate, flat, smooth, apex obtuse, (3-)5-15mm wide, blunt and glaucous (Bhetuwal, 2014)
Scape	20-40(-60)cm, 15-50cm, stout, terete, base covered with leaf bases
Spathe	2-valved, persistent
Umbel	Densely flowered, 2-4cm diameter across, globose, ovate, obtuse and deltoid
Flowering time	June-September
Pedicel	Subequal, slightly shorter than perianth, bracteolate or ebracteolate
Tepal	Tepals dark red-pink to rose, lanceolate, 6-7mm long, exserted
Filament	Longer than tepals, no teeth, connate at the base
Ovary	Subglobose, globose, trigonous with concave nectaries at base
Style	Exserted
Stigma	Capitate
Capsule	Loculicidal

Allium carolinianum

Paired bulbs	Thick emerging leaves	Inflorescences
By Anjula Pandey *et al.*, 2022	By Mark McDonough	By Sherpaworld - Own work, CC BY-SA 3.0, https://commons.wikimedia.org/w/index.php?.curid=33060941

Allium consanguineum Kunth

Subgenus: *Polyprason.*

Section: Orioprason.

Introduction and botany

Traits	*A. consanguineum Kunth*
Chromosome number	2n=16
Habitat; altitudes	Near moraine, sandy soil along river in open fields or in rock crevices (2440-3050m elevations).
World distribution	India, Pakistan
Distribution India	Western Himalaya (Jammu & Kashmir, and Himachal Pradesh)
Cultivation/wild	Wild
Propagation	Seed and bulb
Closest identical relative	Often confused with *A. stracheyi* but for flower colour and plant habit
Plant height (cm)	35-70
Pseudostem-length	Developed
Bulbs	Cylindrical to ovoid on creeping rootstock; outer tunic fibrous, coriaceous, brown, striate
Roots	Creeping rootstock
Leaf	3-5, linear to lanceolate, falcate, flattened, not fistular, arising 2-7cm above the base, apex obtuse; leaf bases sheathing, glabrous; sheath about 1/4 the length of the scape.
Scape	Stout, terete
Spathe	2-valved, elongated-ovate
Pedicel *vs.* perigone	Equal or longer than flowers
Umbel	Hemispherical, densely flowered and often congested, 14-42 flowers
Flowering time	June –September
Pedicel vs. perianth	Equal or longer than flowers
Flower colour	Greenish/yellowish/pinkish-white
Tepal	Ovate, oblong, acute to obtuse, margin inrolled, pouched at base, no teeth, golden yellow or pink
Tepal mid-vein	Prominent
Anther	Oblong
Filament	Exserted, simple filiform, adherent and adnate to tepals at base, entire, linear with broad bases, free, long exserted
Ovary	Globose-subglobose to obvoid, trigonos, trilobed, sulcate at base, style as long as or longer than the filaments, slender, exsrted
Capsule	Light brownish-yellow, globose, loculicidal
Seed no./locule	2
Odour when crushed	Sweet onion-mild garlic
Uses	Seasoning spice
Source: *Anjula Pandey et al., 2021a, 2022.*	

Allium fasciculatum Rendle

Subgenus: *Amerallium.*

Section: *Bromatorrhiza* (Wikispecies, 2023).

Vernacular names: Sikkimese: Shanu Dungdung; Gurung (Manang): Nosyante; Nepali: Lasunesaag; Dolpa Nepali: Faran; Tibetan and Kham: Gokpa, Lhungok, Rig ok, Rung kar, Rukkar (Bhetuwal, 2014).

Synonyms: *Allium gageanum* W.W. Sm. (Tropicos.org, 2023).

Chromosome number: 2n=16.

Introduction and botany

Allium fasciculatum, commonly called as Shanu Dungdung in Sikkim, grows at elevations of 2200–5400 m in the Himalayas of India (Sikkim and Nagaland), Bhutan, Nepal, Tibet, and the Chinese provinces of Qinghai and Sichuan (FOC, 2023b; Anjula Pandey *et al.,* 2022; Hara *et al.,* 1978; Karthikeyan *et al.,* 1989; Noltie, 1994). First report on cultivation was published by Anjula Pandey *et al.,* 2022. The species resembles much with *A. hookeri*, distinguished by characteristically fasciculated roots and tepals (coiled base at maturity). The plant is propagated by bulbs. *Allium fasciculatum* is used as vegetable, in soup, salad and for medicinal purpose.

Allium fasciculatum	***Allium fasciculatum***
Photo courtesy from http://www.efloras.org/object_page.aspx?object_id=87644&flora_id=800 via Bhetuwal (2014)	**By Sherpaworld - Own work, CC BY-SA 3.0, https://commons.wikimedia.org/w/index.php?curid=33060940.**

Allium fasciculatum an annual herb growing up to 40cm tall with tuberous, short but thick roots, and fibrous, solitary or clustered, cylindric bulbs (Anjula Pandey *et al.,* 2022; FOC, 2023b). Tunic is pale brown and fibrous. Linear leaves are 2-5mm wide with obscure midvein, usually longer than lateral scapes that are up to 40cm long, terete, and covered with leaf sheaths for 1/4-2/5 its length. Persistent or deciduous spathe is 1- or 2-valved. Umbels are spherical with large numbers of white to greenish and sweet-scented flowers which continue appearing from July to September. Its fruit is an obovoid, trilobed, light brownish-yellow capsule having papery pericarp and 2-seeded locules. Seeds are ovoid-obovoid, concavo-convex and reticulated on margins (Anjula Pandey *et al.*, 2022; FOC, 2023b).

Allium humile Kunth

Introduction and botany

Major characteristics of this species are tabulated below:

Characteristics	***Allium humile* Kunth**
Subgenus	Reticulatobulbosa
Section	Reticulatobulbosa
Synonyms	*Allium govanianum* Wall. ex Baker, *A. gowanianum* Wall. ex Baker, *A. nivale* Jacquem. ex Regel, *A. nivale* Jacquem. ex Hook. f. & Thomson
Common names	Small alpine onion, Ornamental allium, Dhunar, Dhungr/dhungar, Jambu, Jimbu, Pharan, Koch, Ladum, Laiku, Dodoo and Doodli
Habitat	Along running water, moist meadows, hilly slopes or rocky areas
Altitude	3500-4500m
World distribution	India, Pakistan
Distribution India	Himachal Pradesh, Jammu & Kashmir and Uttarakhand
Status occurrence	Rare
Cultivated/Wild	Wild
Propagation	Seed and bulb; multiplies by extending horizontally in centrifugal pattern
Plant height(cm)	8-30
Bulb-no.	Solitary/clustered
Bulb-shape	Cylindrical
Bulb-tunic colour (at maturity)	Brown
Tunic-type	Reticulate, fibrous
Leaf-no., colour	4-6; green
Leaf-shape/type	Linear, flat, obtuse, glabrous
Spathe	Bifid, ovate or lanceolate
Umbel-shape, no.of flowers	Hemispherical, 9-20 flowered
Flowering	May-July
Flower-shape	Broadly campanulate, stellate
Flower-colour	White or yellowish
Tepal- size	5-9 x 1.5-3mm, acute, often margins inrolled
Tepal-shape	Elliptic or lanccolate
Tepal midline colour	Yellow-green
Mid-line prominence	Finely nerved
Anther-filament	Subulate, 1.5-3.5 x 0.5-1mm, unequal, connate up to 1mm at base forming a cup and adnate to tepals, no teeth

Characteristics	***Allium humile* Kunth**
Anther-lobe	Oblong
Tepals after maturity	Margins inrolled
Ovary	Sub-globose, 1-1.5 x 0.7-2mm, lobes inconspicuous
Style	Filiform, 1-1.2mm long
Capsules	Sub-globose or broadly obovoid, 3 x 3-5mm
Locules	1- seeded
Seed	Oblong, 3 x 2mm, brownish black
Leaf-aroma	Odour of garlic
Source: *Anjula Pandey et al., 2021a, 2022.*	

Allium oreoprasum Schrenk

Vernacular name: Gurung (Manang): Lungho; Nepali: Jimbu (Bhetuwal, 2014).

Chromosome number: 2n=16 (Ogura *et al.*, 2007).

Introduction and botany

The native range of *Allium oreoprasum* is from C. Asia to W. Tibet (POWO, 2023) where it can be found growing at elevations of 1200–2700m (en.wikipedia.org/wiki/Allium_oreoprasum).

Allium oreoprasum produces narrowly ovoid, cylindrical and clustered bulbs that are 0.5-1cm in diameter. Pseudostem tunic is yellowish-brown and reticulate. Narrowly linear leaves are 1-3mm wide and shorter than scape that is 11-30(40) cm tall and covered with leaf sheaths only at base. Spathe is 1- or 2-valved and persistent. Umbel is fascicled to hemispheric and few flowered (Bhetuwal, 2014). Tepals are white or pale red with a dark purple midvein (FOC, 2022). The plant is edible. Its paste is used to treat cough and cold headache, tonsillitis, stomachache, intestinal worms and high altitude sickness (Bhattarai *et al.*, 2006).

Allium oreoprasum

Photo courtesy from http://allium.wikidot.com/allium-o

Allium pratti C.H.Wright (Pratt onion)

Subgenus: *Anguinum* (wikimedia.org, 2023).

Synonyms: *Allium cannifolium* H.Lev., *A.* ellipticum Wall., *A. prattii* var. *ellipticum* F.T.Wang & Tang, *A. prattii* var. *vinicolor* F.T.Wang & Tang, *A. victorialis* var. *angustifolium* Hook.f. (WFO, 2023).

Vernacular names: Sherpa (Gorkha): Ruk pa; Nepali: Ban lasun (Bhetuwal, 2014).

Chromosome number: 2n=16, 32 (FOC, 2023c).

Introduction and botany

Allium prattii is a wild Asian *Allium* found at elevations of 2000–5000m in shady areas, under thickets, at damp places and at sloppy and meadowlands in the Himalayan regions of NE India (Assam, Sikkim), Nepal, Bhutan, Tibet and central and southern China (Bajracharya *et al.*, 2013; TemperatePlantsDatabase, 2023).

The plant is harvested from the wild for local use as a food and as a medicine. It is sometimes sold as a food in local markets often in the form of the dried and powdered leaves (TemperatePlantsDatabase, 2023).

Allium prattii is a perennial herb growing to a height of up to 60cm producing clustered or sometimes solitary and sub-cylindric bulb; a scape that is round in cross-section; and 2 (rarely 3) flat, sub-opposite, linear-lanceolate or very narrowly elliptic leaves that are 0.5-7cm broad and slightly shorter than the scape (Bajracharya *et al.*, 2013; TemperatePlantsDatabase, 2023; Bhetuwal, 2014; en.wikipedia.org/wiki/Allium_prattii; FOC, 2023c). Flowering and fruiting occur from June to September; flowers being several, red or purple and arranged in a hemispheric umbel.

Allium prattii

Photo courtesy from Ms. Prabina Rana, Nepal Academy of Science and Technology (Bhetuwal, 2014)

Allium roylei Stearn

Introduction and botany

The plant is harvested from the wild for local use as a food and medicine. The plant has been used in breeding programmes to confer downy mildew resistance to the onion (*Allium cepa*) and is known to be partially resistant to onion leaf blight.

Major characteristics of this species are tabulated below:

Characteristics	*Allium roylei Stearn*
Subgenus	Polyprason
Section	Oreiorason
Chromosome number	2n=16
Synonyms	*Allium lilacinum* Royle ex Regel
Common names	Bazun (Kashmir)
Habitat	On moist rocky slopes (2000-3200m)
World distribution	India, Afghanistan, Pakistan
Distribution India	Western Himalaya (Jammu & Kashmir)
Cultivated/Wild	Wild and cultivated
Cultivation	Cultivated in kitchen garden
Propagation	Seed and bulb
Closest identical relative	Close relative of cultivated onion; crossable
Plant height (cm)	22-40
Bulb	Ovoid, 2-3cm long, outer tunic membranous and reddish-brown
Stem	Stem 3-4mm diameter
Leaves	Basally and 1-3, linear, 4-20 x 0.1-0.2cm, grooved (not fistular)
Umbel	Hemispherical, 2-3cm across, 15- 40 flowered.
Spathe	Bifid
Fowers	Campanulate; pedicels 7-12mm long; bracteates
Flowering	July-September
Tepals	Oval lanceolate, 6-8mm long, reddish to pale-red, mid-vein present, acute
Stamens and pistil	Exserted, anthers oblong, filaments longer than the tepals, inner with teeth at the base on either side
Anthers	Oblong
Filaments	Longer than the tepals
Ovary	Globose-ovoid, trilobed
Style	Filiform, exserted

Characteristics	*Allium roylei Stearn*
Capsules	Oval-globose, deeply trilobed, pericarp thin and papery, light yellowish brown
Seed	brownish black, orbicular, concavo-convex, and wrinkled on convex surface
Leaf-aroma	Like onion
Uses	Used as seasoning spices in Kishtwar region (Jammu & Kashmir)

	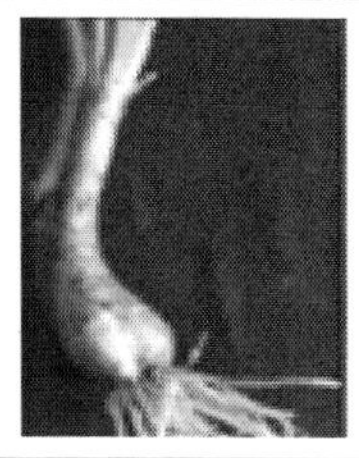			
Field view	**Bulb and root**	**Inflorescences**		**Fully opened inflorescence**

Source: *Anjula Pandey et al., 2021a, 2022.*

References

Abuajah, C.I., A.C. Ogbonna and C.M. Osuji, 2015. Functional components and medicinal properties of food: a review. *J. Food Sci. Tech.,* **52**(5):2522-9. doi: 10.1007/s13197-014-1396-5. //updatepublishing.com/journal/index.php/rip/article/view/5486/4895.

Agoramoorthy, G., M. Chandrasekaran, V. Venkatesalu and M.J. Hsu, 2007. Antibacterial and antifungal activities of fatty acid methyl ester of the blind-your-eye mangrove from India. Braz. *J. Microbiol.,* **38**(4):739-42. doi: 10.1590/S1517-83822007000400028. //updatepublishing.com/journal/index.php/rip/article/view/5486/4895.

Alamgir, A.N.M. 2017. Origin, definition, scope and area, subject matter, importance, and history of development of pharmacognosy. https://www.sciencedirect.com/science/article/pii/S2667142522000811. doi:10.1007/978-3-319-63862-1_2.

Anjula Pandey, R. Pandey, K.S. Negi and J. Radhamani, 2008. Realizing value of genetic resources of Allium in India. *Gen. Res. and Crop Evol.,* **55**(7): 985-94.

Anjula Pandey, K. Madhav Rai, Pavan Kumar Malav and S. Rajkumar, 2021a. *Allium negianum* (Amaryllidaceae): a new species under subg. Rhizirideum from Uttarakhand Himalaya, India. *PhytoKeys,* **183**: 77-93. https://phytokeys.pensoft.net/issue/3467/.

Anjula Pandey, P.K. Malav, K.M. Rai and S.P. Ahlawat, 2021b. Neodomesticates of the Himalayan allium spices (*Allium species*) in Uttarakhand, India and studies on eco-geography and morphology. *Gen. Res.* and *Crop Evol.*, **68**. 10.1007/s10722-021-01164-x. researchgate.net/publication/350708665.

Anjula Pandey, Pavan Kumar Malav, K.M. Rai and S.P. Ahlawat, 2022. Genus *Allium* L. of the Indian Region: A Field Guide for Germplasm Collection and Identification. ICAR-National Bureau of Plant Genetic Resources, New Delhi, India, 109p + i-vi. http://www.nbpgr.ernet.in/Downloadfile.aspx?EntryId=9197.

Arshad, M.S., M. Sohaib, M. Nadeem, F. Saeed, A. Imran, A. Javed, Z. Amjad and S.M. Batoo, 2017. Status and trends of nutraceuticals from onion and onion by-products: A critical review. *Cogent. Food Agric.*, 3. Article 1280254, 10.1080/23311932.2017.1280254. https://www.sciencedirect.com/science/article/pii/S2667142522000811.

Bajracharya, G.B., R. Bhetuwal Bishwa and K.C. Rajendra, 2013. An Unexplored, Wild and Consumable Vegetable *Allium prattii*: Nutritional Value, Cytotoxicity and Antibacterial Potentiality. *J. Nepal Chem. Soc.*, **31**: 1-8.

Bhandari, J., B. Muhammad, P. Thapa and B.G. Shrestha, 2017. Study of phytochemical, anti-microbial, anti-oxidant, and anti-cancer properties of *Allium wallichii*. BMC *Complement Altern. Med.*, **17**(1):102. doi: 10.1186/s12906-017-1622-6. PMID: 28178952; PMCID: PMC5299666. https://www.ncbi.nlm.nih.gov/pmc/articles/PMC5299666/.

Bhattarai, S., R.P. Chaudhary and R.S.L. Taylor, 2006. Ethnomedicinal plants used by the people of Manang District, Central Nepal. *J. Ethnobiol. Ethnomed.*, **2**:41 (retrieved from DOI:10.1186/1746-4269-2-41). www.researchgate.net/profile/Bishwa-Bhetuwal/publication/355049736_Nutritional_Value_Cytotoxicity_Antibacterial_Potentiality_and_Phytochemicals_of_Allium_Pratii_C_H_Wright_Ex-Hemsley/links/ 615b90b749da3d5b113a0230/Nutritional-Value-Cytotoxicity-Antibacterial-Potentiality-and-Phytochemicals-of-Allium-Pratii-C-H-Wright-Ex-Hemsley.pdf.

Bhetuwal, B.R. 2014. Nutritional Value, Cytotoxicity, Antibacterial Potentiality and Phytochemicals of *Allium Prattii* C. H. Wright Ex Hemsley. M.Sc. Dissertation submitted as a partial fulfillment of the requirement for the Master of Science degree in chemistry. Department of Chemistry, Tri-Chandra Multiple Campus Institute of Science and Technology, Tribhuvan University, Ghantaghar, Kathmandu, Nepal. www.researchgate.net/profile/Bishwa-Bhetuwal/publication/355049736_Nutritional_Value

Carballo, A.I., A.L. Martinez, M.E. Gonzalez-Trujano, F. Pellicer, R. Ventura-Martinez, M.I. Diaz-Reval and F.J. Lopez-Munoz, 2010. Antinociceptive activity of *Annona diversifolia* Saff. leaf extracts and palmitone as a bioactive compound. *Pharmacol. Biochem.* Behav., **95**(1):6-12. doi: 10.1016/j.pbb.2009.11.017. //updatepublishing.com/journal/index.php/rip/article/view/5486/4895.

Chandrasekaran, R., Kalaivani Thiagarajan, Amrita Mukherjee, Mariadoss Alphonse,

Kanagam Nachiappan, Parthibanraja, A., Kusum Chauhan, Laxman Singh Kandari, Lakhpat Singh Rawat and Rakesh Kumar Maikhuri, 2020. Large scale cultivation and commercialization opportunities and constraints of *Allium stracheyi* Baker – an endangered medicinal plant of Western Himalaya, India. *Eco. Env. & Cons.*, **26**: S26-S32. http://www.envirobiotechjournals.com/EEC/Vol26OctSuppl20/EEC-5.pdf.

Chhetri, D.B. 1999. Diversity of Medicinal and Aromatic Plants in Manang, Central Nepal with Emphasis on Ecology and Essential oil Variation of Jatamashi (*Natdostachys grandiflora* DC). M. Sc. Thesis. Central Department of Botany, Tribhuvan University, Kathmandu, Nepal. www.researchgate.net/profile/Bishwa-Bhetuwal/publication/355049736_Nutritional_Value_Cytotoxicity_Antibacterial_Potentiality_and_Phytochemicals_of_Allium_Pratii_C_H_Wright_Ex-Hemsley/links/615b90b749da3d5b113a0230/Nutritional-Value-Cytotoxicity-Antibacterial-Potentiality-and-Phytochemicals-of-Allium-Pratii-C-H-Wright-Ex-Hemsley.pdf.

Cundell, D. 2014. Herbal phytochemicals as immunomodulators. *Curr. Immunol.* Rev., **10**: 64-81. 10.2174/1573395510666140915213156.https://www.sciencedirect.com/science/article/pii/S266714252200081.

Dasgupta, S. 2006. Fascicles of Flora of India (Fascicle 23) (ed. Singh, N.P. and M. Sanjappa). Botanical Survey of India, pp. 1-48. https://updatepublishing.com/journal/index.php/rip/article/view/5486/4895

Dobhal, R. 2015. *Allium stracheyi* Baker. In: Spices of Uttarakhand. Uttarakhand St. Council for Science & Technology, Uttarakhand, pp. 45. https://updatepublishing.com/journal/index.php/rip/article/view/5486/4895

Ebadi, P., M.H. Karimi and Z. Amirghofran, 2014. Plant components for immune modulation targeting dendritic cells: implication for therapy. *Immunotherapy,* 6: 1037-53. 10.2217/imt.14.77. https://www.sciencedirect.com/science/article/pii/S2667142522000811.

en.wikipedia.org/wiki/Allium_carolinianum.

en.wikipedia.org/wiki/Allium_oreoprasum

en.wikipedia.org/wiki/Allium_prattii

en.wikipedia.org/wiki/Allium_przewalskianum.

Etewa, S.E. and S.M. Abaza, 2011. Herbal medicine and parasitic diseases. *Herb. Med. Parasites,* **4**: 3-14. https://www.sciencedirect.com/science/article/pii/S2667142522000811.

Farooqui, T. and A.A. Farooqui, 2017. Neuroprotective Effects of Garlic in Model Systems of Neurodegenerative Diseases. Elsevier Inc., Holland. 10.1016/B978-0-12-811959-4.00016-X. https://www.sciencedirect.com/science/article/pii/S2667142522000811.

FOC (Flora of China), 2002. *Allium carolinianum* DC. @ efloras.org" eFlora. Missouri Botanical Garden, St. Louis, MO & Harvard University Herbaria, Cambridge, MA., 2002. Web. Accessed February 2018. http://www.worldfloraonline.org/taxon/wfo-0000755908.

FOC (Flora of China), 2022. Vol. 24, page 180. *Allium oreoprasum* Schrenk. http://www.efloras.org/florataxon.aspx?flora_id=2&taxon_id=200027509. Retrieved 31 December, 2022.

FOC (Flora of China), 2023a. Vol. 24, 183p. *Allium przewalskianum*. http://www.efloras.org/florataxon.aspx?flora_id=2&taxon_id=200027523. https://en.wikipedia.org/wiki/Allium_carolinianum.

FOC (Flora of China), 2023b. Vol 24, 175p. cu gen jiu *Allium fasciculatum* 16 Feb 2023 <http://www.efloras.org/florataxon.aspx?flora_id=2&taxon_id=200027475>

FOC, 2023c. *Allium prattii* C. H. Wright ex Hemsley, tai bai shan cong. Flora of China, Vol. 24, p. 174. http://www.efloras.org/florataxon.aspx?flora_id=2&taxon_id=200027521. Retrieved on Retrieved 20 February 2023.

FOI (Flowers of India), 2015. Ladakh Onion, *Allium przewalskianum.* Archived from the original on 2015-05-31. Retrieved 2015-04-14. http://flowersofindia.net/catalog/slides/Ladakh%20Onion.html.

Gautam, S. S., Prashant Singh, Manindra Mohan and S. Zafar Haider, 2019. *Allium stracheyi* (Baker), the endangered and traditional medicinally important herb of Uttarakhand Himalaya, India: a review. *Research in Pharmacy*, 9:01-05. https://updatepublishing.com/journal/index.php/rip/article/view/5486/4895. doi:10.25081/rip.2019.v9.5486.

Ghimire, S.K., Y.C. Lama, T.N. Gurung, and Y. Aumeeruddy-Thomas, 2000. Conservation of Plant Resources, Community Development and Training in Applied Ethnobotany at Shey-Phoksundo National Park and its Buffer Zone, Dolpa, Kathmandu, WWF Nepal. Third Year. Report Series No. 40. www.researchgate.net/profile/Bishwa-Bhetuwal/publication/355049736_Nutritional_Value_Cytotoxicity. _Antibacterial_Potentiality_and_Phytochemicals_of_Allium_Pratii_C_H_Wright_Ex-Hemsley/links/ 615b90b749da3d5b113a0230/Nutritional-Value-Cytotoxicity-Antibacterial-Potentiality-and-Phytochemicals-of-Allium-Pratii-C-H-Wright-Ex-Hemsley.pdf.

Ghimire, S.K., Y.C. Lama, G.R. Tripathi, S. Schmitt and T.A. Thomas, 2001. Conservation of Plant Resources, Community Development and Training in Applied Ethnobotany at Shey-Phoksundo National Park and its Buffer Zone, Dolpa. *Report Series No. 41*. WWF Kathmandu, Nepal. www.researchgate.net/profile/Bishwa-Bhetuwal/publication/355049736_Nutritional_Value_Cytotoxicity _Antibacterial_Potentiality_and_Phytochemicals_of_Allium_Pratii_C_H_Wright_Ex-Hemsley/links/ 615b90b749da3d5b113a0230/Nutritional-Value-Cytotoxicity-Antibacterial-Potentiality-and-Phytochemicals-of-Allium-Pratii-C-H-Wright-Ex-Hemsley.pdf.

Gonzalez-Trujano, M.E., A.L. Martinez, A. Reyes-Romirez, B. Reyes-Trejo and A. Navarrete, 2006. Palmitone isolated from *Annona diversifolia* induces an anxiolytic-like effect in mice. *Planta Med.*, **72**(8):703-7. doi: 10.1055/s-2006-931598. //updatepublishing.com/journal/index.php/rip/article/view/5486/4895.

Gregrova, A., H. Cizkova, I. Bulantova, A. Rajchl and M. Voldrich, 2013. Characteristics of garlic of the Czech origin. *Czech J. Food Sci.*, **31**: 581-8. 10.17221/539/2012-cjfs. https://www.sciencedirect.com/science/article/pii/S2667142522000811.

Hanelt, P. 2001. Alliaceae. In: Mansfeld's Encyclopedia of Agriculture and Horticultural Crops (ed. Hanelt, P.), vol. 4(3); 2001. Springer-Verlag, Vienna. ISBN: 3-540-41017-1. https://updatepublishing.com/journal/index.php/rip/article/view/5486/4895

Hara, H., W.T. Stearn and H.J. Williams, 1978. *An Enumeration of the Flowering Plants of Nepal,* **1**: 1-154. Trustees of British Museum, London. Retrieved from https://en.wikipedia.org/wiki/Allium_fasciculatum.

Helen O.N., A.O. Job and O. Oghenebrorhie, 2020. Antibiotic and antihelminthic effects of garlic bulbs (*Allium sativum*) and ginger rhizomes (*Zingiber officinale*) as growth promoters on broiler chickens in the tropics. *Adv. Anim. Verterin. Sci.*, **8**: 1019-27. https://www.sciencedirect.com/science/article/pii/S2667142522000811.

HMG. 1982. Wild edible plants of Nepal. Department of Medicinal Plants. His Majesty's Government of Nepal, Kathmandu, Nepal. http://www.forestrynepal.org/images/thesis/MSc_RCNepal.pdf.

IUCN, 2000. National Register of Medicinal Plants. Kathmandu, Nepal. https://en.wikipedia.org/wiki/Jimbu#cite_ref-auto1_4-0.

IUCN, 2004. National Register of Medicinal and Aromatic Plants. IUCN - The World Conservation Union, Kathmandu, Nepal. http://www.forestrynepal.org/images/thesis/MSc_RCNepal.pdf.

Joshi, S. 2016. Studies on supercritical fluid extraction of organic compounds from *Allium stracheyi* Baker. Ph.D. Thesis, Aligarh Muslim University, Aligarh, India. https://updatepublishing.com/journal/index.php/rip/article/view/5486/4895

Joshi, A.R. and J.M. Edington, 1990. The use of medicinal plants by two village communities in the central development region of Nepal. *Econ.* Bot., **44**:71–83.

Kamal, R. and G.L. Sharma, 1984. Steroidal sapogenins from bulbs of *Allium wallichii* Kunth. *Pharmazie,* **39**(10):707–8. https://www.ncbi.nlm.nih.gov/pmc/articles/PMC5299666.

Karthikeyan, S., S.K. Jain, M.P. Nayar and M. Sanjappa, 1989. Florae Indicae Enumeratio: Monocotyledonae: pp. 1-435. Botanical Survey of India, Calcutta. Retrieved from https://en.wikipedia.org/wiki/Allium_fasciculatum.

Kattel, A. and J.A. Maga, 1995.Volatile compounds from dried Jimbu (*Allium wallichii*). *Dev. Food Sci.*, **37**:919–28. doi: 10.1016/S0167-4501(06)80207-4. https://www.ncbi.nlm.nih.gov/pmc/articles/PMC5299666.

Kim, S.J., W.S. Chung, S.S. Kim, S.G. Ko, and J.Y. Um, 2011. Antiinflammatory effect of *ldenlandia diffusa* and its constituent, hentriacontane, through suppression of caspase-1 activation in mouse peritoneal macrophages. *Phytother. Res.*, **25**(11):1537-46. doi: 10.1002/ptr.3443. //updatepublishing.com/journal/index.php/rip/article/view/5486/4895.

Lama, Y.C., S.K. Ghimere and Y. Aumeeruddy-Thomas, 2001. Medicinal Plants in Dolpa: Amchis Knowledge and Conservation. WWF Nepal Program and People and Plant Initiative, Kathmandu, Nepal. www.researchgate.net/profile/Bishwa-Bhetuwal/publication/355049736_Nutritional_Value_Cytotoxicity_Antibacterial_Potentiality_and_Phytochemicals_of_Allium_Pratii_C_H_Wright_Ex-Hemsley/links/ 615b90b749da3d5b113a0230/Nutritional-Value-Cytotoxicity-Antibacterial-Potentiality-and-Phytochemicals-of-Allium-Pratii-C-H-Wright-Ex-Hemsley.pdf.

Nepal, R.C. 2006. Status, Use and Management of Jimbu (*Allium* spp.): A Case Study from Upper Mustang, Nepal. M.Sc. thesis submitted in partial fulfilment of the requirements for the award of the degree of Master of Science in Tropical Ecology and Management of Natural Resources at the Norwegian University of Life Sciences. http://www.forestrynepal.org/images/thesis/MSc_RCNepal.pdf.

Noltie, H.J. 1994. Flora of Bhutan, **3**(1): 1-456. Royal Botanic Gardens, Edinburgh. Retrieved from https://en.wikipedia.org/wiki/Allium_fasciculatum.

Ogura, H., Kondo Katsuhiko, Tan Dun-yan, Mao Jian-feng, Funamoto Tsuneo, Guo Shiguang, Zhang Daming and Hong De-yuan, 2007. A Chromosome Study on Seven Wild *Allium* Species in the Tianshan Mountains, Xinjiang, China. J. Japan. Bot., **82**: 34–40. http://www.jjbotany.com/pdf/JJB_082_034_040.pdf.

Pacificbulbsociety.org/pbswiki/index.php/Allium_carolinianum

Pande, K.K., Lata Pande, Anshulika Upadhyaya, Bharat Pande and Atul Pujari, 2015. Identification and comparison of volatile constituents of Himalayan phytofood; *Allium auriculatum* Kunth and its antimicrobial activity. Intern. *J. Life Sci.*, **4**:3: 168-72. http://www.crdeepjournal.org/wp-content/uploads/2015/07/Vol-4-3-3-IJLS.pdf.

POWO (Plants of the World Online), 2023. *Allium oreoprasum* Schrenk. Royal Botanic Gardens, Kew. https://powo.science.kew.org/taxon/urn:lsid:ipni.org:names:528542-1. Retrieved 16 February, 2023.

Ranjan, S., V.S. Jadon and S. Gupta, 2014. Studies on the anti-inflammatory and analgesic activities of *Allium stracheyi*, a medicinal plant of alpine Himalayas of Uttarakhand. *BDDD*, **1**:27-34. https://www.sciencedirect.com/science/article/pii/S2667142522000811.

Rawat, G.S., B.S. Adhikari, U.K. Tiwari, S. Chandola and N. Raut, 2016. Medicinal plants of Garhwal region Uttarakhand: a baseline on the status and distribution. Wildlife Institute of India and Uttarakhand Forest Development Corporation, Dehradun, India. https://updatepublishing.com/journal/index.php/rip/article/view/5486/4895.

Regel, Eduard August von, 1875. *Trudy Imperatorskago S.-Peterburgskago Botaničeskago Sada*, **3**(2): 164 (in Latin). https://en.wikipedia.org/wiki/Allium_przewalskianum.

Shah, N.C. 2014. Status of cultivated & wild Allium species in India: a review. *The Scitech.*, **1**(9):28-36. https://updatepublishing.com/journal/index.php/rip/article/view/5486/4895

Shrestha, K.K., J.P. Sah and S.K. Ghimere, 1995. Ecology Exploitation Trend and Conservation of Potential High Altitude Medicinal Plants in Gyasumdo Valley, Manang. Annapurna Conservation Area Project (ACAP), King Mahendra Trust for Nature Conservation (KMTNC), Kathmandu, Nepal. Bhetuwal/publication/355049736_Nutritional_Value_Cytotoxicity_Antibacterial_Potentiality_and_Phytochemicals_of_Allium_Pratii_C_H_Wright_Ex-Hemsley/links/615b90b749da3d5b113a0230/Nutritional-Value-Cytotoxicity-Antibacterial-Potentiality-and-Phytochemicals-of-Allium-Pratii-C-H-Wright-Ex-Hemsley.pdf. www.researchgate.net/profile/Bishwa-

Shrestha, T.B. and R. M. Joshi, 1996. Rare, endemic and endangered plants of Nepal. WWF Nepal Program, Kathmandu. http://www.forestrynepal.org/images/thesis/MSc_RCNepal.pdf.

Stearn, W.T. 1960. *Allium* and *Milula* in the Central and Eastern Himalaya. *Bull. Brit. Mus.* (Natural History), **2**:159-91. http://www.forestrynepal.org/images/thesis/MSc_RCNepal.pdf.

TemperatePlantsDatabase, 2023. Temperate Plants Database, Ken Fern. temperate.theferns.info. 2023-02-21. <temperate.theferns.info/plant/Allium+prattii>

Tiwari, U.K., S.J. Adams, S.N. Begum, K.V. Krishnamurthy, K. Ravikumar and V. Padma, 2014. Studies on two Himalayan species of traditional medicinal value: *Allium wallichii* and *Allium stracheyi*. *Not Sci Biol.*, **6**(2):149-54. https://updatepublishing.com/journal/index.php/rip/article/view/5486/4895.

Tropicos.org. Missouri Botanical Garden. 16 Feb 2023. http://www.tropicos.org/Name/18406618

WFO (The World flora Online), 2023. *Allium prattii* C.H.Wright. http://www.worldfloraonline.org/taxon/wfo-0000757055. Retrieved 20 February 2023.

WFOD (World Flora Online data), 2022. http://www.worldfloraonline.org/organisation/Amaryllidaceae; http://www.worldfloraonline.org/taxon/wfo-0000755908.

wikimedia.org, 2023. https://species.wikimedia.org/wiki/Allium_prattii. Retrieved

20 February 2023.

Wikispecies, 2023. https://species.wikimedia.org/wiki/Allium_fasciculatum. Retrieved on 12,February, 2023.

Yamanishi, T., Y. Mikumo and Y. Obata, 1956. Synthetic studies on aroma of foods concerning isovaleraldehyde. *Natural science report of the Ochanomizu University*, **7**(1):101-10. http://www.crdeepjournal.org/wp-content/uploads/2015/07/Vol-4-3-3-IJLS.pdf

11

Other Wild Alliums

Chinese onion (*Allium chinense* G.Don.)

Subgenus: *Cepa.*

Section: *Sacculiferum* (Pandey *et al.,* 2022; Wikispecies, 2023).

Other common names: Chinese onion, Chinese scallion, Japanese scallion (Anjula Pandey *et al.,* 2022; FOI, 2023); in Japanese: Rakkyo; Hindi: Bada lahsun, Chhota lahsun (Anjula Pandey *et al.,* 2022).

Synonyms: *Allium bakeri* Regel; *A. bodinieri* H.Lev. & Vaniot; *A. exsertum* Baker 1874, illegitimate homonym not G. Don 1827; *A. martini* H.Lev. & Vaniot; *A. splendens* Miq. 1867, illegitimate homonym not Willd. 1830; *Caloscordum exsertum* Herb. (POWO, 2023)

Introduction and origin

Chinese onion, distantly related to common onion (Anjula Pandey *et al.,* 2022), has native range from S. China to Hainan (POWO, 2023) and naturalized and cultivated in Japan, Korea, Malaysia, N&S America, Cuba, and India (Anjula Pandey *et al.,* 2022; USDA, 2023). In India, it grows at altitudes of 1200-2000m in Uttarakhand and NEH regions, especially Meghalaya, Arunachal Pradesh and Nagaland where it is cultivated frequently as vegetable. Its leaves and bulbs are harvested and sold in the market of NE Himalayan region and the surplus one is stored for next season though it is rarely cultivated in Uttarakhand (Anjula Pandey *et al.,* 2022).

This onion is always propagated vegetatively through bulbs which are planted after storing these for 1-2 months to overcome dormancy (Chamroy, 2018). While planting the plants are usually spaced 10-15×10-15cm.

Botany

Pandey et al. (2022) reported Chinese onion possessing 2n=2x=16, and 4x=32 chromosokme numbers. The plant grows to a height of 16-50cm with clustered bulbs that are gregarious, elongated *cum* narrowly ovoid, 5-6.5 x 0.5-2cm having membranous and white or tinged-lilac bulb tunic. The slender and terete leaves are equal to scape, up to 3.5mm wide, 3-5 angled, fistulose, sometimes coiled and glabrous. Scape is 22-40cm long, terete and arising lateral. The 2-valved spathe is persistent, ovoid or deltoid. The 7-40 flowered umbel is hemispherical and laxly flowered (10-14). Flowers are broadly campanulate and long pedicellate. Tepals are purple, convex, hooded, tubercled, coherent up to 1mm at base, obtuse or acute, mid-vein prominent, perianth pale-purple to light-purple; tepal elliptic to sub-orbicular, outer shorter than inner, anthers oblong, dorsifixed in the middle or little below, filaments equal, ovary obovoid-globose, trilobed, sessile, surface uneven; style exserted, filiform, often very short and broader above (very prominent in open flower). Flowering occurs from July to October. Seed description is not known.

Chinese onion- flower	Scape	Fresh harvest

Source: Pandey *et al.,*2022

Uses

The leaves and bulbs of Chinese onion can be eaten raw or cooked as vegetables. Bulbs can be pickled and served as a side dish. The plant is traditionally used in China against various ailments.

References

Anjula Pandey, Pavan Kumar Malav, K.M. Rai and S.P. Ahlawat, 2022. Genus *Allium* L. of the Indian Region: A Field Guide for Germplasm Collection and Identification.

ICAR-National Bureau of Plant Genetic Resources, New Delhi, India, 109p + i-vi. http://www.nbpgr.ernet.in/Downloadfile.aspx?EntryId=9197.

Chamroy, T. 2018. *Allium chinense*-Lesser Known Potential Plant. Biotech Articles. https://biotecharticles.com/Agriculture-Article/Allium-Chinense-Lesser-Known-Potential-Plant-4323.html.

FOI (Flowers of India), 2023. Chinese Onion.. https://www.flowersofindia.net/catalog/slides/Chinese%20Onion.html. Retrieved 16 February, 2023.

POWO (Plants of the World Online), 2023. *Allium chinense* G.Don. Royal Botanic Gardens, Kew. https://powo.science.kew.org/taxon/urn:lsid:ipni.org:names:527818-1. Retrieved 16 February, 2023.

USDA, 2023. Agricultural Research Service. National Germplasm Resources Laboratory. Germplasm Resources Information Network (GRIN Taxonomy), Beltsville, Maryland. URL: http://npgsweb.ars-grin.gov/gringlobal/taxon/taxonomydetail?id=101417. Accessed 16 February 2023.

Wikispecies, 2023. https://species.wikimedia.org/wiki/Allium_ chinense. Retrieved on 12,February, 2023.

Cliff onion (*Allium stellatum* Nutt. ex Ker Gawl.)

Subgenus: *Amerallium.*

Section: *Lophioprason.*

Subsection: *Cernua* (Wikispecies, 2023).

Other common names: Autumn onion, Glade onion, Prairie onion (Illinoiswildflowers, 2023; NatureServe, 2023),

Synonyms: *Hexonychia stellatum* (Nutt. ex Ker Gawl.) Salisb.; *Stelmesus stellatus* (Nutt. ex Ker Gawl.) Raf. (POWO, 2023).

Introduction and origin

The Latin name *stellatum* is for "starry" and refers to the umbels. This species was described for science by John Bellenden Ker Gawler in 1813 (Kantrud, 1995; Nuttal, 1813). Cliff onion (Illinoiswildflowers, 2023) is a North American wild species that has native range from Central & SE. Canada to Central & E. Central U.S.A. (POWO, 2023). The entire plant has a strong onion-like smell when cut or bruised, and the bulbs and leaves of this plant were once used in cooking (stews) or eaten raw, although this plant is not generally considered to be of culinary value today (MBG, 2023).

Cliff onion will slowly cluster at the base, or reseed itself under favourable condition.

Cliff onion

By Mason Brock (Masebrock) - Own work, Public Domain, commons. wikimedia.org/w/index.php?curid=33331649

Botany

It has been reported with 2n=14 chromosome numbers. Cliff onion is an herbaceous bulbous perennial very similar to nodding onion in size, culture and general appearance except the floral umbel of the latter species nodding downward rather than standing erect (MBG, 2023). The plant forms tufts of slender but solid leaves and stems (PlantDatabase, 2023). Leaves and flower scapes rise directly from the bulbs (MBG, 2023). The scape typically grows to a height of 30-45(50)cm and is stiffer than the leaves that are flat, narrow, grass-like and up to 30cm long (Illinoiswildflowers, 2023; MBG, 2023). The green leaves appear in spring and die back in summer as the rounded umbel (5-8cm across) of small (6mm or slightly more) pink to purple or white flowers form at the end of the scape (PlantDatabase, 2023; Illinoiswildflowers, 2023), rising slightly above the foliage (MBG, 2023). Each starry, bell-shaped (MBG, 2023) flower has 3 petals and 3 sepals that flare outward (Illinoiswildflowers, 2023). The root system consists of a bulb with unusually deep secondary roots, possibly for survival during dry weather (Illinoiswildflowers, 2023). If conditions are favourable, the cliff onion will slowly cluster at the base or reseed itself (Illinoiswildflowers, 2023).

Uses

The bulbs of wild onions have a very strong flavour which can be eaten raw or parboiled. Their strong taste can be reduced by parboiling and discarding the water (PlantDatabase, 2023). Bulbs were also used by early Americans as cough/cold remedies and as insect repellants (MBG, 2023).

References

Illinoiswildflowers, 2023. Cliff Onion. https://www.illinoiswildflowers.info/prairie/plantx/cl_onionx.htm. Retrieved 30 January, 2023.

Kantrud, H.A. 1995. Pink Wild Onion (*Allium stellatum*). Native *Wildflowers of the North Dakota Grasslands*. Northern Prairie Wildlife Research Center. Archived from the original on 2014-08-13. Retrieved on 19 February, 2023 via https://en.wikipedia.org/wiki/Allium_stellatum.

MBG (Missouri Botanical Garden), 2023. Allium stellatum. www.missouribotanicalgarden.org/PlantFinder/PlantFinderDetails.aspx?taxonid=281848. Retrieved on 19 February, 2023 via https://en.wikipedia.org/wiki/Allium_stellatum.

NatureServe, 2023. *Allium stellatum* (Glade Onion). NatureServe Explorer. https://explorer.natureserve.org/Taxon/ELEMENT_GLOBAL.2.161018/Allium_stellatum. Retrieved 30 January, 2023.

Nuttall, T. ex Ker Gawler and John Bellenden, 1813 Botanical Register 38: plate 1576. https://en.wikipedia.org/wiki/Allium_stellatum.

PlantDatabase, 2023. *Allium stellatum*. Lady Bird Johnson Wildflower Center, University of Texas at Austin. https://www.wildflower.org/plants/result.php?id_plant=ALST. Retrieved 2010-03-24.

POWO, 2023. *Allium stellatum* Nutt. ex Ker Gawl. Plants of the World Online. Royal Botanic Gardens, Kew. https://powo.science.kew.org/taxon/urn:lsid:ipni.org:names:528906-1. Retrieved 30 January, 2023.

Wikispecies, 2023. https://species.wikimedia.org/wiki/Allium_stellatum. Retrieved on 12,February, 2023.

Dark blue garlic (*Allium cyaneum* Regel)

Subgenus: *Rhizirideum.*

Section: *Rhizirideum.*

Synonyms: *Allium hugonianum* Rendle, *A. purdomii* W.W.Sm., *A. szechuanicum* F.T.Wang & Tang, *A. tui* F.T.Wang & Tang (POWO, 2023).

Introduction and origin

This garlic is also a Chinese *Allium* naturally found growing at elevations from 2100 to 5000m on forest edges, slopes and meadows as occurring in Tibet, some provinces of China and possibly in Korea (Kang *et al.*, 2016). It is propagated through seeds when just ripe or in the spring, or through division of offsets in spring and autumn. It also spreads through short rhizomes which make the clumps (RHS, 2021). It is perfect for growing as part of a rockery, in gravel garden or for container display as it is quite ornamental too (gardenersworld.com, 2023).

Botany

It has tetraploid chromosome number, *i.e.* 2n=32 (FOC, 2023). *Allium cyaneum* is a clump-forming perennial herb growing up to 25cm tall with bright green grass-like foliage, bearing open nodding clusters of bell-shaped deep purple-blue flowers above the foliage (gardenersworld.com, 2023; RHS, 2021). Leaves may be shorter or longer than scape (FOC, 2023). Flowering and fruiting occur from August to October. It has cylindrical *cum* clustered bulbs (en.wikipedia.org/wiki/Allium_cyaneum) that are 0.2-0.4(-0.6)cm in diameter (FOC, 2023).

Allium cyaneum

By Daderot - Own work, CC0, https://commons.wikimedia.org/w/index.php?curid=15383253

Uses

Its leaves and scapes are edible and are occasionally consumed by local people as a spice after drying (Kang *et al.*, 2016).

References

en.wikipedia.org/wiki/Allium_cyaneum

FOC (Flora of China), 2023. Allium cyaneum Regel. Flora of China, vol. 24: 20. http://www.efloras.org/florataxon.aspx?flora_id=2&taxon_id=200027467. Accessed on 15 February, 2023

gardenersworld.com, 2023. https://www.gardenersworld.com/plants/allium-cyaneum/. Retrieved 21 February, 2023.

Kang, Jin, Kang Yongxiang, Ji Xiaolian, Guo Quanping, Jacques Guillaume, Pietras Marcin, Łuczaj Nasim, Li Dengwu and Łuczaj Łukasz, 2016. Wild food plants and fungi used in the mycophilous Tibetan community of Zhagana (Tewo County, Gansu, China). *J. Ethnobiol. Ethnomed.*, **12** (21): 21. doi:10.1186/s13002-016-0094-y. PMC 4890536. PMID 27251364.

POWO, 2023. Allium cyaneum Regel. Plants of the World Online. Royal Botanic Gardens, Kew. https://powo.science.kew.org/taxon/urn:lsid:ipni.org:names:527902-1. Retrieved 10 February, 2023.

RHS, 2021. Allium cyaneum. https://www.rhs.org.uk/plants/821/i-allium-cyaneum-i/details. Retrieved 4 June 2021.

Drummond's onion (*Allium drummondii* Regel)

Subgenus: *Amerallium.*

Section: *Amerallium* (Wikispecies, 2023).

Other common names: Prairie onion and Wild garlic (en.wikipedia.org/wiki/Allium_drummondii).

Synonyms: *Allium drummondii* f. *asexuale* Ownbey, *A. helleri* Small, *A. nuttallii* S.Watson, *A. reticulatum* var. *nuttallii* (S.Watson) M.E.Jones (POWO, 2023).

Introduction and origin

Drummond's onion (Wikispecies, 2023) is a North American species of *Allium* native to Arkansas, Kansas, NE Mexico, Nebraska, New Mexico, Oklahoma and Texas (POWO, 2023) where it can be seen growing at elevations from sea level to 1600m, i.e. in the plains, hills and prairies, particularly in limestone soils (FNA, 2023). It is considered invasive in some regions.

The bulbs of Drummond's onion contain a considerable amount of inulin which is a non-reducing sugar that humans cannot digest. Therefore, for human consumption these are required to be thoroughly heated for a long period in order to convert the inulin into digestible sugars (Bailey and Bailey, 1976).

Botany

Drummond's onion

By abbamouse - Flickr: Drummond's Onion (Allium Drummondii), CC BY-SA 2.0, https://commons.wikimedia.org/w/index.php?curid=14951077

This possesses diploid (2n=14) as well as tetraploid (2n=28) chromosome numbers. Drummond's onion is a perennial herb comprising 1-5 ovoid bulbs that are without basal bulbels and of the dimension of 1-1.8 × 0.7–1.5cm. The solid and flat 2-5 persistent leaves that remain green at anthesis are 10–30cm × 1–3(–5)mm size.

Solitary and persistent scape is erect, terete and 10–30cm × 1–3mm. The hemispheric-globose umbel is erect and usually has 10-25 ovate to lanceolate flowers that are white, pink or red and rarely greenish yellow. The flowers are rarely replaced by bulbils (FNA, 2023)

Uses

It is harvested by Native Americans for its small edible bulbs (Bailey and Bailey, 1976).

References

Bailey, L.H. and E.Z. Bailey, 1976. Hortus Third, pp. i–xiv. MacMillan, New York.

en.wikipedia.org/wiki/Allium_drummondii

FNA, 2023. *Allium drummondii* Regel. Flora of North America. FNA vol. 26: 226, 239. http://www.efloras.org/florataxon.aspx?flora_id=1&taxon_id=242101355. Retrieved 30 January, 2023.

POWO, 2023. *Allium drummondii* Regel. Plants of the World Online. Royal Botanic Gardens, Kew. https://powo.science.kew.org/taxon/urn:lsid:ipni.org:names:527963-1. Retrieved 10 February, 2023.

Wikispecies, 2023. https://species.wikimedia.org/wiki/Allium_drummondii. Retrieved on 12,February, 2023.

Korean wild chive (*Allium monanthum* Maxim.)

Subgenus: *Microscordum* (Wikispecies, 2023).

Synonyms: *Allium biflorum* Nakai and *Allium monanthum* var. *floribundum* Z.J. Zhong & X.T. Huang (POWO, 2023).

Introduction and origin

Korean wild chive (KNA, 2015) is native to Russian Far East to Korea, Japan and NE China (Hebei, Heilongjiang, Jilin, Liaoning) (POWO, 2023; FOC, 2023; Ohwi, 1984; Denisov, 2008; Choi and Oh, 2011). The plant has a mild onion flavour and is harvested from the wild for local use as a food. It reproduces both by seeds and bulbs.

Botany

As per FOC (2023), it has diploid (2n=16) to tetraploid (2n=32) chromosome numbers. *Allium monanthum* is unusual in the genus as it is usually dioecious (male and female flowers on separate plants) though rarely hermaphrodite or gynomonoecious (Noda and Kawano, 1988). The plant bears a solitary, globose and very small (0.5-1cm diameter) bulb, along with 1 or 2 flat and narrow leaves that are longer than scape. The slender and terete scape is relatively short for the genus and it is 3-8mm wide and rarely more than 10cm in length. The white, pink or red flowers being 2-4 (or sometimes even 5) usually appearing in small umbels in May in male plants and only 1 in female plants (FOC, 2023; Choi and Oh, 2011; Maximowicz, 1886).

Korean wild chives

By Moosimi - http://moosimi.tistory.com/6, CC BY 4.0, https://commons.wikimedia.org/w/index.php?curid=59821241

Uses

Korean wild chives can be eaten raw, used in soups, cooked as spring vegetables, may be pickled, *etc*. It makes a good last minute addition to stews and other dishes.

References

Choi, H.J. and B.U. Oh, 2011. A partial revision of *Allium* (Amaryllidaceae) in Korea and north-eastern China. *Bot. J. Linn. Soc.*, **167**: 153-211.

Denisov, N. 2008. Addition to Vascular Flora of the Kozlov Island (Peter the Great Gulf, Japanese sea). *Turczaninowia,* **11**(4): 29-42. https://en.wikipedia.org/wiki/Allium_monanthum.

FOC, 2023. *Allium monanthum* Maxim. Flora of China, vol. 24: 202. Accessed on 15 February, 2023 via http://www.efloras.org/florataxon.aspx?flora_id=3&taxon_id=200027503.

KNA, 2015. Pocheon: Korea National Arboretum, p. 347. ISBN 978-89-97450-98-5. Archived from the original (PDF) on 25 May 2017. Retrieved 6 December 2016 – via Korea Forest Service.

Maximowicz, C.J. 1886. *Bulletin de l'Academie Imperiale des Sciences de St-Petersbourg*, **31**(1): 109. https://en.wikipedia.org/wiki/Allium_monanthum.

Noda, S. and Kawano Shoichi, 1988. The Biology of *Allium monanthum* (Liliaceae) I. Polyploid Complex and Variations in Karyotype. *Pl. Spec. Biol.*, **3** (1): 13–26. doi:10.1111/j.1442-1984.1988.tb00167.x. ISSN 1442-1984. https://en.wikipedia.org/wiki/Allium_monanthum#cite_note-2.

Ohwi, J. 1984. Flora of Japan (in English), 1-1067. Smithsonian Institution, Washington, D.C. Retrieved on 15 February, 2023 from https://en.wikipedia.org/wiki/Allium_monanthum.

POWO, 2023: *Allium monanthum* Maxim. Plants of the World Online. Royal Botanic Gardens, Kew. https://powo.science.kew.org/taxon/urn:lsid:ipni.org:names:528433-1. Retrieved 30 January, 2023.

Wikispecies, 2023. https://species.wikimedia.org/wiki/Allium_monanthum. Retrieved on 12,February, 2023.

Nodding onion (*Allium cernuum* Roth)

Subgenus: *Amerallium.*

Section: *Lophioprason.*

Subsection: *Cernua* (Wikispecies, 2023).

Other common names: New Mexican nodding onion (PFAF, 2023), Lady's leek (Burnie, 1999).

Synonyms: *Allium alatum* Schreb. ex Roth; *A. allegheniense* Small; *A. cernuum* f. *alba* J.K.Henry; *A. cernuum* f. *obtusum* Cockerell; *A. cernuum* subsp. *neomexicanum* (Rydb.) Traub & Ownbey; *A. cernuum* subsp. *obtusum* (Cockerell) Traub & Ownbey; *A. cernuum* var. *neomexicanum* (Rydb.) J.F.Macbr.; *A. cernuum* var. *obtusum* (Cockerell) Cockerell; *A. neomexicanum* Rydb.; *A. nutans* Schult. & Schult.f.; *A. oxyphilum* Wherry; *A. recurvatum* Rydb.; *A. tricorne* Poir.; *Calliprena cernua* (Roth) Salisb.; *Cepa cernua* (Roth) Moench; *Gynodon cernuus* (Roth) Raf.; *Gynodon elliotii* Raf.; *Gynodon rupestris* Raf.

Introduction and origin

Nodding onion (Illinoiswildflowers, 2023) can be found growing in deciduous woodlands and open grasslands in its native range of W&S Canada to NE Mexico though now it is hardly found in its range (POWO, 2023; Coffin and

Pfannmuller, 1988). In Great Britain it is an introduced species. In the southern part of its range in North America it is limited to mountainous habitats, and in other parts of its North American range where it is limited to local and disjunct population and absent from North Dakota and most of the states of Great Plains and intermountain region of the U.S. (Coffin and Pfannmuller, 1988). In gardens it is grown for its distinctive nodding flowers that are white, pink or maroon. It is winter hardy in U.S. hardiness zones of 3–9 (Burnie, 1999). Though it can be grown by seeds but is easier to establish through transplanting the bulbs or offsets.

Nodding onion is easy to identify because of the nodding habit of its floral umbels which may be an evolutionary adaptation to protect nectar from rain and to restrict insect visitors to bees, as other insects are more reluctant to hang upside down while attempting to feed on nectar or pollen (Illinoiswildflowers, 2023).

Botany

The species is diploid with 2n=14 chromosomes. It is a perennial herb comprising 3-5 linear and erect basal leaves; one or sometimes more light green to light reddish purple *cum* glabrous floral stalks (Illinoiswildflowers, 2023; FNA, 2023) terminating in downward nodding hemispheric umbels of 8-35 flowers (FNA, 2023) that are white or pink, pedicellate and emerging out from a pair of sack-like deciduous membranes (Illinoiswildflowers, 2023); and a cluster of 2-5 or more oblong, elongate bulbs that are 1–3cm long, 8–15mm wide and often basally short-rhizomatous (rhizome not stout or iris-like) (FNA, 2023). The greyish or brownish outer coats enclose 1 or more bulbs (FNA, 2023).

Leaves of Nodding onion	**Nodding onion (pink flowered)**	**Nodding onion (white flowered)**

By Illinoiswildflowers (2023)	**By ghislain118 http://www.fleurs-des-montagnes.net - Own work, CC BY-SA 3.0, https://commons.wikimedia.org/w/index.php?curid=15906332**	**By Illinoiswildflowers (2023)**

Basal leaves some 10–25cm long and 1–6mm wide are channelled to broadly V-shaped in cross-section either with entire margin or denticulate (FNA, 2023).

Its little mid-vein keeled leaves are solid and flattened throughout rather than being hollow similar to those in non-native 'field garlic' (*A. vineale*), chives (*A. schoenoprasum*) and cultivated onions (*A. cepa*) having parallel venation (Illinoiswildflowers, 2023). Solid, ridged or terete, up to 50cm long and 1–3mm across scapes are persistent, particularly distally, sometimes flattened and narrowly winged and abruptly recurved near apex (FNA, 2023).

The hemispheric umbels are persistent, cernuous and loose; and the spathe bracts persistent, 2, 3-veined, lanceolate, ± equal, apex-acuminate and beakless. The campanulate flowers with yellow anthers *cum* pollen are 4–6mm across and bloom in July and August. The flowers mature into spherical crested fruits which later on split open to reveal dark dull or shiny seeds (FNA, 2023).

Uses

Similar to common onions, the nodding onion also has strong onion flavour and is edible so often used in cooking (Bailey and Bailey, 1976).

References

Bailey L.H.; Bailey E.Z. (1976). Hortus Third. *New York: MacMillan*. pp. i–xiv, 1–1290.

Burnie Geoffrey, 1999. Botanica: The Illustrated A-Z of Over 10,000 Garden Plants. *Welcome Rain*. p. 75. ISBN 0760716420.

Coffin Barbara and Pfannmuller Lee, 1988. Minnesota's Endangered Flora and Fauna. *U of Minnesota Press.* p. 47. ISBN 978-0-8166-1689-3.

FNA, 2023. Flora of North America @ efloras.org. http://www.efloras.org/flora_page.aspx?flora_id=1. Retrieved 18 February, 2023.

Illinoiswildflowers, 2023. Nodding Onion. https://www.illinoiswildflowers.info/prairie/plantx/nod_onionx.htm. Retrieved 30 January, 2023.

PFAF, 2023. Plants For A Future. https://pfaf.org/User/plant.aspx?latinname=Allium+cernuum. Retrieved 19 February, 2023.

POWO, 2023. *Allium cernuum* Roth. Plants of the World Online. Royal Botanic Gardens, Kew. https://powo.science.kew.org/taxon/urn:lsid:ipni.org:names:527801-1. Retrieved 30 January, 2023.

Wikispecies, 2023. https://species.wikimedia.org/wiki/Allium_cernuum. Retrieved on 12,February, 2023.

Round-headed leek (*Allium rotundum* L. 1762)

Subgenus: *Allium*.

Section: *Allium* (Wikispecies, 2023).

Common names: Purple-flowered garlic (Wikipedia, 2023), Wild garlic, Wild leek, Kormen, Sirmo and Çatlangus (Tasci and Koca, 2016).

Synonyms: *Allium ampeloprasum* subsp. *paterfamilias* (Boiss.) K.Richt.; *A. ampeloprasum* Thunb. 1794, illegitimate homonym not L. 1753; *A. ampeloprasum* var. *paterfamilias* (Boiss.) Nyman; *A. baumannianum* K.Koch; *A. cambiasii* De Not.; *A. cilicicum* Boiss.; *A. descendens* Pall. ex Schult. & Schult.f. 1830, illegitimate homonym not L. 1753; *A. erectum* G.Don; *A. gracilescens* Sommier & Levier; *A. jajlae* var. *baidarense* Seregin; *A. jajlae* Vved.; *A. multiflorum* Kunth 1843, illegitimate homonym not Desf. 1798; *A. paterfamilias* Boiss.; *A. porphyroprasum* Heldr. & Sart. ex Boiss.; *A. preslianum* Schult. & Schult.f.; *A. rotundifolium* Lumn. ex Steud.; *A. rubellum* C.Presl; *A. rubicundum* Niven ex G.Don; *A. scariosum* Jan ex Schult. & Schult.f.; *A. scorodoprasum* subsp. *jajlae* (Vved.) Stearn; *A. scorodoprasum* subsp. *rotundum* (L.) Stearn; *A. scorodoprasum* subsp. *waldsteinii* (G.Don) Stearn; *A. tmoleum* O.Schwarz; *A. waldsteinianum* Schult. & Schult.f.; *A. waldsteinii* G.Don; *Porrum polyanthum* Fourr.; *P. rotundum* (L.) Rchb. (WFO, 2023).

Introduction and origin

Round-headed leek is a wild allium native of Central Europe to NW Iran and Morocco though naturalized in the United States (Michigan), the Canary Islands and South Australia (POWO, 2022). The species grows in disturbed habitats such as roadsides, cultivated fields, *etc.* (FNA, 2023), in the calcareous areas, clayey slopes, lawns, rural areas, beaches and sands between 0 and 1400m above sea level (Tasci and Koca, 2016). It is propagated by seed when ripe and by dividing the clumps in spring.

Botany

Tzanoudakis (2000) reported *Allium rotundum* containing 2n=4x=32 chromosomes. It is a perennial herb producing up to 50 or more egg-shaped bulbs, each up to 1.5cm long. The outer coat surrounds several very dark purple bulbils. The leaves 15–40cm × 2–5(–7)mm numbering 2-5 are solid, flat or channeled and carinate with entire margins. The solitary and erect scape is 25–90cm × 1–4mm bearing globose and quite dense umbel comprising 80–200 bell-shaped flowers that are campanulate and up to 7mm across. Erect and oblong to ovate tepals are purple in colour though sometimes with white margins. Anthers are yellow or purple and pollen is light yellow to white. Flowering usually occurs from May to July (FNA, 2023).

Uses

Leaves and bulbs both can be eaten raw or cooked and used as spices in cooking. The bulbs are believed to have antioxidant, antifungal, antibacterial and antiviral effects. Its various pharmacological activities such as antimicrobial, antihypercholesterolemic, antihyperlipidemic, antihyperpertansive, hepato-protective, diuretic and antitumour have been reported. In folk medicine, it is used to combat diabetes and in order to increase visual accuracy. Tasci and Koca (2016) reported the traditional consumption of *Allium scorodoprosum* L. ssp. *rotundum* in Turkey and its impacts on health (Tasci and Koca, 2016).

Aroma components

Koca *et al.* (2016) reported that aroma components of white and green parts of kormen were analyzed by Headspace/gas chromatography/mass spectrometry (HS/GC/MS). Trimethylene oxide, 1-propanethiol, methyl-thiirane, allyl mercaptan, methyl isocyanide, hexanal, 1,2-dithiolane and diallyl disulphide were identified in the green kormen. Major volatile components detected in kormen were trimethylene oxide (7.12±2.06%) and 1-propanethiol (8.11±2.63%), which are not found in garlic.

References

FNA, 2023. *Allium rotundum* Linnaeus. Flora of North America, 26:244. http://www.efloras.org/florataxon.aspx?flora_id=1&taxon_id=242101396. Accessed on: 14 Feb 2023.

Koca, I., B. Karadeniz and B. Tekguler, 2016. Aroma components of green kormen (*Allium scorodoprasum* L. *spp. rotundum*) and garlic (*Allium sativum*) plants. *Acta Hortic.*, No. 1143, pp. 291-6. DOI: 10.17660/ActaHortic.2016.1143.40. doi.org/10.17660/ActaHortic.2016.1143.40; www.actahort.org/books/1143/1143_40.htm.

POWO, 2022: *Allium rotundum* L. Plants of the World Online. Royal Botanic Gardens, Kew. https://powo.science.kew.org/taxon/urn:lsid:ipni.org:names:528750-1. Retrieved 10 December, 2022.

Tasci, B. and I. Koca, 2016. Use of *Allium scorodoprasum* L. subsp. *rotundum as food. Acta Hortic.*, No. 1143, pp. 153-8. DOI: 10.17660/ActaHortic.2016.1143.22. https://doi.org/10.17660/ActaHortic.2016.1143.22. https://www.actahort.org/books/1143/1143_22.htm.

Tzanoudakis, D. 2000. The genus *Allium* in Cyprus: a preliminary cytotaxonomical study. *Bocconea*, **11**: 105-15. https://flora-of-cyprus.eu/cdm_dataportal/taxon/f659abf8-a7e5-4650-adab-3f2644073e29.

WFO, 2023. Allium rotundum L. Published on the Internet; http://www.worldfloraonline.org/taxon/wfo-0000757183. Accessed on: 14 Feb 2023.

Wikipedia, 2023. en.wikipedia.org/wiki/Allium_rotundum. Retrieved 30 February, 2023.

Wikispecies, 2023. https://species.wikimedia.org/wiki/Allium_rotundum. Retrieved on 12, February, 2023.

Snowdrop onion (*Allium galanthum* Kar. & Kir.)

Subgenus: *Cepa.*

Section: *Cepa* (Wikispecies, 2023).

Synonym: *Allium pseudocepa* Schrenk (POWO, 2023).

Introduction and origin

Snowdrop onion is an Asian species of wild alliums that grows on gravelly and stony slopes from the low mountains to the middle mountain zone (AgroAtlas, 2023) at elevations of 500-1500m (FOC, 2023). The native range of this species is from SW Siberia to Central Asia and Mongolia (POWO, 2023). The plant is usually harvested from the wild for food by the local population.

The plants can be propagated by seeds and bulbs.

Botany

Through FOC (2023), it has been found containing diploid chromosome number of 2n=16. Snowdrop onion is a perennial bulbous plant having clusters of conical-cylindrical, thin-skinned and un-split bulbs with red-brown tunics (AgroAtlas, 2023) having 1.5-3cm diameter (FOC, 2023). Cylindrical leaves are 2 (rarely 3) in number, fistular, 3-10mm wide (AgroAtlas, 2023), 1/2-2/3 as long as scape (FOC, 2023), which is 20-50cm tall, terete, solid and covered with leaf sheaths only at base (AgroAtlas, 2023). The flowering and fruiting occurs from August to October (FOC, 2023). Umbel is hemispheric or more frequently globular, polyanthous and more or less dense; tepals are white with a barely visible vein, 4-5mm long, equal in size, oblong and obtuse; and the filaments are scarcely longer than the perianth (AgroAtlas, 2023).

Snowdrop onion

By Flora of central Asia Flora of Tien Shan - https://tien-shan.myspecies.info/taxonomy/term/6419, CC BY-SA 4.0, https://commons.wikimedia.org/w/index.php?curid=124535717

Uses

Snowdrop onion is a source of vitamins and an excellent table onion. It can be cooked as vegetable similar to those of common onion.

References

AgroAtlas, 2023. *Allium galanthum* Kar. et Kir. - Snowdrop onion. Economic Plants and their Diseases, Pests and Weeds. *Interactive Agricultural Ecological Atlas of Russia and Neighboring Countries.* http://www.agroatlas.ru/en/content/related/Allium_galanthum/. Retrieved on 12, February, 2023.

FOC, 2023. *Allium galanthum* Karelin & Kirilov. http://www.efloras.org/florataxon.aspx?flora_id=2&taxon_id=200027481. Retrieved 30 January, 2023.

POWO, 2023. *Allium galanthum* Kar. & Kir. Plants of the World Online. Royal Botanic Gardens, Kew. https://powo.science.kew.org/taxon/urn:lsid:ipni.org:names:528080-1. Retrieved 30 January, 2023.

Wikispecies, 2023. https://species.wikimedia.org/wiki/Allium_galanthum. Retrieved on 12, February, 2023.

Three-cornered leek (*Allium triquetrum* L.)

Subgenus: *Amerallium.*

Section: *Briseis* (en.wiktionary.org, 2023).

Orher common names: Threecorner leek and Triquetrous garlic (en.wiktionary.org, 2023)

Synonyms: *Allium medium* G.Don, *A. opizii* Wolfner, *A. triquetrum* f. *normale* (L.) Maire & Weiller, *A. triquetrum* var. *bulbiferum* Batt. & Trab., *A. triquetrum* var. *typicum* (L.) Regel, *Briseis triquetra* (L.) Salisb. (POWO, 2023)

Introduction and origin

The specific epithet *triquetrum* refers to the 3-cornered shape of the flower stalks (Hyam and Pankhurst, 1995). This is native to the W&C Mediterranean basin, *viz.* Algeria, Baleares, Corse, France, Italy, Morocco, Portugal, Sardegna, Sicilia, Spain and Tunisia (POWO, 2023). *Allium triquetrum* is a garden escape, introduced from SW Europe, and is potentially a noxious weed (FNA, 2023).

Botany

Through FNA (2023), it has been found containing 2n=18 chromosome numbers. Three-corner leek produces 5–20+ ovoid bulbs that are 1–2 × 1–2cm and not basally clustered on stout primary rhizome as being absent. Basally sheathing 2-3 solid, flat, unfalcate and green leaves especially at anthesis are linearly narrow with 15–50cm × 3–15mm dimension (FNA, 2023). The leaves have a distinct onion smell when crushed (en.wikipedia.org/wiki/Allium_triquetrum). Clustered and persistent 1-4 scapes are erect, solid, sharply 3-angled, 10–40cm × 1–10mm dimention; each scape producing an umbel of 3-15 loose flowers in March-April or May (FNA, 2023). White tepals with strong green lines are 1–1.8cm long (Clive, 1997).

Threecorner leek

By Meneerke bloem - Own work, CC BY-SA 3.0, https://commons.wikimedia.org/w/index.php?curid=15063706

Uses

All the parts of this plant whether fresh or cooked are edible. The plant has a subtle flavour like leek or spring onion (en.wikipedia.org/wiki/Allium_triquetrum).

References

Clive, A.S. 1997. Allium L. – Onions. New Flora of the British Isles (2nd ed.). Cambridge University Press, pp. 936–41. ISBN 978-0-521-58935-2.

en.wikipedia.org/wiki/Allium_triquetrum

en.wiktionary.org, 2023. Allium triquetrum. https://species.wikimedia.org/wiki/Allium_triquetrum. Retrieved 30 January, 2023.

FNA, 2023. Allium triquetrum L. Flora of North America, vol. 26: 224, 230, 257, 258. http://www.efloras.org/florataxon.aspx?flora_id=1&taxon_id=242101411. Retrieved 30 January, 2023.

Hyam, R. and R.J. Pankhurst, 1995. Plants and Their Names: A Concise Dictionary. Oxford University Press, p. 18. ISBN 978-0-19-866189-4.

POWO, 2023. Allium triquetrum L. Plants of the World Online. Royal Botanic Gardens, Kew. https://powo.science.kew.org/taxon/urn:lsid:ipni.org:names:529047-1. Retrieved 30 January, 2023.

Twincrest onion (*Allium bisceptrum* S.Watson)

Subgenus: *Amerallium* (en.wiktionary.org, 2023).

Other common name: Aspen onion.

Synonyms: *Allium bisceptrum* var. *palmeri* (S.Watson) Cronquist, *Allium bisceptrum* var. *utahense* M.E.Jones, *Allium palmeri* S.Watson (POWO, 2023).

Introduction and origin

Twincrest onion is native of West U.S.A. to West New Mexico (POWO, 2023). It has been recorded growing at elevations of 1100--3000m under damp and shady conditions or less commonly on open slopes in mountains in Arizona, California, Idaho, Nevada, New Mexico, Oregon and Utah (FNA, 2023). The two recognized varieties of *Allium bisceptrum* cannot be distinguished confidently at their overlapping ranges of central Utah and central Nevada, except by chromosome number (FNA, 2023).

Twincrest onion

By Gary A. Monroe - http://plants.usda.gov/java/profile?symbol=ALBIB&photoID=albib_001_avp.tif, Public Domain, https://commons.wikimedia.org/w/index.php?curid=15232840

Twincrest onion can be propagated by seeds in the early fall, or by bulbs and bulblets which are collected from mature plants and stored in dry place to be planted in the late fall when the first frost comes. Daughter bulblets may be planted next to mature bulbs to keep the dense pattern (en.wikipedia.org/wiki/Allium_bisceptrum).

Botany

As per FNA (2023) report, this onion is diploid (2n=14) to tetraploid (2n=28). The twincrest onion grows to a height of 10-40cm with solitary or clustered (1–7+), round and egg-shaped bulbs that are 1–2 × 0.6–1.8cm in dimension and have a strong odour. Their long, flat, 8–30cm × 1–13mm leaves are persistant, remaining green at anthesis and usually appear in pairs of two. The persistent, solitary or clustered (1-3) scape are erect, solid, terete, 10–30(–40)cm × 1–5mm. Flowering usually occurs from May to July. The erect umbel comprises 15–40 flowers with varying shades of purple or pink. Each flower head contains six sharply pointed tepals and often with a darker stripe. In this case the plants remain green for longer duration than for most other onions which on crushing or scratching emit an odour when scratched (FNA, 2023).

Allium bisceptrum flower

By Gary A. Monroe - http://plants.usda.gov/java/profile?symbol=ALBIB&photoID=albib_003_ahp.tif, Public Domain, https://commons.wikimedia.org/w/index.php?curid=15232758

Uses

The leaves and greens of this plant are used locally as a spice for flavouring of foods. The young leaves and stems that are only few inches above the ground are harvested after the last snow melts in early spring. Its leaves are similar to chives and can be used in various culinary forms whether eaten raw or stir-fried with meats and/or other vegetables (en.wikipedia.org/wiki/Allium_bisceptrum).

References

en.wikipedia.org/wiki/Allium_bisceptrum

en.wiktionary.org, 2023. *Allium bisceptrum*. https://en.wiktionary.org/wiki/Allium_bisceptrum. Retrieved 30 January, 2023.

FNA, 2023. *Allium bisceptrum* S. Watson. Flora of North America, vol. 26: 230, 256, 257. http://www.efloras.org/florataxon.aspx?flora_id=1&taxon_id=242101335. Retrieved 30 January, 2023.

POWO, 2023. *Allium bisceptrum* S.Watson. Plants of the World Online. Royal Botanic Gardens, Kew. https://powo.science.kew.org/taxon/urn:lsid:ipni.org:names:527688-1. Retrieved 30 January, 2023.

Twinleaf/Kellogg's onion (*Allium anceps* Kellogg)

Subgenus: *Amerallium*.

Section: *Lophioprason* (Wikispecies, 2023).

Introduction and origin

Twinleaf onion is a wild onion having native range from South Oregon to

Nevada ~ California, Idaho and Oregon (POWO, 2023), where it grows on usually barren heavy clay soils at elevations of 1200-1600m (FNA, 2023).

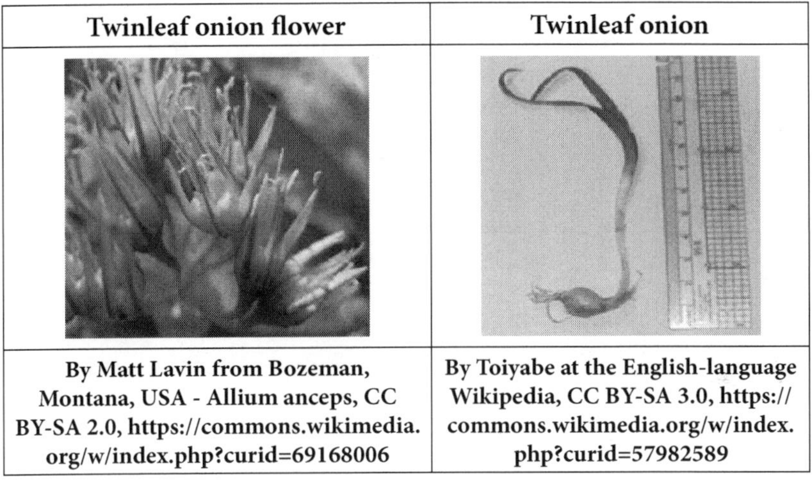

Twinleaf onion flower	Twinleaf onion
By Matt Lavin from Bozeman, Montana, USA - Allium anceps, CC BY-SA 2.0, https://commons.wikimedia.org/w/index.php?curid=69168006	By Toiyabe at the English-language Wikipedia, CC BY-SA 3.0, https://commons.wikimedia.org/w/index.php?curid=57982589

Botany

As described through FNA (2023), it has diploid chromosome number as 2n=14. Twinleaf onion is a perennial herb having up to 5 ovoid bulbs that are 1.5–2cm long and 1.2–2cm wide. One or sometimes more bulbs are enclosed in the brown or yellow-brown outer coat. The two flat, smooth-edged, sickle-shaped leaves are 7- 26 cm long by 4-6 mm wide. Solitary, solid and erect scape is flattened, narrowly winged and of 10-15 × 1–3mm dimension. Once the seeds become mature the scape dies and breaks off, usually along with the leaves. Umbel, with two spathes at the base bears 15-35 star-shaped flowers. Each flower is roughly 1cm wide and has six greenish-veined pink tepals (FNA, 2023).

Uses

The bulbs are edible and a food source for the Northern Paiute.

References

FNA, 2023. *Allium anceps* Kellogg. Flora of North America. FNA vol. 26, page 232 and 270. http://www.efloras.org/florataxon.aspx?flora_id=1&taxon_id=242101332. Retrieved 30 January, 2023.

POWO, 2023. *Allium anceps* Kellogg. Plants of the World Online. Royal Botanic Gardens, Kew. https://powo.science.kew.org/taxon/urn:lsid:ipni.org:names:527590-1. Retrieved 30 January, 2023.

Wikispecies, 2023. https://species.wikimedia.org/wiki/Allium_anceps. Retrieved on 12, February, 2023.

Colour Plates

pg. 2 pg. 2 pg. 36

pg. 39 pg. 39

pg. 41 pg. 44

pg. 48
pg. 51
pg. 68
pg. 98
pg. 98
pg. 99
pg. 99
pg. 99
pg. 100

pg. 100

pg. 102

pg. 106

pg. 107

pg. 109

pg. 110

pg. 111

pg. 112

pg. 113

pg. 115

pg. 115 pg. 116

pg. 118 pg. 119 pg. 120

pg. 120 pg. 121

pg. 122 pg. 151 pg. 152

pg. 154 pg. 155

pg. 155

pg. 156

pg. 160

pg. 166

pg. 198

pg. 246

pg. 247

pg. 249 pg. 250 pg. 251

pg. 252 pg. 256

pg. 256

pg. 266 pg. 267 pg. 271

pg. 273 pg. 274 pg. 279

pg. 289 pg. 292 pg. 293

pg. 293 pg. 302

pg. 307 pg. 308

pg. 323 pg. 324 pg. 329

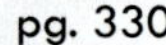

pg. 330

pg. 331

pg. 331

pg. 343

pg. 345

pg. 346

pg. 354

pg. 356

pg. 358

pg. 360

pg. 370

pg. 374

pg. 346

pg. 380

pg. 386

pg. 387

pg. 389

pg. 392

pg. 393

pg. 394

pg. 404 pg. 406

pg. 108 pg. 409 pg. 411

pg. 413 pg. 413 pg. 418 pg. 419

pg. 421 pg. 422 pg. 423 pg. 423

Index
Scientific Names

Common Name

Cepa group, 5, 39, 42, 43

Genus, Sub Genus, Section

Section